AF581681

Reliability of Mechatronic Systems and Machine Elements: Testing and Validation

Editors

Thomas Gwosch
Sven Matthiesen

MDPI • Basel • Beijing • Wuhan • Barcelona • Belgrade • Manchester • Tokyo • Cluj • Tianjin

Editors
Thomas Gwosch
Karlsruhe Institute of
Technology (KIT)
Germany

Sven Matthiesen
Karlsruhe Institute of
Technology (KIT)
Germany

Editorial Office
MDPI
St. Alban-Anlage 66
4052 Basel, Switzerland

This is a reprint of articles from the Special Issue published online in the open access journal *Machines* (ISSN 2075-1702) (available at: https://www.mdpi.com/journal/machines/special_issues/RMSME).

For citation purposes, cite each article independently as indicated on the article page online and as indicated below:

LastName, A.A.; LastName, B.B.; LastName, C.C. Article Title. *Journal Name* **Year**, *Volume Number*, Page Range.

ISBN 978-3-0365-7476-9 (Hbk)
ISBN 978-3-0365-7477-6 (PDF)

Cover image courtesy of Thomas Gwosch

Contents

About the Editors

Thomas Gwosch

Thomas Gwosch (Dr.-Ing.) graduated as a mechanical engineer (M.Sc.) at the Karlsruhe Institute of Technology (KIT) in 2014 and received his doctoral degree (Dr.-Ing.) in 2019. He worked for 3 years as a postdoc at the Institute of Product Engineering at the Karlsruhe Institute of Technology (KIT). In 2022, he received a research fellowship from the German Research Foundation (DFG) and continues his research at the ETH Zurich. Thomas Gwosch has many years of research experience in testing and validation of mechatronic systems with a focus on measurement and control. His focus is on the early stages of the product development process with the design of methods for frontloading in the engineering of mechatronic systems.

Sven Matthiesen

Sven Matthiesen (Univ.-Prof. Dr.-Ing.) is a full professor at the Institute for Product Engineering at the Karlsruhe Institute of Technology (KIT). Based on his extensive practical experience in the development of power tools and the management of multidisciplinary development areas, Sven Matthiesen's research focuses on the research and development of methods and processes to support the product development of human–machine systems. Prof. Matthiesen conducts research on human–machine systems, mechatronic machine elements, and system reliability. In addition to the validation and testing of mechatronic systems, the use of AI methods, and the exploration of machine elements in a system context, the interaction of technology with humans is the focus of his research. Another focus of his research is design methodology, which deals with the exploration of the mental design process and the development of design methodologies.

Preface to "Reliability of Mechatronic Systems and Machine Elements: Testing and Validation"

Advanced and complex systems require technologies and methodologies to develop mechatronic systems in an efficient and effective way. In addition to the design of these systems, testing and validation activities are a key challenge in product development. These activities ensure the functionality and reliability of the system.

The papers in this Special Issue are intended to support the analysis and design of advanced mechatronic systems and are aimed at scientists and engineers in the fields of product development, reliability engineering, dynamics and control engineering.

The guest editors would like to express their sincere appreciation to all the authors who contributed to this Special Issue.

Thomas Gwosch and Sven Matthiesen
Editors

 MDPI

Editorial

Reliability of Mechatronic Systems and Machine Elements: Testing and Validation

Thomas Gwosch * and Sven Matthiesen

Institute of Product Engineering, Karlsruhe Institute of Technology, 76131 Karlsruhe, Germany
* Correspondence: thomas.gwosch@kit.edu

1. Reliability of Complex Mechatronic Systems

The design of reliable systems is a key challenge in product engineering [1]. Functional reliability means the ability of a system to fulfill its required functionality under specified conditions for a specified period of time [2], ensuring the performance and robustness in operation of a system. In case of complex systems, system reliability is particularly challenging due to strong interactions between the subsystems of a single technical system and its environment. It is therefore important to be aware of future applications of the system in order to take them into account during product development. This can be enabled by application data using integrated sensors and validation, verification and testing (VVT) strategies.

At the system level, investigation of functionality and system modeling as well as simulation techniques are needed. Strategies for the design and operation of mechatronic systems also require optimization methods and control approaches.

On the topic of machine elements, research is needed in sensor technologies and signal processing for machine diagnostics and prognosis. For mechatronic systems, there are approaches for integrating sensors into common machine elements for intelligent mechatronic systems [3,4].

In addition, for the system design and the development of intelligent machine elements, research on VVT methods for complex mechatronic systems is needed. The role of testing and validation is a key activity in industry, but is often underestimated in research. VVT is the knowledge base for successful engineering, both for the next generation of products and for proving the functionality and reliability of the current product design. Therefore, research in the field of VVT is essential to manage the complexity of mechatronic systems in the future. This Special Issue (SI), entitled "Reliability of Mechatronic Systems and Machine Elements: Testing and Validation", makes a valuable contribution to research in this field.

The Special Issue presents original research on reliability strategies and design for reliability, machine elements for intelligent mechatronic systems, and the role of testing and validation during the product development of these systems.

Citation: Gwosch, T.; Matthiesen, S. Reliability of Mechatronic Systems and Machine Elements: Testing and Validation. *Machines* **2023**, *11*, 317. https://doi.org/10.3390/machines11030317

Received: 13 February 2023
Accepted: 20 February 2023
Published: 21 February 2023

2. Reliability Strategies and Design for Reliability

From the perspective of the design of complex mechatronic systems, there is a need for operating strategies and design methods that force system reliability and robust design. This concerns the topic areas of maintenance strategies, design optimization and control strategies.

With a focus on reliable and robust mechatronic systems, methodical approaches and system modeling for machines and dynamic systems are needed to solve key challenges in product engineering.

This SI includes contributions on preventive maintenance strategies and system reliability, robust and optimized systems, and system control for powertrains and power systems. Uncertainty in manufacturing is considered in some of the contributing articles.

In [5], the selection of a preventive maintenance strategy for industrial machines is proposed based on a multi-criteria decision method. As a result, a multidimensional matrix is shown to determine the best preventive maintenance strategy for the components of a multistage machine.

In [6], a method for the optimization of an additive manufacturing process is shown. The developed method, which is based on a physically informed machine learning approach, helps to reduce the uncertainty of mechanical properties and enhance the quality of printed parts, increasing system reliability.

In [7], an optimization method for brushless direct current (BLDC) motors that takes manufacturing uncertainty into account is presented. The authors demonstrated a reduction in the motor failure rate and performance variation by the proposed method.

In [8], a new control strategy for heating portable fuel cell systems is presented. This study investigates reliable operation under subfreezing conditions and evaluates system efficiency. The results were verified by simulation and experiments on a test bench.

In [9], a new position control of an electrohydraulic actuator (EHA) for use in early product development stages is presented. Based on a multicriteria optimization of the control in simulations, system performance was investigated in simulation and verified on a test bench. The study shows a good prediction of system behavior for the optimized control.

3. Machine Elements for Intelligent Mechatronic Systems

Machine diagnostics and prognosis are key factors for successful and innovative mechatronic products in future. Product innovation requires knowledge about the state of the system. This requires sensor technologies as well as prognostic methods for state estimation.

In order to manufacture intelligent mechatronic systems, sensor integration into system components is especially important. Therefore, research on the prognosis and health prediction of machine elements is necessary. This SI includes contributions to the diagnosis of gears, the investigation of bearing damage, research on the estimation of external excitations and loads, and a scaling approach for drive components.

In [10], the dynamic characteristics of high-contact-ratio spur gear bearing systems are investigated. The vibrational behavior of the system is evaluated based on a proposed dynamic model with tooth spalling defects. The dynamic model is verified by experimental results.

In [11], a scaling model for drive components is investigated. A functional investigation of the drive system was conducted on a X-in-the-Loop test bench based on the coupling of a geometrically scaled drive component and powertrain. This study demonstrates a new method for scaled component tests.

In [12], a model of a two-tooth difference swing-rod movable teeth transmission system is proposed. The influence of external excitation and system load on the system dynamics is investigated. The simulation results were verified by experiments on a test bench.

In [13], a study of electrical bearing damage is presented. The authors propose a metric to evaluate damage progression using surface property data. The experimental data show a suitable approach for quantifying electrical bearing damage.

In [14], the degradation effects of thin-film sensors are investigated. The investigation includes reliability measurements after high loads, which can occur after the application of a critical component load. Based on the results, a sensor data fusion method is proposed. The presented results can be used for system design using these sensors in machine elements.

In [15], sensor integration in an industrial gear is shown. Based on a simulation, a gear modification was proposed and the effect on the load carrying capacity was investigated. The approach shows the trade-off between gear performance and sensor integration used in intelligent systems.

4. The Role of Verification, Validation and Testing of Mechatronic Systems

The papers in this SI present new approaches and related research to address the reliability challenges of mechatronic systems and machine elements. From a product engineering perspective, the role of VVT is a key aspect of reliable and robust systems.

The increasing complexity of mechatronic systems encourages research of new technologies and methods, especially in the field of VVT. In this context, current challenges are related to research on new reliability approaches, but also on machine elements to be used in intelligent mechatronic systems [3,4]. Especially in human–machine systems, interactions with humans and the environment are highly sophisticated [16,17].

Methods and system models are needed to study and simulate the strong interactions between the subsystems of these complex systems. It is important to appreciate the value of VVT activities in research and their relevance for successful product engineering in the future.

Acknowledgments: We thank all our colleagues who are interested in these research topics and submitted their research for our SI. A special thanks is due to all reviewers for their excellent work and efforts to maintain the high quality of all contributions.

Conflicts of Interest: The authors declare no conflict of interest.

References

1. O'Connor, P.D.T. *Test Engineering: A Concise Guide to Cost-Effective Design, Development and Manufacture*; Wiley: New York, NY, USA, 2001; ISBN 0471498823.
2. Birolini, A. *Reliability Engineering*; Springer: Berlin/Heidelberg, Germany, 2004; ISBN 978-3-662-05411-6.
3. Vorwerk-Handing, G.; Gwosch, T.; Schork, S.; Kirchner, E.; Matthiesen, S. Classification and examples of next generation machine elements. *Forsch Ingenieurwes* **2020**, *84*, 21–32. [CrossRef]
4. Stücheli, M.; Meboldt, M. Mechatronic Machine Elements: On Their Relevance in Cyber-Physical Systems. In *Smart Product Engineering*; Abramovici, M., Stark, R., Eds.; Springer: Berlin/Heidelberg, Germany, 2013; pp. 263–272. ISBN 978-3-642-30816-1.
5. García, F.J.Á.; Salgado, D.R. Analysis of the Influence of Component Type and Operating Condition on the Selection of Preventive Maintenance Strategy in Multistage Industrial Machines: A Case Study. *Machines* **2022**, *10*, 385. [CrossRef]
6. Wenzel, S.; Slomski-Vetter, E.; Melz, T. Optimizing System Reliability in Additive Manufacturing Using Physics-Informed Machine Learning. *Machines* **2022**, *10*, 525. [CrossRef]
7. Jeon, K.; Yoo, D.; Park, J.; Lee, K.-D.; Lee, J.-J.; Kim, C.-W. Reliability-Based Robust Design Optimization for Maximizing the Output Torque of Brushless Direct Current (BLDC) Motors Considering Manufacturing Uncertainty. *Machines* **2022**, *10*, 797. [CrossRef]
8. Zimprich, S.; Dávila-Portals, D.; Matthiesen, S.; Gwosch, T. New Control Strategy for Heating Portable Fuel Cell Power Systems for Energy-Efficient and Reliable Operation. *Machines* **2022**, *10*, 1159. [CrossRef]
9. Dörr, M.; Leitenberger, F.; Wolter, K.; Matthiesen, S.; Gwosch, T. Model-Based Control Design of an EHA Position Control Based on Multicriteria Optimization. *Machines* **2022**, *10*, 1190. [CrossRef]
10. Cheng, Z.; Huang, K.; Xiong, Y.; Sang, M. Dynamic Analysis of a High-Contact-Ratio Spur Gear System with Localized Spalling and Experimental Validation. *Machines* **2022**, *10*, 154. [CrossRef]
11. Steck, M.; Matthiesen, S.; Gwosch, T. Functional Investigation of Geometrically Scaled Drive Components by X-in-the-Loop Testing with Scaled Prototypes. *Machines* **2022**, *10*, 165. [CrossRef]
12. Wei, R.; Yi, Y.; Wu, M.; Chen, M.; Jin, H. Analysis of Load Inhomogeneity of Two-Tooth Difference Swing-Rod Movable Teeth Transmission System under External Excitation. *Machines* **2022**, *10*, 502. [CrossRef]
13. Harder, A.; Zaiat, A.; Becker-Dombrowsky, F.M.; Puchtler, S.; Kirchner, E. Investigation of the Voltage-Induced Damage Progression on the Raceway Surfaces of Thrust Ball Bearings. *Machines* **2022**, *10*, 832. [CrossRef]
14. Ottermann, R.; Steppeler, T.; Dencker, F.; Wurz, M.C. Degeneration Effects of Thin-Film Sensors after Critical Load Conditions of Machine Components. *Machines* **2022**, *10*, 870. [CrossRef]
15. Bonaiti, L.; Knoll, E.; Otto, M.; Gorla, C.; Stahl, K. The Effect of Sensor Integration on the Load Carrying Capacity of Gears. *Machines* **2022**, *10*, 888. [CrossRef]

16. Amaya-Toral, R.M.; Piña-Monarrez, M.R.; Reyes-Martínez, R.M.; de La Riva-Rodríguez, J.; Poblano-Ojinaga, E.R.; Sánchez-Leal, J.; Arredondo-Soto, K.C. Human–Machine Systems Reliability: A Series–Parallel Approach for Evaluation and Improvement in the Field of Machine Tools. *Appl. Sci.* **2022**, *12*, 1681. [CrossRef]
17. Gwosch, T.; Dörr, M.; Matthiesen, S. Modeling of Human-Machine Systems for System Reliability Testing: Investigation of the User Impact on the Load of the Machine. In Proceedings of the 2021 Stuttgarter Symposium für Produktentwicklung, SSP 2021, Stuttgart, Germany, 20 May 2021; Fraunhofer-Institut für Arbeitswirtschaft und Organisation IAO: Stuttgart, Germany, 2021; pp. 1–10.

MDPI

Article

Analysis of the Influence of Component Type and Operating Condition on the Selection of Preventive Maintenance Strategy in Multistage Industrial Machines: A Case Study

Francisco Javier Álvarez García [1,*] and David Rodríguez Salgado [2]

1 Department of Mechanical, Energy and Materials Engineering, University of Extremadura, C/Sta. Teresa de Jornet 38, 06800 Mérida, Spain
2 Department of Mechanical, Energy and Materials Engineering, University of Extremadura, Avda. Elvas s/n, 06006 Badajoz, Spain; drs@unex.es
* Correspondence: fjag@unex.es

Abstract: The study of industrial multistage component's reliability, availability and efficiency poses a constant challenge for the manufacturing industry. Components that suffer wear and tear must be replaced according to the times recommended by the manufacturers and users of the machines. This paper studies the influence of the individual maintenance values of Main Time To Repair (MTTR), Time To Provisioning (TTPR) and Time Lost Production (TLP) of each component, including the type of component and operation conditions as variables that can influence deciding on the best preventive maintenance strategy for each component. The comparison between different preventive maintenance strategies, Preventive Programming Maintenance (PPM) and Improve Preventive Programming Maintenance (IPPM) provide very interesting efficiency and availability results in the components. A case study is evaluated using PPM and IPPM strategies checking the improvement in availability and efficiency of the components. However, the improvement of stock cost of components by adopting IPPM strategy supposes the search of another more optimal solution. This paper concludes with the creation of a multidimensional matrix, for that purpose, to select the best preventive maintenance strategy (PPM, IPPM or interval between PPM and IPPM) for each component of the multistage machine based on its operating conditions, type of component and individual maintenance times. The authors consider this matrix can be used by other industrial manufacturing multistage machines to decide on the best maintenance strategy for their components.

Keywords: maintenance strategies; preventive maintenance; operation condition; type of component; global operation condition; multistage industrial machine; thermoforming

Citation: García, F.J.Á.; Salgado, D.R. Analysis of the Influence of Component Type and Operating Condition on the Selection of Preventive Maintenance Strategy in Multistage Industrial Machines: A Case Study. *Machines* **2022**, *10*, 385. https://doi.org/10.3390/machines10050385

Academic Editors: Sven Matthiesen and Thomas Gwosch

Received: 3 March 2022
Accepted: 16 May 2022
Published: 17 May 2022

1. Introduction

The study of the reliability, availability and efficiency of industrial multistage components poses a constant challenge for the manufacturing industry.

Manufacturing processes can be studied by adopting combinations of different machines that work in a coordinated way, either in series or parallel. The machines can be single stage machines, i.e., machines in charge of one phase of production, or multistage machines, which develop several phases of the production process.

Maintenance strategies must be different for each type of process and machine used. With processes based on a series-parallel combination of single stage machines, an unexpected stoppage caused by the failure of a component does not necessarily stop production altogether, but it may decrease the value of production capacity. With processes based on multistage machines, except for redundancy of these multistage machines, a stop caused by a component failure can cause the entire production process to stop.

Because of this condition in multistage machines, the components, their operating condition and their reliability for the performance of the work they must carry out must be carefully studied. The preventive maintenance strategy [1] is one of the most popular strategies in the industry. According to Colledani [2], equipment availability, product quality and system productivity are strongly related. Moreover, Colledani stated that preventive maintenance policies significantly affect the completion time of a batch [3].

Cheng-Hung [4] proposed a Dynamic Dispatch and Preventive Maintenance model (DDPM) that considers dispatching-dependent deterioration and machine health-dependent production rates for C, a dynamic decision model. Colledoni [5,6] cited in his works that opportunistic maintenance affected the performance in multistage manufacturing systems. Similarly, Xiaojun [7] proposed opportunistic preventive maintenance for Serial-Parallel Multistage Manufacturing Systems (SP-MMSs).

Recent works of Azimpoor [8] reveal that a machine's lifetime is divided into two stages in a failure process, showing defect arrival and then failure arrival. So, the maintenance schedule can be a combination of orders to repair and inspect machines. In this way, Ruiz Hernández [9] believed that poor maintenance could not reinstate the machine to an "as-new" status and this had to be considered when designing maintenance policies. Additionally, Guanghan [10] cited four degradations stages of multistage machines: normal stage, slow degradation, fast degradations and fail.

An effective maintenance policy typically seeks high-quality mechanical reliability, and the minimum possible maintenance cost [11–15]. Xiaojun [16] studied the Condition-Based Maintenance (CBM) policy in multistage manufacturing systems and the positive effects on the quality of the machine work with the most appropriate preventive maintenance decisions. Qipeng [17] proposed using Multistage Stochastic Mixed-Integer Programming (MSMIP) to seek optimal operations regarding maintenance outage scheduling of the machine.

Yingsai [18] studied preventive maintenance based on a policy to improve operation efficiency by modelling an algorithm to obtain the optimal parameters to ascertain the frequencies of inspections and maintenance. Similarly, Grossmann [19] concluded that a Markov decision-making process model is an interesting framework for modelling the stochastic dynamic decision-making process of condition-based maintenance.

Qing [20] proposed preventive maintenance based on quality rework loops for detecting random machine failures. Qiuhua [21] proposed using a constraint to a two-stage assembly flow shop against a fixed preventive maintenance time, using the Weibull probability distribution to calculate the optimal maintenance interval. This ensures the production flow is continuous and ensures the reliability of the machine.

1.1. Preventive Programming Maintenance

Preventive programming maintenance is used in most manufacturing industries. The work of Jun-Hee [22] proportionated preventive maintenance scheduling to minimise the risk of failure in a single-process machine. Other studies by Taghipour [23] and Duffuaa [24] developed models based on integrating the maintenance schedule into the production to improve the machine's quality and performance.

The study of the availability to show the performance level of a multistage system was developed by Arvanitoyannis [25]. Ahmadi [26] used Reliability-Centred Maintenance (RCM) based on condition-based maintenance to decide which maintenance action must be undertaken. Zhen [27] studied the health index to obtain and measure the reliability of a complex production process to reflect the in-time operation state of the production process.

Jiři [28] studied the losses in production and analysed the priority in the corrective and preventive maintenance as the fastest return to the normal activity of the machine. He also used the main time between failure and main time to failure and other delayed times to minimise the cost and improve the availability of the machine. Liberopoulos [29] analysed all the times involved in main time to repair in a single-parallel multistage machine. He also proposed the use of time lost for production as an indicator of the availability of the process.

1.2. Improved Preventive Programming Maintenance

The improved preventive programming is based on the PPM strategy. This strategy minimises the TTPR of all the components by improving the safety of own stocks to reduce the MTTR and improve efficiency and availability ratios. Ren [30] analysed the product-service system (PSS) as an important challenge to providers to perform preventive maintenance based on historical data combined with real-time operational data.

Gharbi [31,32] analysed the effects of joint production and preventive maintenance controls for manufacturing systems, using a make-to-stock strategy and age-based on preventive maintenance, minimising the inventory cost according to unreliable manufacturing periods.

Hongbing [33] studied the optimisation of preventive maintenance by a joinder of maintenance and production considering the maintenance costs, processing costs and completion rewards using the Markov model to form a decision process.

García and Salgado [34] studied the modelling of a multistage machine and preventive maintenance strategies to improve the machine's efficiency and availability.

Ferreira [35] introduced the reactive and proactive concepts to evaluate the components obsolescence and then a new Key Performance Indicators (KPI_S) for a matrix decision for industrial maintenance evaluation.

This paper studies a real case based on a MultiStage Thermoforming Machine (MSTM).

The objective is focused on the selection of the most appropriate preventive maintenance strategy for the components of the studied machine. For that purpose, the preventive maintenance strategies PPM and IPPM are studied, and their results compared. Initially the machine works with PPM strategy. Looking for the improvement of efficiency and availability, IPPM strategy is proposed for use. Results show that a combination of different maintenance strategies is more interesting from a cost point of view. In the aim to reach the objective, the authors propose a methodology for selecting PPM or IPPM strategy for the components depending on the location of the component and new indicators, defined in Step 5 in Section 2. This research proposes a n dimensional matrix for that purpose.

2. Materials and Methods

The work carried out in this article is based on the analysis of one year working of the MSTM. The results obtained with PPM and IPPM strategies are different, so a multicriteria decision method is studied for selecting the appropriate preventive maintenance strategy for different components in the same machine. The result of this multicriteria analysis is the multidimensional matrix proposed and adopted for the machine.

The methodology used in this research and its ordered executed steps, is as follow:

- Step One: First, the MSTM is selected as case study. Then the thermoforming multistage machine is characterised and subsequently all the components are identified and classified by component type. See Section 3.1;
- Step Two: Definition of the concept Global Operation Condition (GOC_i) as an interesting parameter to propose a maintenance strategy. See Section 3.2;
- Step Three: Definition of maintenance times for each component an efficiency and availability definition. See Section 3.3;
- Step Four: Study and collection the individual maintenance times for each component in the MSTM. Evaluation of the results of applying the PPM and IPPM strategies in the same machine in MTTR, TLP, efficiency and availability terms. Evaluation for each component type. See Sections 3.4 and 3.4.1 for PMM strategy and 3.4.2 for IPPM strategy. The results provided by the Section 3.4 are not part for the results of this research, due to the fact that the objective of this research is the selection of the appropriate preventive maintenance strategy for each component and the results of this section are the efficiency and availability values for both strategies;
- Step Five: Definition of Key Performance Indicators (KPIs) as the result of proposed expressions based on maintenance times defined and studied in step four. See Section 3.5.1;

- Step Six: Proposal of a multidimensional matrix for evaluating the maintenance strategy suggested for each component in the same MSTM as a combination of a Global Operation Condition (GOCi), key performance indicators and type of component. See Section 3.5.2;

  Results, discussion, conclusions, and futures research are shown in Sections 4–6. This paper is organised following the outlined steps.

Other Considerations

If the analysed machine will present a lot of unexpected failures in different types of components, the authors suggest using FMECA analysis to achieve a better design and manufacture of the multistage machine.

The definition of the maintenance strategies is lined up to EN 13306:2001. Also, the new KPIs proposed are not the same that technical groups in EN 15341:2007 but allow new and interesting results.

3. Case Studied

3.1. Definition of A Multistage Thermoforming Machine and List of Studied Components

Thermoforming and tub-filling machines are one case among the many that exist. This study covers this type of machine. Figure 1 shows the MSTM and the components' placement.

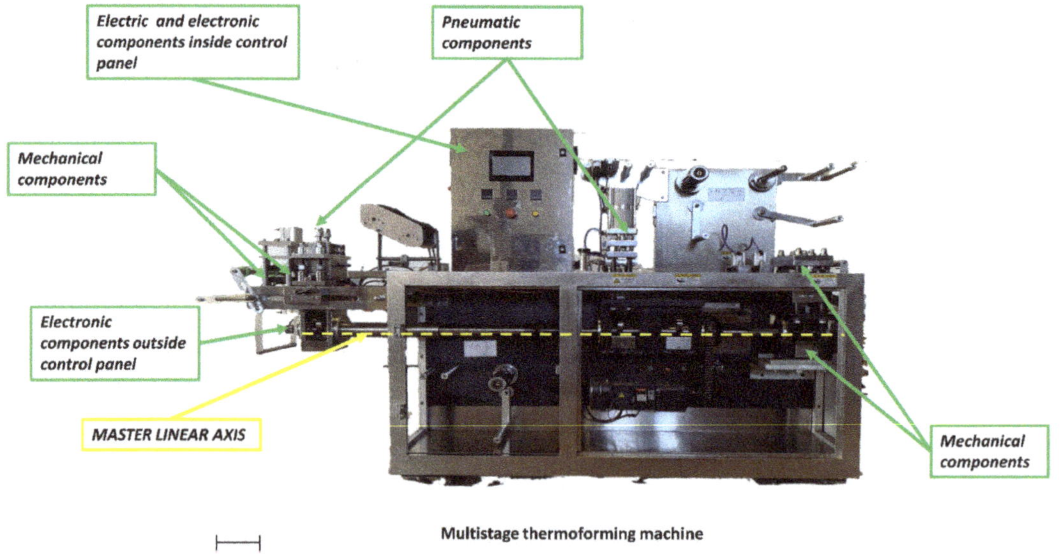

Figure 1. A multistage thermoforming machine of 6 terrines per cycle and its type of components.

These machines comprise several steps, from managing the polymer film, the container and the lid to the dosage and final cut.

The cycle time in this machine is 4 s, during which six terrines are manufactured. Standard operation requires the constant coordination of all steps since a failure in one of them means the global failure and loss of the ongoing production.

There is a master linear axis in the lower part of the machine from the thermal conditioner of the polymer for the container thermoformer to the cutter for finished tubs, which ensures the coordinated operation of the entire machine (see Figure 1).

A structural, fixed part is usually not subject to wear and tear but must be protected against corrosion and meet health and food operation conditions. This multistage machine has many component types. The classification of the components is as follows:

- Electrical components;
- Electronic components;
- Mechanical components;
- Pneumatic components.

The assignment of the type of component has been made by applying the following criteria:

- Electrical components are all those that works in alternating voltage and current;
- Electronic components are all those that need analog signals of voltage or current to work. Those components that use electronic control cards and power electronics equipment in their operating principle, such as thyristors or insulated gate bipolar transistors, are also included;
- Mechanical components are all those that move or are actuated abruptly. The peristaltic pump is included in this group since its drive is carried out by a servomotor that controls the proper dosage. Additionally, the thermocouple sensor is considered as a mechanical component due to its location is inside of the mechanical base for thermoforming creating tub (see step 3 in Figure 2) and inside of the mechanical base for the thermal adhesion (see step 6 in Figure 2). Both mechanical bases are in constant movement;
- Pneumatic components are all those that require pressurised air for their operation.

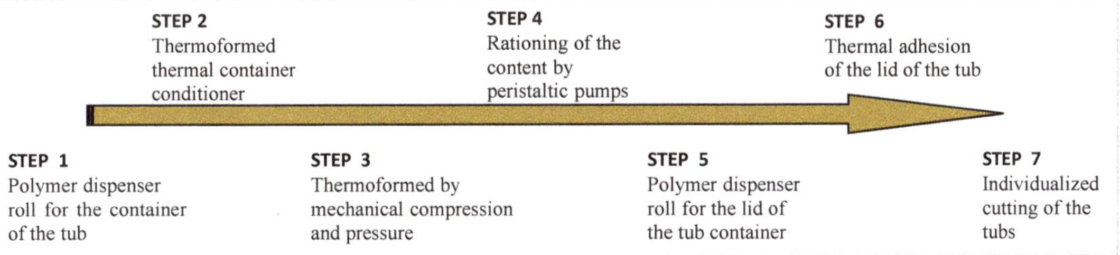

Figure 2. Subprocess in the studied MSTM.

3.1.1. Electrical Components

They are usually inside the control panel, but those that entail human-machine interface (HMI) are on the outer face of the door control panel. Table 1 shows the list of electrical components in this MSTM, with the failure source and event.

Table 1. Electrical components in the studied MSTM with failure source and event.

Component	Failure Source	Failure Event
Master power switch	Ambient condition, Power supplier event	Stop
Plug-in relay	Ambient condition, Power supplier event, Unexpected hit	Malfunction
Command and signalling	Ambient condition, Power supplier event	Stop
Safety limit switch	Ambient condition, Power supplier event, Unexpected hit	Stop

3.1.2. Electronic Components

Electrical components are usually inside the control panel, e.g., the programming logic controller (PLC) and solid-state relays, but some components can be on the outer face of the control panel, as with the electrical components that have human-machine interactions. Sensors are typically distributed around the machine and are subject to degrading and unexpected hits. Table 2 shows the list of electronic components in this MSTM, with the failure source and event.

Table 2. Electronic components in the studied MSTM with failure source and event.

Component	Failure Source	Failure Event
PLC	Ambient condition, Power supplier event	Stop
HMI	Ambient condition, Power supplier event	Stop
Chromatic sensor	Ambient condition, Power supplier event	Stop
Safety relay	Ambient condition, Power supplier event	Stop
Temperature controller	Ambient condition, Power supplier event, Unexpected hit	Stop
Solid-state relay	Ambient condition, Power supplier event	Stop
Frecuency inverter	Ambient condition, Power supplier event	Malfunction
Pressure sensor	Pressure failure, Global fatigue	Malfunction
Servo drive peristaltic pump	Ambient condition, Power supplier event	Stop
Absolute encoder	Ambient condition, Power supplier event, Unexpected hit	Malfunction

3.1.3. Mechanical Components

Some are subject to movement, degrading and unexpected hits. They are selected with fatigue-resistant materials but may be damaged by wear, environmental conditions, and unexpected hits. Table 3 shows the list of mechanical components in this MSTM, with the failure source and event.

Table 3. Mechanical components in the studied MSTM with failure source and event.

Component	Failure Source	Failure Event
Safety button	Ambient condition, Power supplier event	Stop
Thermal resistance	Ambient condition, Power supplier event	Malfunction
Thermocouple sensor	Global fatigue	Malfunction
Motor belt	Ambient condition, Power supplier event	Stop
Bronze cap	Global fatigue	Malfunction
Linear axis	Global fatigue	Malfunction
Linear bearing	Global fatigue	Malfunction
Peristaltic pump	Ambient condition, Power supplier event	Stop
Terrine cutter	Global fatigue	Malfunction

3.1.4. Pneumatic Components

These components are distributed all over the machine and are subject to degrading and unexpected hits. Table 4 shows the list of pneumatic components in this MSTM, with the failure source and event.

Table 4. Pneumatic components in the studied MSTM with failure source and event.

Component	Failure Source	Failure Event
Pneumatic valve	Ambient condition, Power supplier event, Global fatigue, Pressure failure	Malfunction
Pneumatic cylinder	Ambient condition, Power supplier event, Global fatigue, Pressure failure, Failure pneumatic valve	Malfunction

3.2. *Operation Conditions*

Operation conditions are different depending on the situation and type de component selected. In this study, the operation conditions assessed are:

- Work temperature;
- Work Humidity, studied by Ingress Protection rating (IP) according to IEC 62262 [36];
- Impact Protection rating (IK) according to IEC 62262 [36].

Table 5 shows the classification of operation conditions and three operation stages defined in this study.

Table 5. Type of operation conditions for temperature, humidity and IK.

Type of Operation Condition	Temperature	Humidity	IK Rating
A	Outdoor and ventilated situation	Indoor with appropriate IP	Indoor and mechanically protected
B	Indoor and ventilated situation	Outdoor with appropriate IP	Outdoor and protected against mechanical shock
C	Indoor and non-ventilated situation	Outdoor with not appropriate IP	Outdoor and not protected against mechanical shock

The authors propose a definition of Global Operation Condition (GOC_i) for each component as a decisive variable to select the appropriate preventive maintenance strategy for the "i" component. This GOC_i is defined by a sequence of three letters (A, B or C) that mention the three operation conditions studied in Step 3 in Section 2 (see Table 5). In general, the structure of a GOC_i is expressed at it follows:

- First letter: Temperature condition (A, B or C);
- Second letter: Humidity condition (A, B or C);
- Third letter: IK rating condition (A, B or C).

So, if we consider, for example, a component with a BAA value of GOC_i, it means the following:

- First letter B: Indoor and ventilated situation;
- Second letter A: Indoor with appropriate IP;
- Third letter A: Indoor and mechanically protected.

3.3. *Expressions Proposed for the Preventive Maintenance Study*

As stated in Section 1, a failure of most components should lead to a global failure in this type of machine. So, if the critical scenario is studied, the MSTM will present a global failure if a component fails.

The times studied for the failures [28,29] are:

- TTRP: Time to replace a component;
- TTC: Time to configure;
- TTMA: Time to mechanical adjustment;
- TTPR: Time to provisioning;
- MTTR: Mean time to repair;
- MTTF: Mean time to failure;
- MTBF: Mean time between failure;
- TTLR: Line restart time, defined by expert knowledge;
- TLP: Time lost production.

MTTR (1), TLP (2), MTBF (3), efficiency (4) and availability (5) can be calculated with these equations. Efficiency and availability are used as indicators of success in preventive maintenance.

$$\text{MTTR} = \text{TTRP} + \text{TTC} + \text{TTMA} + \text{TTPR} \tag{1}$$

$$\text{TLP} = \text{MTTR} + \text{TTLR} \tag{2}$$

$$\text{MTBF} = \text{MTTR} + \text{MTTF} \tag{3}$$

$$\text{Efficiency} = 1 - \frac{\text{TLP}}{\text{MTTR} + \text{MTTF}} \tag{4}$$

$$\text{Availability} = \frac{\text{MTBF}}{\text{MTBF} + \text{MTTR}} \tag{5}$$

3.4. Preventive Maintenance Strategies for Multistage Thermoforming Machines

The maintenance strategies studied for this MSTM are PPM and IPPM. The efficiency and availability results improve by applying the IPPM strategy. However, applying the IPPM strategy for all the components is not the best scenario for the end user because the stock costs increase. The value of TTLR for both strategies is set at 14,400 s given by the user experience of the machine.

3.4.1. Preventive Programming Maintenance

This strategy uses its own times per component (TTPR, TTC, TTMA, TTRP). All the times are obtained for the usage of the machine. The results are shown in a different table for each type of component.

Table 6 shows the electrical components times and the value of efficiency and availability calculated with Equations (4) and (5).

Table 6. Electrical components times in seconds. Efficiency and availability calculated in % with PPM strategy.

Component	MTTR	TTPR	MTTF	TLP	Efficiency	Availability
Master power switch	14,400	10,800	9,999,999	28,800	99.71%	99.86%
Plug-in relay	14,400	10,800	4,999,999.5	28,800	99.43%	99.71%
Command and signalling	14,400	10,800	4,999,999.5	28,800	99.43%	99.71%
Safety limit switch	14,400	10,800	9,999,999	28,800	99.71%	99.86%

Table 7 shows the electronic components times and the value of efficiency and availability calculated with Equations (4) and (5).

Table 7. Electronic components times in seconds. Efficiency and availability calculated in % with PPM strategy.

Component	MTTR	TTPR	MTTF	TLP	Efficiency	Availability
PLC	435,600	345,600	9,999,999	450,000	95.69%	95.99%
HMI	435,600	345,600	9,999,999	450,000	95.69%	95.99%
Chromatic sensor	176,520	172,800	4,999,999.5	190,920	96.31%	96.70%
Safety relay	14,400	10,800	9,999,999	28,800	99.71%	99.86%
Temperature controller	435,600	345,600	9,999,999	450,000	95.69%	95.99%
Solid-state relay	176,400	172,800	4,999,999.5	190,800	96.31%	96.70%
Frecuency inverter	435,600	345,600	9,999,999	450,000	95.69%	95.99%
Pressure sensor	176,700	172,800	4,999,999.5	191,100	96.31%	96.70%
Servo drive peristaltic pump	435,600	345,600	9,999,999	450,000	95.69%	95.99%
Absolute encoder	360,000	172,800	4,999,999.5	374,400	93.01%	93.71%

Table 8 shows the mechanical components times and the value of efficiency and availability calculated with Equations (4) and (5).

Table 8. Mechanical components times in seconds. Efficiency and availability calculated in % with PPM strategy.

Component	MTTR	TTPR	MTTF	TLP	Efficiency	Availability
Safety button	14,400	10,800	9,999,999	28,800	99.71%	99.86%
Thermal resistance	25,500	10,800	3,700,800	39,900	98.93%	99.32%
Thermocouple sensor	14,700	10,800	3,700,800	29,100	99.22%	99.61%
Motor belt	187,200	172,800	4,999,999.5	201,600	96.11%	96.52%
Bronze cap	288,000	172,800	7,750,000	302,400	96.24%	96.54%
Linear axis	288,000	172,800	7,625,000	302,400	96.18%	96.49%
Linear bearing	288,000	172,800	7,500,000	302,400	96.12%	96.43%
Peristaltic pump	547,200	518,400	4,999,999.5	561,600	89.88%	91.02%
Terrine cutter	288,000	172,800	9,999,999	302,400	97.06%	97.28%

Finally, Table 9 shows the pneumatic components times and the efficiency and availability value calculated with Equations (4) and (5).

Table 9. Pneumatic components times in seconds. Efficiency and availability calculated in % with PPM strategy.

Component	MTTR	TTPR	MTTF	TLP	Efficiency	Availability
Pneumatic valve	176,400	172,800	9,999,999	190,800	98.13%	98.30%
Pneumatic cylinder	176,400	172,800	9,999,999	190,800	98.13%	98.30%

Figure 3 shows a comparative average ratio for efficiency and availability in the PPM strategy by component type.

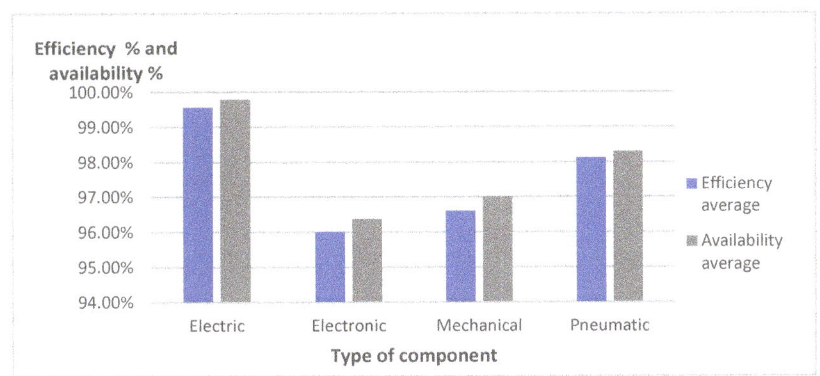

Figure 3. Average efficiency and availability values in % by type of component with the PPM strategy.

Electrical components have the higher average values of efficiency and availability, with a maximum value of 99.71% in efficiency and availability.

Figure 4 also shows the maximum and minimum values for each type of component. Here, the minimum efficiency and availability values of electronic and mechanical components suggest using another maintenance strategy. The machine efficiency and availability levels depend on the efficiency and availability of all the components. So, the objective of the maintenance strategy is to achieve higher efficiency and availability for all the components and not for just many components.

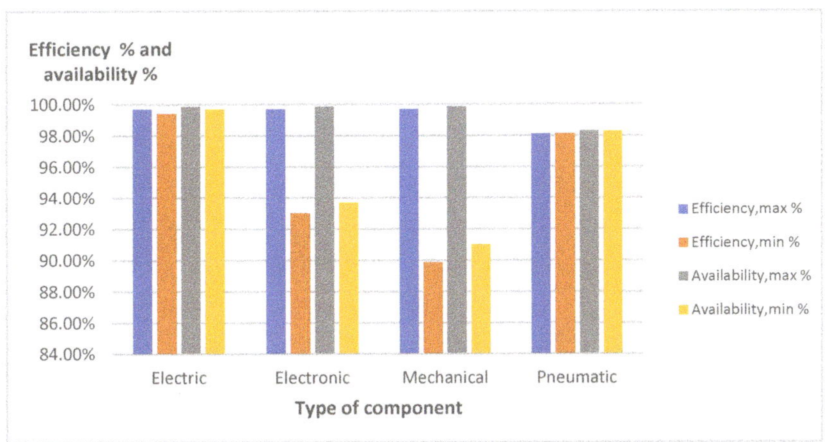

Figure 4. Maximum and minimum efficiency and availability values in % by type of component with the PPM strategy.

3.4.2. Improve Preventive Programming Maintenance

The IPPM strategy is based on reducing the TTPR time for all the components by increasing the security stocks for the components. The TTPR value in this strategy is a residual value consisting in the transport and picking time to machine of the component waiting for in the safety stock.

Table 10 shows the electrical components times and the efficiency and availability values calculated with Equations (4) and (5).

Table 10. Electrical component times in seconds. Efficiency and availability calculated in % with the IPPM strategy.

Component	MTTR	TTPR	MTTF	TLP	Efficiency	Availability
Master power switch	3900	300	9,999,999	18,300	99.82%	99.96%
Plug-in relay	3900	300	4,999,999.5	18,300	99.63%	99.92%
Command and signalling	3900	300	4,999,999.5	18,300	99.63%	99.92%
Safety limit switch	3900	300	9,999,999	18,300	99.82%	99.96%

Table 11 shows the electronic components times and the efficiency and availability values calculated with Equations (4) and (5).

Table 11. Electronic component times in seconds. Efficiency and availability calculated in % with IPPM strategy.

Component	MTTR	TTPR	MTTF	TLP	Efficiency	Availability
PLC	90,300	300	9,999,999	104,700	98.96%	99.11%
HMI	90,300	300	9,999,999	104,700	98.96%	99.11%
Chromatic sensor	4020	300	4,999,999.5	18,420	99.63%	99.92%
Safety relay	3900	300	9,999,999	18,300	99.82%	99.96%
Temperature controller	90,300	300	9,999,999	104,700	98.96%	99.11%
Solid-state relay	3900	300	4,999,999.5	18,300	99.63%	99.92%
Frecuency inverter	90,300	300	9,999,999	104,700	98.96%	99.11%
Pressure sensor	4200	300	4,999,999.5	18,600	99.63%	99.92%
Servo drive peristaltic pump	90,300	300	9,999,999	104,700	98.96%	99.11%
Absolute encoder	187,500	300	4,999,999.5	201,900	96.11%	96.51%

Table 12 shows the mechanical components times and the efficiency and availability values calculated with Equations (4) and (5).

Table 12. Mechanical component times in seconds. Efficiency and availability calculated in % with the IPPM strategy.

Component	MTTR	TTPR	MTTF	TLP	Efficiency	Availability
Safety button	3900	300	9,999,999	18,300	99.82%	99.96%
Thermal resistance	15,000	300	3,700,800	29,400	99.21%	99.60%
Thermocouple sensor	4200	300	3,700,800	18,600	99.50%	99.89%
Motor belt	14,700	300	4,999,999.5	29,100	99.42%	99.71%
Bronze cap	115,500	300	7,750,000	129,900	98.35%	98.55%
Linear axis	115,500	300	7,625,000	129,000	98.32%	98.53%
Linear bearing	115,500	300	7,500,000	129,900	98.29%	98.51%
Peristaltic pump	29,100	300	4,999,999.5	43,500	99.14%	99.42%
Terrine cutter	115,500	300	9,999,999	129,900	98.72%	98.87%

Finally, Table 13 shows the pneumatic components times and the efficiency and availability values calculated with Equations (4) and (5).

Table 13. Pneumatic component times in seconds. Efficiency and availability calculated in % with the IPPM strategy.

Component	MTTR	TTPR	MTTF	TLP	Efficiency	Availability
Pneumatic valve	3900	300	9,999,999	18,300	99.82%	99.96%
Pneumatic cylinder	3900	300	9,999,999	18,300	99.82%	99.96%

As with Figures 3 and 5 shows a comparative average ratio for efficiency and availability in the IPPM strategy by component type.

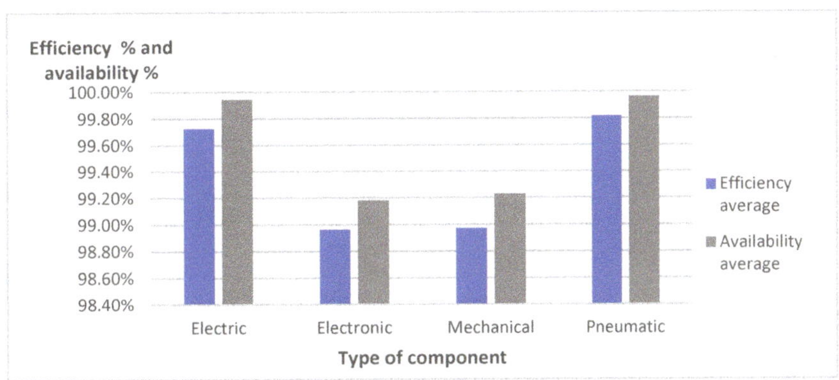

Figure 5. Average efficiency and availability values in % by component type with the IPPM strategy.

When comparing Figures 3 and 5, the efficiency and availability increase their values in all components, especially in pneumatic components. Electronic and mechanical components also increase their values. Figure 6 show the minimum and maximum new efficiency and availability values when applying the IPPM strategy.

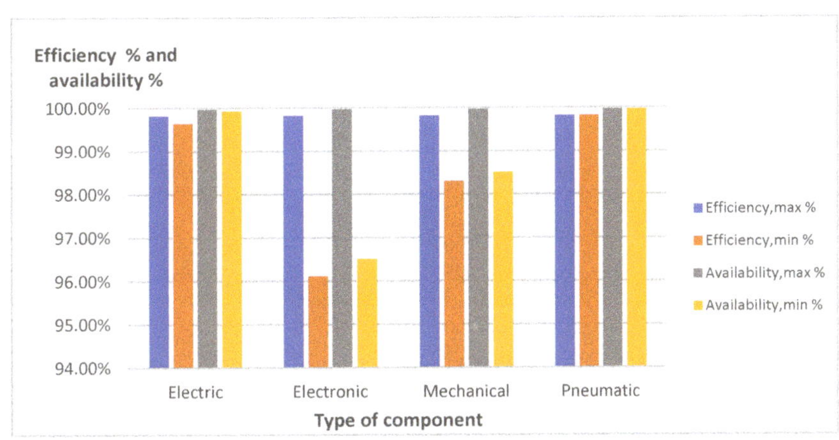

Figure 6. Maximum and minimum efficiency and availability values in % by component type with the IPPM strategy.

A simple comparison between Figures 4 and 6 show the increase obtained in pneumatic components by applying the IPPM strategy. Their efficiency and availability levels improve to above 99.82% values. Electronic and mechanical components also improve their minimum efficiency and availability values by 3.09% and 2.81% for electronic components and 8.42% and 7.48% for mechanical components. Increasing the components' minimum efficiency and availability values allows for improving MSTM efficiency and availability globally.

3.5. Selection of Preventive Maintenance Strategies

The results obtained by applying the PPM and IPPM strategies show very interesting efficiency and availability values, but in the same machine, as is well-known, all the components do not have to use the same maintenance strategy. In this section, the authors propose a singular study that allows for selecting the best strategy for each component. This selection depends on parameters such as type of component, operation condition and own time values of all the components.

3.5.1. Parameter for the Selection under Study

An analysis of the results obtained when applying PPM and IPPM shows that component type is an important parameter for deciding on preventive maintenance strategy. For example, applying IPPM in electrical components does not show a remarkable increase, so the conclusion could be not applying IPPM in electrical components.

The operation conditions in Table 5 denote the relevance of the adequate operation conditions for all the components by selecting the appropriate tolerance to temperature and an appropriate value of IP and IK ratings. A high-quality component working in inadequate operation conditions, compared with their datasheet, could entail unexpected failures and a decrease in efficiency and availability of the MSTM. Each component may have an individual condition, shown by three letters A, B or C for this study.

Attending to the definition of each operation condition, it is easy to understand that many combinations of operation conditions cannot exist simultaneously. For example, a component cannot be located indoors and outdoors, so combinations such as AAA are impossible.

Individual maintenance times of all the components allow for knowing the influence of the MMTR and TLP times in the efficiency and availability of the MSTM, so for this study, these KPIs are used:

$$KPI_1 = (MTTR - TTPR)/MTTR \quad (6)$$

$$KPI_2 = TTPR/TLP \quad (7)$$

Equation (6) states the influence of the TTPR in MTTR. For this KPI_1 singular value is used at 25% result.

Equation (7) also shows the influence of the TTPR in the TLP (see Equation (2)). A Higher value of TTPR greater than TLR could entail a considerable stop time in the MSTM and the assumption of undesirable opportunity costs. In this KPI_2, the singular value used is 70% result.

3.5.2. N-Dimensional Matrix for Preventive Maintenance Selection

Using the three parameters explained in the previous subsection, an n-dimensional matrix is proposed to select the appropriate maintenance strategy for the component, where *n* is set in five dimensions:

- Operation condition;
- Type of component;
- Value of Equation (6);
- Value of Equation (7);
- Combination of Equations (6) and (7) values.

To understand and use this 5-dimension matrix, the authors propose a plane conversion in which the operation condition is the column, and the rest of the conditions are fixed in mixed lines, as shown in Table 14.

Table 14. N-dimensional matrix for preventive maintenance selection in MSTM.

Type of Component	KPI	Operation Condition					
		ABB	ABC	ACB	ACC	BAA	CAA
Electrical	$KPI_1 > 25\%$ $KPI_2 < 70\%$	II	III	II	III	II	II
	$KPI_1 < 25\%$	I	II	II	II	I	II
	$KPI_2 > 70\%$	III	III	III	III	III	III
Electronic	$KPI_1 > 25\%$ $KPI_2 < 70\%$	II	III	III	III	II	III
	$KPI_1 < 25\%$	II	III	II	III	II	III
	$KPI_2 > 70\%$	III	III	III	III	III	III

Table 14. *Cont.*

Type of Component	KPI	Operation Condition					
		ABB	ABC	ACB	ACC	BAA	CAA
Mechanical	$KPI_1 > 25\%$ $KPI_2 < 70\%$	II	III	II	III	II	III
	$KPI_1 < 25\%$	I	III	I	III	I	III
	$KPI_2 > 70\%$	III	III	III	III	III	III
Pneumatic	$KPI_1 > 25\%$ $KPI_2 < 70\%$	II	III	III	III	III	III
	$KPI_1 < 25\%$	II	III	II	III	III	III
	$KPI_2 > 70\%$	III	III	III	III	III	III

Selection I indicates PPM strategy; selection III indicates IPPM strategy, and; selection II shows an intermediate situation between PPM and IPPM strategies, where there is a special consideration of the necessary constant for analysing the TTPR of each component. Selection II is called "Interval PPM to IPPM strategy" in this paper.

4. Results

Applying the n-dimensional matrix for the preventive maintenance strategy in the components of a multistage thermoforming machine allows for improving efficiency and availability. Table 15 shows the application of the n-matrix in this study.

Table 15. Results and comparison of the application of n-dimensional matrix for maintenance strategy decision.

Component	PPM Strategy		IPPM Strategy		N-MATRIX PROPOSAL		
	Efficiency	Availability	Efficiency	Availability	Efficiency	Availability	Maintenance Strategy
Master power switch	99.71%	99.86%	99.82%	99.96%	99.71%	99.86%	PPM
PLC	95.69%	95.99%	98.96%	99.11%	98.96%	99.11%	IPPM
HMI	95.69%	95.99%	98.96%	99.11%	98.96%	99.11%	IPPM
Chromatic sensor	96.31%	96.70%	99.63%	99.92%	99.63%	99.92%	IPPM
Plug-in relay	99.43%	99.71%	99.63%	99.92%	99.43%	99.71%	PPM
Command and signalling	99.43%	99.71%	99.63%	99.92%	99.43%	99.71%	PPM
Safety limit switch	99.71%	99.86%	99.82%	99.96%	99.71%	99.86%	PPM
Safety relay	99.71%	99.86%	99.82%	99.96%	99.71%	99.86%	PPM
Safety button	99.71%	99.86%	99.82%	99.96%	99.71%	99.86%	PPM
Temperature controller	95.69%	95.99%	98.96%	99.11%	98.96%	99.11%	IPPM
Solid-state relay	96.31%	96.70%	99.63%	99.92%	99.63%	99.92%	IPPM
Thermal resistance	98.93%	99.32%	99.21%	99.60%	98.93%	99.32%	Interval PPM to IPPM
Thermocouple sensor	99.22%	99.61%	99.50%	99.89%	99.22%	99.61%	Interval PPM to IPPM
Frequency inverter	95.69%	95.99%	98.96%	99.11%	98.96%	99.11%	IPPM
Motor belt	96.11%	96.52%	99.42%	99.71%	99.42%	99.71%	IPPM
Bronze cap	96.24%	96.54%	98.35%	98.55%	96.24%	96.54%	PPM to IPPM Interval
Linear axis	96.18%	96.49%	98.32%	98.53%	96.18%	96.49%	PPM to PPM Interval
Linear bearing	96.12%	96.43%	98.29%	98.51%	96.12%	96.43%	PPM to PPM Interval
Pneumatic valve	98.13%	98.30%	99.82%	99.96%	99.82%	99.96%	IPPM
Pneumatic cylinder	98.13%	98.30%	99.82%	99.96%	99.82%	99.96%	IPPM
Pressure sensor	96.31%	96.70%	99.63%	99.92%	99.63%	99.92%	IPPM
Servo drive peristaltic pump	95.69%	95.99%	98.96%	99.11%	98.96%	99.11%	IPPM
Peristaltic pump	89.88%	91.02%	99.14%	99.42%	99.14%	99.42%	IPPM
Terrine cutter	97.06%	97.28%	98.72%	98.87%	97.06%	97.28%	PPM to IPPM Interval
Absolute encoder	93.01%	93.71%	96.11%	96.51%	93.01%	93.71%	PPM to IPPM Interval

The comparison of combined average values for all the components by efficiency and availability is shown in Figure 7.

The average values show that a mixed preventive maintenance strategy for the components in a machine is an efficient solution to reach reasonable values of efficiency and availability. The IPPM strategy provides better results for all the components but can increase the maintenance cost for the whole machine, due to PPM strategy does not need stock of components and IPPM strategy that needs stock for all the components (see the beginning of the Section 3.4.2.).

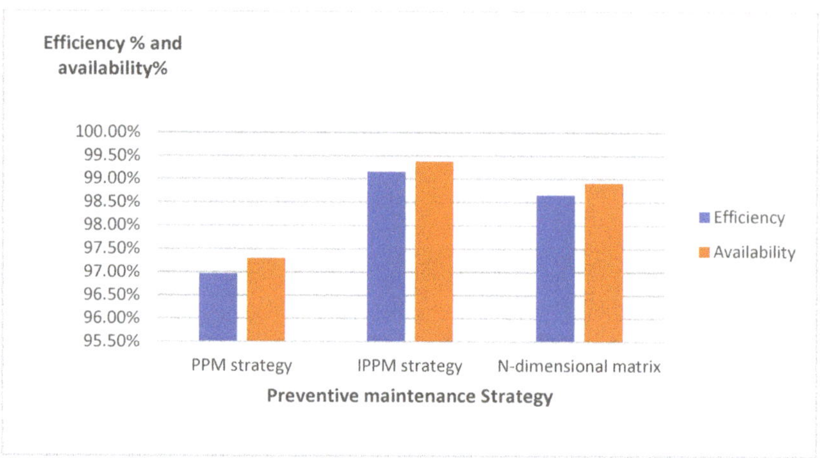

Figure 7. Comparison of mixed average values in applying the PPM strategy, the IPPM strategy and the n-dimensional matrix preventive maintenance proposal.

5. Results Discussion

The results obtained showed an optimisation procedure to select the appropriate maintenance strategy for different components in a multistage machine. Table 16 shows the efficiency and availability improvements comparing PPM and IPPM strategies for type of component.

Table 16. Efficiency and availability improvements comparing PPM and IPPM strategies for type of component.

Type of Component	Efficiency Maximum %	Efficiency Minimum %	Availability Maximum %	Availability Minimum %	Efficiency Average	Availability Average
Electrical	0.10%	0.21%	0.10%	0.21%	0.16%	0.16%
Electronic	0.10%	3.09%	0.10%	2.81%	2.95%	2.82%
Mechanical	0.10%	8.42%	0.10%	7.48%	2.37%	2.22%
Pneumatic	1.69%	1.69%	1.66%	1.66%	1.69%	1.66%

For Electrical components, the possibility to not use IPPM strategy selected for a component that only needs PPM strategy is an improve of the global maintenance strategy for the machine. See maximum, minimum and average compared values of efficiency and availability in electrical components.

In the case of Electronic and Mechanical components, the IPPM application improves the efficiency and availability values; this is also demonstrated in Pneumatic components, with minor improvement values in efficiency.

The application of n dimensional matrix can reduce maintenance costs. In this machine, the application of different strategies for each component supposes a change in the number of components that require stock due to their strategy adopted. Table 15 shows the maintenance strategy proposed for n-matrix. Table 17 shows the number of each component used in the MSTM and then compares the number of those that need stock depending on their maintenance strategy adopted, PPM, IPPM and n-matrix. In the outlined comparison data, the authors use two scenarios with n-matrix:

- Scenario one. The denominated "PPM to IPPM Interval" used in Table 15 is declined to PPM strategy, so the components do not need stock;
- Scenario two. The denominated "PPM to IPPM Interval" used in Table 15 is declined to IPPM strategy, so the components need stock.

Table 17. Comparing the number of type of components that require stock due to the maintenance strategy adopted.

Number of Type Component	Component	Number of Components Who Require Stock due to Their Maintenance Strategy Adopted			
		PPM Strategy	IPPM Strategy	N-Matrix (Scenario One)	N-Matrix (Scenario Two)
1	Master power switch	0	1	0	0
1	PLC	0	1	1	1
1	HMI	0	1	1	1
1	Chromatic sensor	0	1	1	1
3	Plug-in relay	0	3	0	0
1	Command and signalling	0	1	0	0
1	Safety limit switch	0	1	0	0
1	Safety relay	0	1	0	0
1	Safety button	0	1	0	0
4	Temperature controller	0	4	4	4
4	Solid state relay	0	4	4	4
4	Thermal resistance	0	4	0	4
4	Thermocouple sensor	0	4	0	4
2	Frequency inverter	0	2	2	2
2	Motor Belt	0	2	2	2
8	Bronze cap	0	8	0	8
1	Linear axis	0	1	0	1
8	Linear bearing	0	8	0	8
4	Pneumatic valve	0	4	4	4
6	Pneumatic cylinder	0	6	6	6
2	Pressure sensor	0	2	2	2
1	Servo drive peristaltic pump	0	1	1	1
3	Peristaltic pump	0	3	3	3
1	Terrine cutter	0	1	0	1
1	Absolute encoder	0	1	0	1

The application of n-matrix supposes a decrease in number of components that required stock due to the maintenance strategy adopted. According to Table 17, Figure 8 shows the global number of components that require stock due to their maintenance strategy.

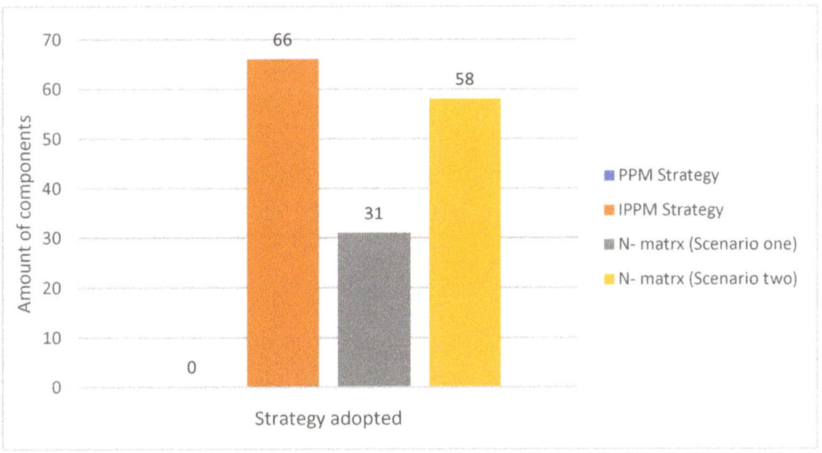

Figure 8. Comparison of the global number of components that require stock due to the maintenance strategy adopted.

Depending on the scenario adopted with n-matrix, the decrease in the number of components that require stock is between 12.12% and 53.03%. It is easy to understand that this decrease will mean decreases in maintenance costs.

Additionally, as Figure 7 shows, the efficiency and availability ratios are maintained in higher values. For more understanding, the authors consider relevant the analysis of efficiency and availability values with the application of PPM, IPPM and n-dimensional matrix strategies as follows:

- The comparison of PPM and n-dimensional matrix strategies, shown in Figure 7, indicates an improve of 1.65% in efficiency and 1.6% in availability.
- The comparison of IPPM and n-dimensional matrix strategies indicates a decrease of 0.51% in efficiency and 0.48% in availability.
- The comparison of PPM and IPPM shows the higher average values, situated at 2.16% in efficiency and 2.08% in availability.

So, the decrease in values of efficiency and availability caused by the application of n-matrix compared with IPPM strategies allows for applying the n-dimensional matrix for the optimal strategy for each component.

The authors consider that the results shown in Table 14 indicate that the KPI_1 and KPI_2, defined in Section 3.5.1, suggest a precise cost study for the best decision making. This study can offer the negative impact of a component failure due to its main time to repair and line restart time, as a new dimension for the proposed matrix.

Table 15 shows a Maintenance Strategy called *PPM to IPPM Interval*. For real case, a strategy PPM or IPPM must be selected for this component, so the authors consider important the cost analysis for this decision.

6. Conclusions

The preventive maintenance strategy for all the components of a multistage machine depends on the individual maintenance times and depends on the operation condition of each type of component. In this way, the definition of the Global Operation Condition (GOCi.) for each component allows for the study of the optimal preventive maintenance strategy used, PPM or IPPM.

The authors consider the application of the methodology used in this research relevant, step by step, for other industrial multistage or single machines, due to the fact that multistage and single machines need an appropriate preventive maintenance strategy for all their components.

Obtaining an n-dimensional matrix to select the best preventive maintenance strategy by type of component allows for maintaining a higher values of efficiency and availability for type of components. A minor decrease compared with IPPM strategy (see Table 16) is offset by the decrease on stock cost. The end users of the industrial multistage or single machines always need information and procedures to applying the appropriate maintenance strategy, so this contribution allows them to fulfill that need.

The location of a component in the machine allows for knowing the Global Operation Condition (GOCi) depending on the individual maintenance times and makes it possible to find the same type of components with different preventive maintenance strategies proposed by the n-dimensional matrix. In this way, the results shown and discussed in this research can be interesting for the industrial machinery manufacturer, by the preliminary study of the optimal operation condition for each component of the machine.

For the preventive maintenance strategy, it is necessary to study the individual maintenance times as it shown in Section 3.3. Additionally, the analysis of the individual values of TTPR, TLP, MTTR for each component allows for calculating the KPI's used by the n-dimensional matrix. These KPIs are different, as shown in EN 15341:2007, but allow new results.

For a more precise decision with the *PPM to IPPM Interval* maintenance strategy proposed by the n-dimensional matrix (see Table 15), it will be interesting to study the cost of all the components as a new dimension of the matrix.

The authors consider the following future research:

With the applied methodology of this research and given the recent increase in materials costs, it is suggested to study the analysis of the impact and variation in the cost of the components to decide the best preventive maintenance strategy, using a new dimension n-dimensional matrix.

Selective study of the suitability of components with high TTPR in multistage machines significantly influences the efficiency and availability of the industrial multistage machine. This study can determinate the maximum TTPR to maintain the adequate preventive maintenance strategy and then suggest the possible change of the component for another with minor TTPR.

Application of n-dimensional matrix in other multistage industrial machines. Results and comparison of the same used methodology in this research.

Adding predictive maintenance strategy in n-dimensional matrix for undefined interval PPM and IPPM maintenance strategies.

Author Contributions: Conceptualisation, F.J.Á.G. and D.R.S.; methodology, F.J.Á.G.; validation, F.J.Á.G. and D.R.S.; formal analysis, F.J.Á.G.; investigation, F.J.Á.G. and D.R.S.; resources, F.J.Á.G.; writing-original draft preparation, F.J.Á.G.; writing-review and editing, F.J.Á.G. and D.R.S.; visualisation, F.J.Á.G.; supervision F.J.Á.G.; project administration, F.J.Á.G. and D.R.S.; funding acquisition, F.J.Á.G. and D.R.S. All authors have read and agreed to the published version of the manuscript.

Funding: This study has been carried out through the Research Project GR-18029 linked to the VI Regional Research and Innovation Plan of the Regional Government of Extremadura.

Institutional Review Board Statement: Not applicable.

Informed Consent Statement: Not applicable.

Data Availability Statement: Not applicable.

Acknowledgments: The authors wish to thank the European Regional Development Fund "Una manera de hacer Europa" for their support towards this research. This study has been carried out through the Research Project GR-21098 linked to the VI Regional Research and Innovation Plan of the Regional Government of Extremadura.

Conflicts of Interest: The authors declare no conflict of interest.

References

1. Hoffmann Souza, M.L.; Da Costa, C.A.; Oliveira Ramos, G.D.; Da Rosa Righi, R. A survey on decision-making based on system reliability in the context of Industry 4.0. *J. Manuf. Syst.* **2020**, *56*, 133–156. [CrossRef]
2. Colledani, M.; Tolio, T. Integrated quality, production logistics and maintenance analysis of multi-stage asynchronous manufacturing systems with degrading machines. *CIRP Ann. Manuf. Technol.* **2012**, *61*, 455–458. [CrossRef]
3. Colledani, M.; Angius, A.; Yemane, A. Impact of condition-based maintenance policies on the service level of multi-stage manufacturing systems. *Control Eng. Pract.* **2018**, *76*, 65–78. [CrossRef]
4. Wu, C.-H.; Yao, Y.-C.; Dauzère-Pérès, S.; Yu, C.-J. Dynamic dispatching and preventive maintenance for parallel machines with dispatching-dependent deterioration. *Comput. Oper. Res.* **2020**, *113*, 104779. [CrossRef]
5. Colledani, M.; Magnanini, M.C.; Tolio, T. Imapct of opportunistic maintenance on manufacturing system performance. *Manuf. Technol.* **2018**, *67*, 499–502. [CrossRef]
6. Colledani, M.; Angius, A.; Silipo, L.; Yemane, A. Impact of Preventive Maintenance on the Service Level of Multi-stage Manufacturing Systems with Degrading Machines. *IFAC PapersOnLine* **2016**, *49*, 568–573. [CrossRef]
7. Lu, B.; Zhou, X. Opportunistic preventive maintenance scheduling for serial-parallel multistage manufacturing systems with multiple streams of deterioration. *Reliab. Eng. Syst. Saf.* **2017**, *168*, 116–127. [CrossRef]
8. Azimpoor, S.; Taghipour, S.; Farmanesh, B.; Sharifi, M. Joint Planning of Production and Inspection od parallel Machines with two-phase of Failure. *Reliab. Eng. Syst. Saf.* **2022**, *217*, 108097. [CrossRef]
9. Ruiz-Hernández, D.; Pinar-Pérez, J.M.; Delgado-Gómez, D. Multi-machine preventive maintenance scheduling with imperfect interventions: A restless bandit approach. *Comput. Oper. Res.* **2020**, *119*, 104927. [CrossRef]
10. Wang, J.; Han, H.; Zhang, Y.; Bai, G. Modelling the varying effects of shocks for a multi-stage degradation process. *Reliab. Eng. Syst. Saf.* **2021**, *215*, 107925. [CrossRef]
11. Farahani, A.; Tohidi, H. Integrated optimization of quality and maintenance. A literature review. *Comput. Ind. Eng.* **2021**, *151*, 16924. [CrossRef]
12. Resaei-Malek, M.; Siadat, A.; Dantan, J.Y.; Tavakkoli-Moghaddam, R. An approximation Approach for a Integrated Part Quality Inspection and Preventive maintenance Planning in a Nonlinear Deteriorating Serial Multi-stage Manufacturing System. *IFAC PapersOnLine* **2018**, *51*, 270–275. [CrossRef]
13. He, Y.; Liu, F.; Cui, J.; Han, X.; Zhao, Y.; Chen, Z.; Zhou, D.; Zhang, A. Reliability-oriented design of integrated model of preventive maintenance and quality control policy with time-between-events control chart. *Comput. Ind. Eng.* **2019**, *129*, 228–238. [CrossRef]

14. Wang, K.; Yin, Y.; Du, S.; Xi, L. Variation management key control characteristics in multistage machining processes considering quality-cost equilibrium. *J. Manuf. Syst.* **2021**, *59*, 441–452. [CrossRef]
15. Bouslah, B.; Gharbi, A.; Pellerin, R. Joint production, quality and maintenance control of two-machine line subject to operation-dependent and quality-dependent failures. *Int. J. Prod. Econ.* **2018**, *195*, 210–226. [CrossRef]
16. Lu, B.; Zhou, X. Quality and reliability oriented maintenance for multistage manufacturing systems to condition monitoring. *J. Manuf. Syst.* **2019**, *52*, 76–85. [CrossRef]
17. Huang, Z.; Zheng, Q.P. A multistage stochastic programming approach for preventive maintenance scheduling of GENCOs with natural gas contract. *Eur. J. Oper. Res.* **2020**, *287*, 1036–1051. [CrossRef]
18. Cao, Y. Modeling the effects of dependence between competing failure processes on the condition-base preventive maintenance policy. *Appl. Math. Model.* **2021**, *99*, 400–417. [CrossRef]
19. Ye, Y.; Grossmann, I.E.; Pinto, J.M.; Ramaswamy, S. Integrated optimization of design, storage sizing, and maintenance policy as a Markov decision process considering varying failure rates. *Comput. Chem. Eng.* **2020**, *142*, 107052. [CrossRef]
20. Cheng, Z.; Qing, C.; Arinez, J. Data-Enabled Modelling and Analysis of Multistage Manufacturing Systems with Quality Rework Loops. *J. Manuf. Syst.* **2020**, *56*, 573–584. [CrossRef]
21. Zhang, Z.; Tang, Q. Integrating flexible preventive maintenance activities into two-stage assembly flow shop scheduling with multiple assembly machines. *Comput. Ind. Eng.* **2021**, *159*, 107493. [CrossRef]
22. Yu, T.-S.; Han, J.-H. Scheduling proporcionate flow shops with preventive machine maintenance. *Int. J. Prod. Econ.* **2021**, *231*, 107874. [CrossRef]
23. Ghaleb, M.; Taghipour, S.; Sharifi, M.; Zolfagharinia, H. Integrated production and maintenance scheduling for a single machine with deterioration-based failures. *Comput. Ind. Eng.* **2020**, *143*, 16432. [CrossRef]
24. Duffuaa, S.; Kolus, A.; Al-Turki, U.; El-Khalifa, A. An integrated model of production scheduling, maintenance and quality for a single machine. *Comput. Ind. Eng.* **2020**, *142*, 106239. [CrossRef]
25. Panagiotis, H.T.; Atvanitoyannis, I.S.; Varzakas, T.H. Reliability and maintainability analysis of cheese (feta) production line in a Greek medium-size company: A case Study. *J. Food Eng.* **2009**, *94*, 233–240. [CrossRef]
26. Rahmati, S.H.A.; Ahmadi, A.; Karimi, B. Developing simulation-based optimization mechanism for a novel stochastic reliability centered maintenance problem. *Trans. E Ind. Eng.* **2018**, *25*, 2788–2806. [CrossRef]
27. Niu, G.-C.; Wang, Y.; Hu, Z.; Zhao, Q.; Hu, D.-M. Application of AHP and EJE in reliability Analysis of Complex production Lines Systems. *Math. Probl. Eng.* **2019**, *2019*, 7238785. [CrossRef]
28. Jiři, D.; Tuhý, T.; Jančíková, Z.K. Method for optimizing maintenance location within the industrial plant. *Int. Sci. J. Logist.* **2019**, *6*, 55–62. [CrossRef]
29. Liberopoulos, G.; Tsarouhas, P. Reliability analysis of an automated pizza production line. *J. Food Eng.* **2005**, *69*, 79–96. [CrossRef]
30. Wang, N.; Ren, S.; Liu, Y.; Yang, M.; Wang, J.; Huisingh, D. An active preventive maintenance approach of complex equipment based on novel product-service system operation mode. *J. Clean. Prod.* **2020**, *177*, 123365. [CrossRef]
31. Gharbi, A.; Kenne, J.-P.; Beit, M. Optimal safety stocks and preventive maintenance periods in unreliable manufacturing systems. *Int. J. Prod. Econ.* **2007**, *107*, 422–434. [CrossRef]
32. Ait El Cadi, A.; Gahrbi, A.; Dhouib, K.; Artiba, A. Joint production and preventive maintenance controls for unreliable and imperfect manufacturing systems. *J. Manuf. Syst.* **2021**, *58*, 263–279. [CrossRef]
33. Yang, H.; Li, W.; Wang, B. Joint optimization of preventive maintenance and production scheduling for multi-stage production systems based on reinforcement learning. *Reliab. Eng. Syst. Saf.* **2021**, *214*, 107713. [CrossRef]
34. García, F.J.Á.; Salgado, D.R. Maintenance Strategies for Industrial Multi-Stage Machines: The study of a Thermoforming Machine. *Sensors* **2021**, *21*, 6809. [CrossRef] [PubMed]
35. Ferreira, S.; Silva, F.J.G.; Casais, R.B.; Pereira, M.T.; Ferreira, L.P. KPI development and obsolescence management in industrial maintenance. *Procedia Manuf.* **2019**, *38*, 1427–1435. [CrossRef]
36. Webstore International Electrotechnical Commission. Available online: https://webstore.iec.ch/publication/64485 (accessed on 21 April 2021).

 machines

MDPI

Article

Optimizing System Reliability in Additive Manufacturing Using Physics-Informed Machine Learning

Sören Wenzel *, Elena Slomski-Vetter and Tobias Melz

Research Group System Reliability, Adaptive Structures, and Machine Acoustics SAM, Technical University of Darmstadt, Otto-Berndt-Str. 2, 64287 Darmstadt, Germany; elena.slomski-vetter@sam.tu-darmstadt.de (E.S.-V.); tobias.melz@sam.tu-darmstadt.de (T.M.)

* Correspondence: soeren.wenzel@tu-darmstadt.de

Abstract: Fused filament fabrication (FFF), an additive manufacturing process, is an emerging technology with issues in the uncertainty of mechanical properties and quality of printed parts. The consideration of all main and interaction effects when changing print parameters is not efficiently feasible, due to existing stochastic dependencies. To address this issue, a machine learning method is developed to increase reliability by optimizing input parameters and predicting system responses. A structure of artificial neural networks (ANN) is proposed that predicts a system response based on input parameters and observations of the system and similar systems. In this way, significant input parameters for a reliable system can be determined. The ANN structure is part of physics-informed machine learning and is pretrained with domain knowledge (DK) to require fewer observations for full training. This includes theoretical knowledge of idealized systems and measured data. New predictions for a system response can be made without retraining but by using further observations from the predicted system. Therefore, the predictions are available in real time, which is a precondition for the use in industrial environments. Finally, the application of the developed method to print bed adhesion in FFF and the increase in system reliability are discussed and evaluated.

Keywords: reliability optimization; physics-informed machine learning; recurrent neural network; knowledge transfer; additive manufacturing; Latin hypercube sampling

Citation: Wenzel, S.; Slomski-Vetter, E.; Melz, T. Optimizing System Reliability in Additive Manufacturing Using Physics-Informed Machine Learning. *Machines* **2022**, *10*, 525. https://doi.org/10.3390/machines10070525

Academic Editors: Sven Matthiesen and Thomas Gwosch

Received: 30 March 2022
Accepted: 24 June 2022
Published: 29 June 2022

1. Introduction

The production of plastic parts by means of fused filament fabrication (FFF) is on its way to becoming established for mass production [1]. The major advantage of mass-produced products from FFF over conventional manufacturing is the greater variability and individuality of the products as well as lower cost for smaller production quantities [1,2]. Furthermore, FFF is an inexpensive and widely used additive manufacturing process [3]. In FFF, a polymer is heated until it reaches a semi-fluid state. Then it is squeezed out of a nozzle, cools down, and becomes solid shortly after. In this way, products are created layer by layer with a 3D printer. The process depends on print parameters that determine, among other things, the number and properties of the layers, speeds, and temperatures. In order to be able to use the FFF process economically in mass production, the machine costs and personnel costs must be reduced [4] as well as the uncertainty in the product quality such as aesthetics, dimensional accuracy, and mechanical properties [5]. Research is conducted to control the uncertainty by using simulations of the whole printing process [6] and by optimizing the printing parameters [7] among others [5]. Three-dimensional printing is a complex process with a minimum of 75 printing parameters and more to be optimized simultaneously. Physically informed machine learning (PIML) combines machine learning using data and physical knowledge in the form of models or constraints to reduce errors of machine learning and physical models [8,9]. PIML has shown promising results in high-dimensional contexts [9]. Through integration of mathematical physics models into machine learning fewer data are needed for the training of the neural network [10].

The method developed in this paper differs from the literature mentioned above by deriving empirical models from domain knowledge (DK), which can be in the form of research results or other sources. These models are then combined with data collected from 3D printers. System reliability is ensured by individually optimizing the input parameters for each 3D printer, paying attention to the disturbance variables.

The summarized contributions of this paper are as follows:

1. Development of a new method using physics-informed machine learning to optimize system reliability in additive manufacturing.
2. Reliability optimization is implemented by using neural networks to predict the system responses and an optimization algorithm with respect to system specific boundary conditions.
3. System behavior of 3D printers is quantified by unsupervised machine learning and is used for predictions of system responses for the knowledge transfer.
4. The new method is designed to be used for a number of 3D printers in an industrial environment in continuous operation and to provide calculation proposals for input parameters in real time.

2. Methods

This section gives an overview of the techniques used by the proposed method for reliability optimization. These methods include Latin hypercube sampling, deep neural networks, recurrent neural networks, which have been widely used and closely studied in the scientific field [11], and PIML, which yields promising results.

2.1. Latin Hypercube Sampling (LHS)

The influence that a parameter has on a target value is called the total effect, which is composed of interaction effects and the main effect. The main effect is the influence of the parameter that depends only on the parameter itself. Interaction effects also depend on other parameters. The order of an interaction indicates how many other parameters the influence depends on. To conduct experiments, the research question must be clarified to determine what type of effects should be identified. For the proposed method, detailed knowledge of the effects is required, and therefore the goal of knowing main and interaction effects is pursued. This is achieved using stochastic simulation. In Latin hypercube sampling (LHS), each parameter is divided into equally sized intervals, with the number of intervals corresponding to the number of experiments performed. Then, the experiments are distributed uniformly among the intervals [12]. This method was extended to generate approximately orthogonal, i.e., particularly balanced, experimental designs [13]. Currently, the method is accepted and widely used, although the number of experiments cannot be expanded without changing the overall number of samples. This method was chosen because it does create a pseudo-random experimental design, but it is also approximately orthogonal. These properties are important for quantifying the behavior of 3D printers, as effects can be determined independently of neural networks.

2.2. Artificial Neural Networks (ANN)

Artificial neural networks (ANN) consist of artificial neurons which in turn are composed of a weighted sum connected to an activation function. Therefore, each artificial neuron has one or more inputs and one output. ANN can be constructed from artificial neurons in layers, where a neuron has all the outputs of the neurons of the previous layer as inputs. This setup is called multilayer perceptron (MLP). The value of the neurons in the first layer of the MLP, which is called the input layer, are the inputs of the MLP. The last layer is called the output layer, as the output from the artificial neurons in this layer are the outputs of the total MLP. The layers in between are called hidden layers; see Figure 1 for a graphical representation of an MLP. An MLP with multiple hidden layers is called a deep neural network (DNN) [14].

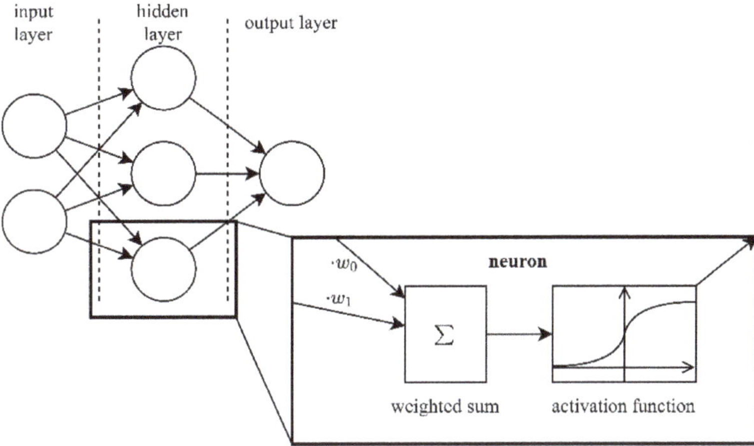

Figure 1. A multilayer perceptron (MLP) consists of neurons organized in layers. Each neuron is composed of a weighted sum of all inputs and an activation function to define the output.

During the training, the weights of the artificial neurons are altered in a way such that the input from the DNN matches the desired output. This is performed by means of an optimization algorithm. Consider a DNN A_1 that is trained by a set of input vectors $\boldsymbol{X} = \{X_1, \ldots, X_n\}$, $X_i \in \mathbb{R}^m$ and a set of output vectors $\boldsymbol{Y} = \{Y_1, \ldots, Y_n\}$, $Y_i \in \mathbb{R}^k$, where each output is assigned to one input $Y_i \to X_i$. The inputs and outputs are combined in the training data $D_{\text{train}}(\boldsymbol{X}, \boldsymbol{Y})$. If A_1 is given an X_i as an input, A_1 can be called to estimate $\hat{Y}_i$, which can be written as $A_1(X_i) = \hat{Y}_i$. To quantify how well the estimations are matching with the desired outputs, a loss function $\mathcal{L}$ is used, e.g., the root mean square error (RMSE), which is defined as

$$\mathcal{L}_{RMSE}(\boldsymbol{Y}, \hat{\boldsymbol{Y}}) = \sqrt{\sum \frac{(Y_i - \hat{Y}_i)^2}{n}} \tag{1}$$

For the initialization, random values are assigned to the weights from the weighted sum in all neurons. Consider multiple DNNs A_j using the same architecture and a random, and thus different, initialization. The training is performed using the same training data $D_{\text{train}}(\boldsymbol{X}, \boldsymbol{Y})$, but the resulting A_j differ. While after a successful training the loss function is minimized and $Y_i \approx \hat{Y}_i$ can be assumed for all A_j, different A_j are explained by using a gradient decent optimization algorithm for the training, which minimizes the loss function and can only find local optima. Therefore, different optima are found [15,16]. ANNs show substantial differences in weights in intermediate and higher-level networks with more than six layers, despite similar performance [17].

A DNN with a special architecture can be used as an encoder–decoder for lossy compression of data. The encoder–decoder (ED) has a small central layer and is trained to reconstruct input data $\boldsymbol{X}$ as output $\boldsymbol{X} = ED(\boldsymbol{X})$, as shown in Figure 2. The values of the small central layer are called compressed feature vector (CFV) $\boldsymbol{Y}$. The CFV is significantly smaller than the input and output. The size of the CFV is a main factor on how well the output can be recreated from the CFV [18]. The encoder–decoder can be split into two DNNs, the encoder E and the decoder D. E uses data $\boldsymbol{X}$ to generate CFV $\boldsymbol{Y} = E(\boldsymbol{X})$ and D uses CFV to estimate data $\hat{\boldsymbol{X}} = D(\boldsymbol{Y})$ [19].

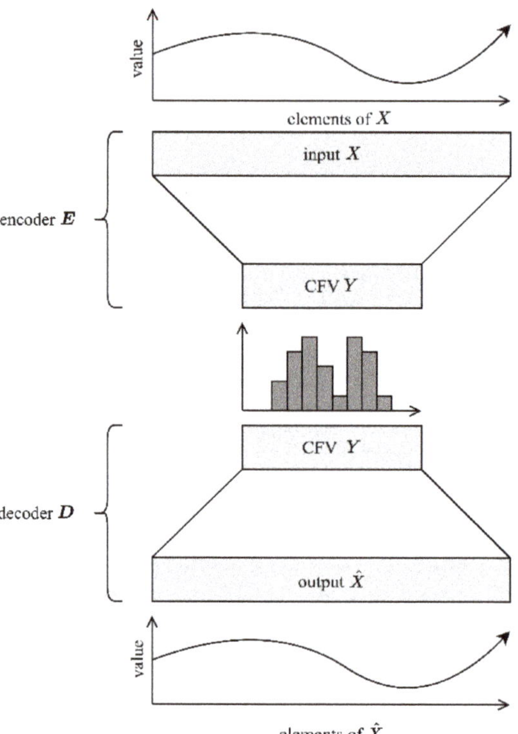

Figure 2. The encoder–decoder (ED) consists of two neural networks trained in unity to output $\hat{X}$ from X while information must pass the compressed feature vector (CFV).

The ED is chosen for the proposed method because it contains a way of unsupervised learning using algorithms for supervised learning. This is intended to allow for ease of connection with the other ANNs that use supervised learning.

2.3. *Recurrent Neural Networks (RNN)*

Recurrent neural networks (RNN) are ANNs that can use information from previous calls. This makes it possible for RNNs to analyze data with respect to past inputs and makes estimations dependent on previous calls. A simple implementation uses an artificial neuron, where the output from the previous call is going to be an additional input for the next estimation. With this implementation, the problem of vanishing gradients occurs. This means that the output from the previous call has to be mostly defining the state of the next call, if the neuron needs to keep its state over a larger number of calls [20]. A type of cell which can solve this problem is the long short-term memory (LSTM) [21]. Here, the cell is replaced by multiple cells with well-defined tasks. One of the cells is the memory cell, where the output is an input for next call with a fixed weight of one. Therefore, the problem of vanishing gradients cannot occur, as the state is transported over various time steps without vanishing or growing exponentially [20]. This property predestines LSTM for the proposed method, since patterns found in the observational data should not be lost due to vanishing gradients.

2.4. *Physics-Informed Machine Learning (PIML)*

The amount of data needed to train a neural network can be reduced by including physical knowledge in machine learning. This physical knowledge includes laws or observations typically found through extensive scientific work in the form of experimentation, modeling, and other measures. This knowledge is called domain knowledge (DK) [22]. The more DK that is used, the fewer the data that are needed to train a neural network [10]. Several possibilities to include DK into neural networks exist. Physical laws can be added to the loss function as an extra term and can therefore penalize unphysical calls during training, called physics-guided neural networks (PGNN) [23]. Kapusuzoglu et al. [10] compared different approaches of PIML in the context of 3D printing. Based on this comparison, the approach shown in Figure 3 was selected for the developed method in this paper. In this approach, neural networks are pretrained with estimates based on DK of inputs and outputs. The network is pretrained to gain knowledge of how certain input parameters must be combined to obtain the desired results. Physical laws are not enforced in further training and are only an initialization for the neural network to learn fundamental correlations [10]. This feature is important to adapt neural networks to a wide range of 3D printers.

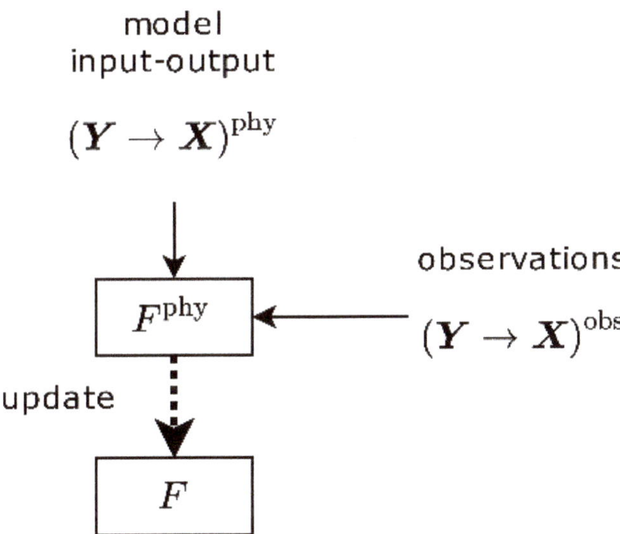

Figure 3. Physical knowledge can be included in machine learning by pretraining a neural network with input and output from a model, based on Figure 4c (Kapusuzoglu et al. [10]).

3. Developed Method

In this paper, the developed method aims at optimizing system reliability with changing requirements by predicting outcome and suggesting input variables. Reliability can be ensured without further training by transferring and enriching DK on conducted observations, though neural networks can only generalize from data that were used to train the neural networks. Predictions from neural networks can only be as informed as DK and observations are.

3.1. Prediction Versus Suggestion

Consider a system S_a which uses a vector of input parameters X to generate a set of target values $Y = S_a(X)$. Essentially, for a reliable operation, it is important that target values Y must be achieved within certain limits. Hence, for an explicit system S_a, an explicit vector X that achieves this target value Y needs to be identified. This could be achieved directly by a neural network suggesting X. As in most cases some parameters in X cannot be set freely, the presented method is developed to solve this problem. To identify an explicit vector X, an optimization algorithm is better suited as it can satisfy constraints in the form of mathematical equations and optimize the target values, see Figure 4. Thus, the aim of this work is to generate predictions $\hat{Y}$ for the system S_a and freely chosen $\boldsymbol{X}$ that can be used in an optimization algorithm to identify an explicit X with a desired estimation for $\hat{Y}$. By using this method, target values and, thus, system reliability can be optimized. Since constraints depend on the application, further considerations can be found in Section 4, where the application to additive manufacturing is discussed.

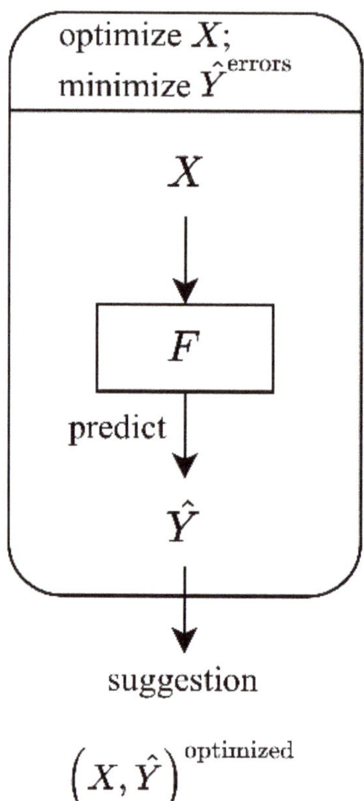

Figure 4. Suggestions are generated using an optimization algorithm with a neural network predicting model output $\hat{Y}$.

3.2. *Predictions of Target Values*

For a system S_a, all observations of S_a should have a direct influence on the predictions $\hat{Y}$. This is ensured using a recurrent neural network R that creates an estimation on the general reactions of S_a based on all observations $\left(\mathbf{Y}^{\text{obs}} \rightarrow \mathbf{X}\right)_a$. In this paper, this is called behavior and is quantified in the behavioral vector B_a, which is estimated by calling the RNN R as follows:

$$\hat{B}_a = R\left(\mathbf{X}, \mathbf{Y}^{\text{obs}}\right). \tag{2}$$

After that, a feed-forward DNN F is called to generate an estimation of the target values:

$$\hat{Y} = F(X, \hat{B}_a). \tag{3}$$

To calculate estimations for different X, only the network F is used, while $\hat{B}_a$ stays the same until new observations $\mathbf{Y}^{\text{obs}} \rightarrow \mathbf{X}$ are made. The whole process of predictions is shown in Figure 5.

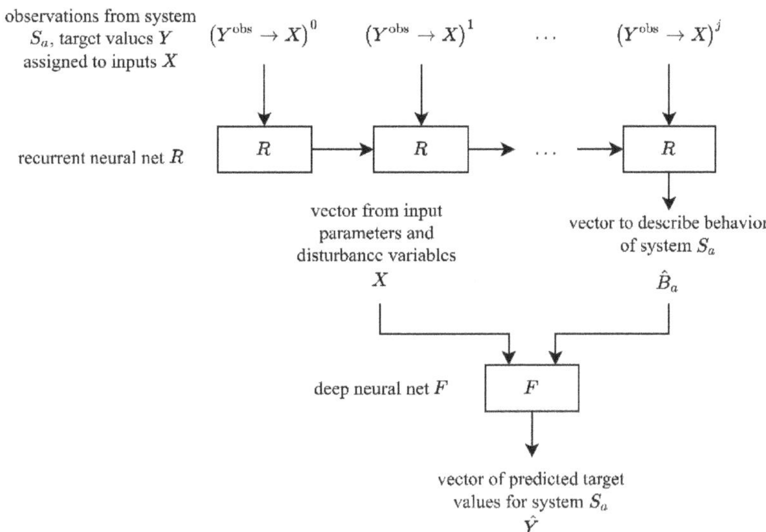

Figure 5. Predictions are made using observations Y^{obs} and a vector X of inputs.

3.3. Behavioral Vector

To define the behavioral vectors B for various known systems, a DNN encoder–decoder is used. The encoder–decoder ED is trained using predictions for the target value $\hat{Y}^{LHC}$ that is estimated using the DNN network F and vectors X^{LHC} that are, respectively, derived from Latin hypercube sampling:

$$\hat{Y}^{LHC} = F\left(X^{LHC}, \hat{B}_a\right) \quad (4)$$

For the training of ED, the prediction for the target values $\hat{Y}^{LHC}$ are used as an input and output. ED encodes the input to a CFV and decodes it afterwards. The CFV is defined as the behavioral vector B, because the CFV contains the necessary information to recover the prediction $\hat{Y}^{LHC}$. After training the DNN ED is split in the encoder DNN E and the decoder DNN D to calculate B.

$$B = E\left(\hat{Y}^{LHC}\right) \quad (5)$$

$$\hat{Y}^{LHC} = D(B) \quad (6)$$

The complete usage of the encoder–decoder ED is illustrated in Figure 6.

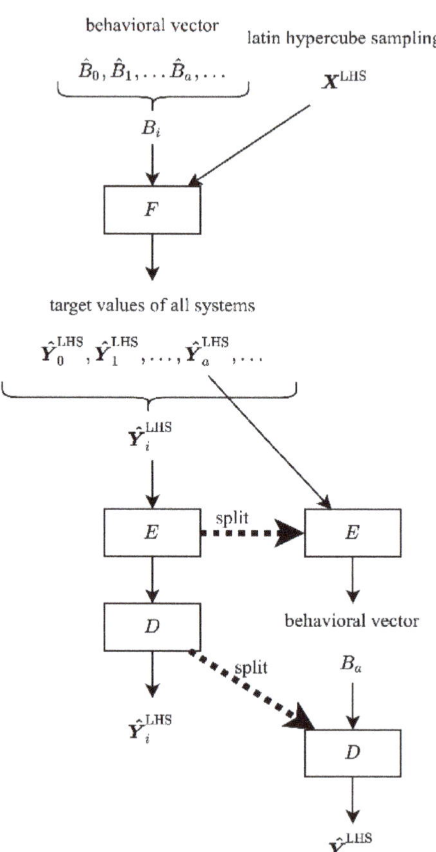

Figure 6. Usage of the encoder–decoder ED. The behavioral vector B is defined from the output of the encoder from Latin hypercube (LHC) sampled target values $\hat{Y}^{\text{LHC}}$.

4. Method Transfer

The developed method is designed to ensure the system reliability in operating 3D printers used for mass production. In this case, disturbance variables, materials, and optimization targets are constantly changing, while in an industrial environment, high demands on quality are placed [24]. In the context of this work, print defects in additive manufacturing mean that the printing process must be canceled, mechanical properties are not met, dimensional accuracy could not be kept, or optical defects are present in the 3D-printed product [1]. Due to these print defects, the printed parts must be reprinted.

The influence of a single parameter on an outcome is called the total effect. The total effect consists of interaction effects and the main effect. The main effect is the influence of the parameter, which depends only on the parameter itself. Interaction effects are also dependent on other parameters. The order of an interaction effect indicates how many other parameters are involved.

In practice, there are guidelines (manufacturer´s instructions) to avoid print defects by changing printing parameters. One larger collection of print defects by Richter [25] names more than 200 total effects of 18 input parameters on 40 print defects while mechanical properties as print defects are not considered. All of these total effects are documented to be monotonic, and the general direction of the changing outcomes of the effects is claimed to be applicable to all FFF 3D printers, regardless of the printing material and printer design. According to Richter, some geometric features seem to have effects on print defects.

Scientific literature also names interaction effects [26] and non-monotonic effects [27] for 3D printers. The summarized results of those studies show that the avoidance of print defects could lead to opposing goals, such as surface roughness and dimensional accuracy, i.e., [26]. In consequence, the reliable operation of 3D printers is a complex task, as in the best case, all effects must be known in order to counteract print defects and to ensure high-quality parts.

As there are presumably more than 200 total effects on more than 40 print defects, an intelligent, automated process must be developed to adjust the print parameters just in time. This should enable the printing system to perform more efficiently concerning time and costs than is the case so far, and ultimately ensures the system reliability. The effects may be non-monotonic, so a simple linear regression is insufficient. Interaction effects make statistical investigation labor-intensive, as rapid screening methods are not sufficient, as they can only be used to determine total effects. The method developed and explained in the previous sections highly encourages knowledge transfer among systems, which is convenient given that the effects are comparable among 3D printers. By using an optimization algorithm, it is ensured that opposing goals could be weighted and individually adjusted to the printed part. The results of the method transfer to 3D-printing are shown in the following.

4.1. Application and Results

To apply the developed method on FFF 3D printers, the general system must be defined. Three-dimensional printers are considered individual systems, and with a substantial change of the 3D printer, a new system is formed. This can be a modification such as a changed print bed or print nozzle. Each system requires printing parameters, printer parameters, and disturbance variables X to produce the outcomes Y. Print parameters are controllable parameters for the print, which can be optimized. Disturbance variables contain all parameters that can be measured or determined, such as ambient temperature or print time since the last maintenance, e.g., printer parameters are specific for the printer, such as, for example, the printer model or the nozzle diameter. Product parameters are, for example, geometric features of the printed product, such as the maximum overhang angle and the contact surface of the product on the print bed. The outcome contains all print defects and some other metrics, such as production time and used material for the final product. Overall, 75 printing parameters, 4 disturbance variables, 8 printer parameters, 9 product parameters, 40 print defects, and 4 print metrics are recorded for each observation of a print process.

The possible occurrence of certain printing defects related to geometry can be derived from the product parameters of the printed product. Combined with known customer requirements, the required outcome of a successful print and the optimization criterion can be created. This criterion is used for the optimization to find print parameters for a successful print and thus a reliable operation of the printer.

4.2. Use of Domain Knowledge (DK)

This paper addresses a well-known and crucial problem, i.e., the print bed adhesion [21], to show the method transfer exemplarily. If the print bed adhesion could be controlled, the printed part can be easily detached from the print bed after the print. This is crucial in order to enable an automation of the printing process and to make this technology more available for mass production.

Following the general initialization process explained in Section 2.4, DK must be defined based on scientific research on the subject. This is performed via various measurements and datasets of the print bed adhesion taken from scientific literature. Two publications suitable for this application are particularly noteworthy. The first publication is written by Kujawa [28], who investigated the influence of the first layer print parameters on the print bed adhesion. The second publication is written by Spoerk et al. [27], who optimized the print bed adhesion by varying print parameters. The problem with using literature data is that not all print parameters and disturbance variables are known for these measurements, so no complete observations (see Figure 7) can be created. In the next paragraphs it will be explained how DK can still be taught to neural networks.

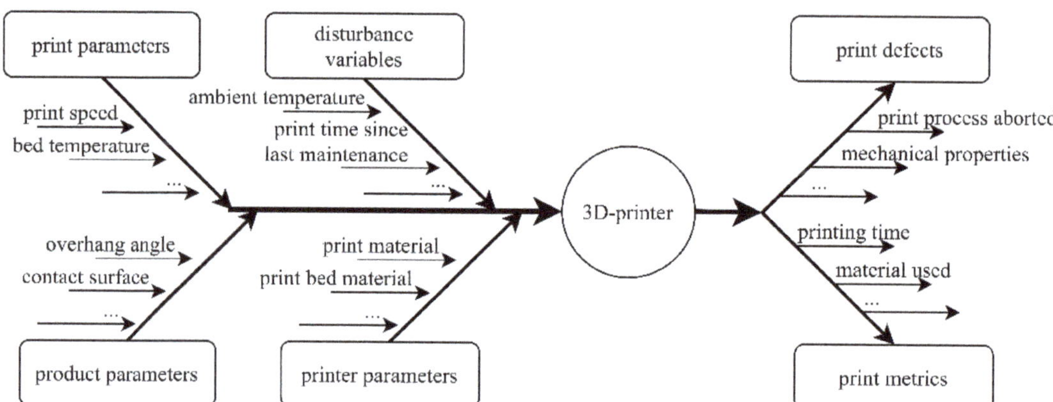

Figure 7. Extended Ishikawa diagram visualizing all relevant input and output parameters for the proposed method from a 3D-printed product in FFF.

In the first publication, by Kujawa [28], only total effects were measured in a one-factor-at-a-time design of experiments. Measurements were repeated, and usable data are given in the publication. To use the data in the developed method, synthetic observations are created from the data. This is achieved using linear interpolation between the measurements. See Figure 8a for a graphical representation.

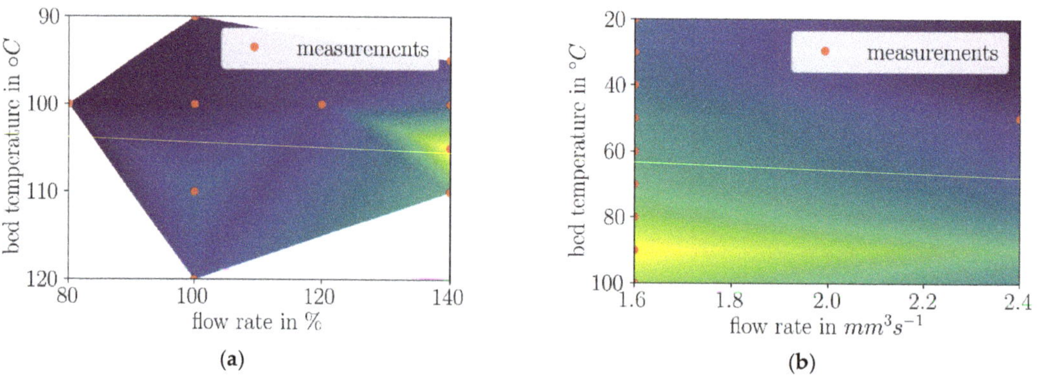

Figure 8. Simple synthetic data generation using measurements for the pretraining of the neural networks are shown: (**a**) linear interpolation using data from Kujawa [28] and (**b**) superposition using data from Spoerk et al. [27].

In the second publication, by Spoerk et al. [27], linear main effects and interaction effects were calculated for three print parameters in a full-factorial design of experiment. In a second step, the printer bed temperature was singularly investigated from 20 °C to 80 °C in steps of 10 °Celsius. For the pretraining of the ANN used in the proposed method, these two designs of experiments are combined using superposition, as can be see in Figure 8b.

The authors of both publications have chosen parameters that are expected to have the biggest effects on the printer bed adhesion. Not-investigated effects, which are to be expected, should be equal to or smaller than the investigated effects. ANNs learn about important effects with DK, resulting in fewer experimental data needed until ANNs are successfully trained [29].

In order to train the first neuronal net F^{phy} using DK, the behavioral vector B_i^{bin} must be set, which is a binary representation of the system number. In this case, it is a vector of the length 1 with values 0 and 1. Synthetic data generation and initialization is carried out as described. Afterwards, FFF 3D printers are used and the observations should be documented in the format shown in Figure 5 in Section 4.1. The observations can then be used to update F^{phy} based on DK to F. Suggestions for print parameters can be generated using a simple gradient-based optimization algorithm with the predictions from F, as described in Section 3.1 and Figure 4.

4.3. Critical Discussion

The aim of the proposed method is to ensure system reliability in operating 3D printers. Thus, systematic errors are avoided by suggesting optimal parameters. A reliable operation is strongly dependent on the suggested parameters and, as a result, strongly dependent on the accuracy of the underlying prediction. With this in mind, the accuracy of the prediction can be used as an indicator for system reliability.

There are also other approaches to ensure the reliable operation of 3D printers. These approaches are the intuitive and the statistical approach that are discussed in the following. In an intuitive or popular approach, which is widely used in the non-scientific literature, the print parameters are only changed when print defects occur. Working print parameters are stored for different materials, conditions, and 3D printers. In this approach, the total effects are used only qualitatively, although most parameters have effects on multiple defects. The result is a trial-and-error analysis that can lead to long setup times due to various print defects, since different print defects can occur after each step [25].

With the newly developed method, setup times should be shortened, because if working as intended, all print defects should be predicted and avoided simultaneously. On the contrary, for the intuitive approach only the total effects are known qualitatively, which gives only an indication of how to adjust print parameters to avoid individual print defects. A reliable operation cannot be ensured, since the known global total effect could be locally incorrect. With the developed method, on the contrary, the print parameters can be adjusted before print defects occur.

In direct comparison of the proposed method with the widely used manual prevention of print defects, the proposed method has the potential to save setup time, to automatically avoid print defects, to ensure the reliability of the printing process, and thus improve the time and cost efficiency of the printing process.

Currently, one disadvantage of the newly developed method is certainly the initialization and updating process. However, since this initialization phase only needs to be performed once, the overall result could be a more efficient and reliable process that ensures the possibility of automating 3D printing in the future.

For example, a DNN can be combined with LHS to predict print defects using statistics. The statistical study of a 3D printer is lengthy but possible, even with more than 90 input parameters and more than 40 output parameters. In comparison, the proposed method uses previously published observations and further DK to predict print defects of a single 3D printer with fewer observations. In a statistical approach, a study is conducted at a particular time and under particular conditions. Any deviation from these conditions can lead to inaccurate predictions and thus a decrease in reliability. This can only be avoided by further investigation or by transferring knowledge from a single 3D printer to all other 3D printers and printing materials. Knowledge transfer could be an essential part of the proposed method and is achieved by creating a behavioral vector from the observations of one 3D printer. This vector is used to make predictions for all 3D printers using the same neural network. As conditions change, the behavior of the 3D printer changes and the behavioral vector could be adjusted by the neural network. Thus, compared to a statistical approach, the proposed method requires fewer observations and no manual statistical analysis is required to transfer knowledge. Another advantage is that no retraining is required to derive new predictions from new observations. As a result, the reliability of 3D printing is increased because real-time knowledge transfer allows real-time response to changing conditions. As it is summarized in Table 1.

Table 1. Comparison among different approaches to specify parameters for the reliable operation of 3D printers.

	Intuitive Approach Using Qualitatively Known Total Effects	Statistical Approach	Proposed Method
Functionality	Total effects give multiple possibilities to eliminate print defects.	Predictions are based on previous observations to avoid print defects.	Predictions are based on observed and quantified behavior to avoid print defects.
Observations needed for setup	None, only DK is used.	Most, because a full sensitivity analysis is needed.	Less than the statistical approach, because DK is used.
Observations needed for reliable print	Eliminating print defects often leads to other print defects.	None.	Enough to quantify behavior.
Transfer of knowledge	Qualitatively known total effects are mostly universal.	Deviations in print process could make all observations obsolete.	Integrated through quantification of behavior.

Other approaches of mitigating print defects exist as well, such as closed-loop control [5] or ANN-based optimizing of process parameters [8,10]. These approaches have in common that they are supposed to eliminate certain or only a few defects at the same time. While these approaches might be better used to eliminate individual defects, it is unclear how these systems work together to eliminate all print defects. FFF 3D printers have a large number of parameters that influence numerous defects. The presented method aims to avoid all defects at the same time by only setting print parameters. Therefore, not all problems could be solved by the proposed method, such as mechanical problems, but it could be combined with other approaches as the 3D printers are not altered in any way.

4.4. Case Study

A case study is conducted to show the effectiveness of the proposed method. On four identical and unmodified Prusa i3 MK3s, 3D printers in a temperature- and humidity-controlled room without direct sunlight, 1273 print bed adhesion measurements are taken. Prior to printing, the print beds are cleaned using an alcohol-based solvent and a fresh paper towel every day, and each printer used one black PLA filament coil from Verbatim for all prints. For the measurements, a basic geometry is printed because our focus lies on print bed adhesion. The print bed is left to cool down to 35 °C after printing and the printed test cubes with an edge length of 10 mm are slowly pulled horizontally from the print bed using a pull arm with a worm gear driven by a stepper motor. The resulting force is recorded at 48 KHz and the maximum value is noted.

These measurements are part of an LHS experimental design, which includes 400 experiments; each comprises a minimum of three measurements, resulting in a total of 1273 print bed adhesion measurements. Within the design of experiments, 75 print parameters are varied and the print bed adhesion is recorded. Parameters which directly influence support structures, overhangs, or bridges are not chosen because of the focus on print bed adhesion. Temperature, relative humidity, prints since last print bed cleaning, and prints since last calibration are recorded as disturbance variables. The test data comprise 80 experiments with 210 measurements. The evaluation in the comparison is derived based on the prediction accuracy of the said test data with the loss-function $\mathcal{L}_{RMSE}$.

A comparison with the intuitive approach is not conducted, since it strongly depends on the person choosing print parameters and the prognosis of the printing defects is not calculated. Therefore, the proposed method is compared with a possible statistical approach from the last subsection. A similar approach is used in the statistical approach as in the proposed method for better comparison. A DNN is trained with the same data and nearly the same architecture as the DNN in the proposed method and with this comparison, the influence of the PIML approach in the proposed method can be determined due to the similarities of the two methods.

In the proposed method, the training of the encoder–decoder failed, arguably because of the low number of 3D printers in the case study, which were also very similar. As a result, the redefinition of the behavioral vector, as described in Section 3.3, was not carried out. In Figure 9, a comparison of RMSE in prognosis of the print bed adhesion is illustrated.

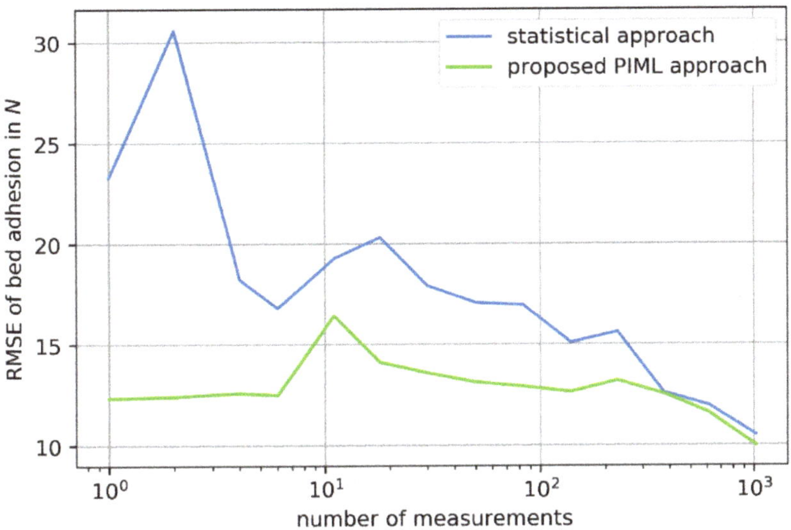

Figure 9. Comparison of prognosis error of print bed adhesion between the proposed PIML approach and a statistical approach using a similar DNN and identical training data and test data.

Around 3000 physical experiments as training data are created for pretraining for the proposed PIML approach, which are used 50 times for training. In addition to the pretraining with the physical experiments of the proposed PIML approach, both approaches used the real measurements 50 times for training. For every set of values in the figure, each approach was carried out and evaluated 10 times. Data and results can be found in the supplementary materials [30] linked below. High variance occurs in the evaluation of the statistical approach, when the number of measurements for training is low. The variance of the evaluated RMSE is mostly 50 to 100 times bigger in the statistical approach when compared to the proposed method. The variance as well as the prediction error is similar, with around 350 or more measurements available for training. The spike around 10 measurements in the proposed method is caused by some consecutive unexpectedly high measurements in the training data. In the case study, the proposed method has a lower RMSE for predicting unknown experiments while knowing only a limited amount of measurements. This indicates that transfer learning from domain knowledge is working, but with an increasing number of measurements, the effect disappears.

5. Conclusions

In this paper, a method for optimizing the system reliability of FFF 3D printers is presented. The developed method optimizes reliability by suggesting optimal print parameters using physics-informed machine learning. Print defects such as insufficient mechanical properties, dimensional accuracy, and aesthetics are avoided simultaneously by optimizing predictions from ANNs. Based on real-world experiments, a behavioral vector is determined for an individual 3D printer which is used to predict the system outcome of a 3D printer. Literature and first experiments on FFF 3D printers suggest that the total effects of the printing parameters on printing defects remain qualitatively the same for different materials and printers. This suggests that for accurate predictions, only a few observations are needed for each individual 3D printer when DK is used. In consequence, accurate predictions ensure the reliable operation of 3D printers by avoiding multiple print defects simultaneously.

For the initialization of the neural networks, DK is used. In the selected literature, effects of print parameters on print defects are investigated. The investigations are used to pretrain the neural network. This way, quantitative effects between print parameters and print defects are learned by the neural network without using real-world observations.

Different approaches for a reliable operation of multiple 3D printer are compared and discussed. The intuitive and popular approach is to use qualitative knowledge of effects to avoid print defects. This can cause the appearance of different print defects in the setup. Consequently, with slowly changing disturbance variables, print defects may occur. When the proposed method is compared to a general statistical approach, the main advantages of the new developed method are the integrated knowledge transfer among different printers and materials as well as less-needed observations by the use of DK. This is shown with the implementation of the proposed method in the form of a case study on the parameter print bed adhesion. Initial results indicate that the transferability of domain knowledge with the proposed method is given with a few experimental results, but with an increasing number of experimental results, this effect decreases.

Three-dimensional printers can behave in unforeseen ways, and to quantify such behavior, future research is investigating a synthetic data generation method using DK in neural networks. The presented approach is extended to all print defects. All significant parameters must be determined and investigated in terms of their influence on the final result. DK must be researched and models for PIML have to be derived for all print defects. It must be defined how those defects can be measured appropriately. Possible combinations with other conventional methods could be researched, such as reverse modelling of the print products and mitigating geometric inaccuracies in the product model. Further research on the effects of the parameters of the proposed approach on the prediction accuracy is needed and will be implemented. Investigations of the capabilities and limitations of the transfer of knowledge will be carried out.

Author Contributions: Conceptualization and methodology, S.W. and E.S.-V.; formal analysis, E.S.-V.; investigation, S.W.; resources, T.M.; writing—original draft preparation, S.W.; writing—review and editing, E.S.-V.; supervision, E.S.-V. and T.M.; project administration and funding acquisition, T.M. All authors have read and agreed to the published version of the manuscript.

Funding: This research received no external funding.

Institutional Review Board Statement: Not applicable.

Informed Consent Statement: Not applicable.

Data Availability Statement: The data presented in this study are openly available in https://figshare.com/articles/software/Print_bed_adhesion/19854967 accessed on 30 March 2022 at 10.6084/m9.figshare. 19854967.v1 reference number [30].

Acknowledgments: We would like to thank Ian Moses Muchungi for conducting the print bed adhesion measurements and Sophie Charlotte Lueth for her help with the programming.

Conflicts of Interest: The authors declare no conflict of interest.

References

1. Shahrubudin, N.; Lee, T.C.; Ramlan, R. An Overview on 3D Printing Technology: Technological, Materials, and Applications. *Procedia Manuf.* **2019**, *35*, 1286–1296. [CrossRef]
2. Ngo, T.D.; Kashani, A.; Imbalzano, G.; Nguyen, K.T.; Hui, D. Additive manufacturing (3D printing): A review of materials, methods, applications and challenges. *Compos. Part B Eng.* **2018**, *143*, 172–196. [CrossRef]
3. Mwema, F.M.; Akinlabi, E.T. Basics of Fused Deposition Modelling (FDM). In *Fused Deposition Modeling: Strategies for Quality Enhancement*; Mwema, F.M., Akinlabi, E.T., Eds.; Springer International Publishing: Cham, Switzerland, 2020; pp. 1–15. ISBN 978-3-030-48259-6.
4. Thomas, D. Costs, Benefits, and Adoption of Additive Manufacturing: A Supply Chain Perspective. *Int. J. Adv. Manuf. Technol.* **2016**, *85*, 1857–1876. [CrossRef] [PubMed]
5. Mercado Rivera, F.J.; Rojas Arciniegas, A.J. Additive manufacturing methods: Techniques, materials, and closed-loop control applications. *Int. J. Adv. Manuf. Technol.* **2020**, *109*, 17–31. [CrossRef]
6. Xia, H.; Lu, J.; Tryggvason, G. Simulations of fused filament fabrication using a front tracking method. *Int. J. Heat Mass Transf.* **2019**, *138*, 1310–1319. [CrossRef]
7. Kapusuzoglu, B.; Sato, M.; Mahadevan, S.; Witherell, P. Process Optimization under Uncertainty for Improving the Bond Quality of Polymer Filaments in Fused Filament Fabrication. *J. Manuf. Sci. Eng.* **2021**, *143*, 021007. [CrossRef]
8. McGowan, E.; Gawade, V.; Guo, W.G. A Physics-Informed Convolutional Neural Network with Custom Loss Functions for Porosity Prediction in Laser Metal Deposition. *Sensors* **2022**, *22*, 494. [CrossRef]
9. Karniadakis, G.E.; Kevrekidis, I.G.; Lu, L.; Perdikaris, P.; Wang, S.; Yang, L. Physics-informed machine learning. *Nat. Rev. Phys.* **2021**, *3*, 422–440. [CrossRef]
10. Kapusuzoglu, B.; Mahadevan, S. Physics-Informed and Hybrid Machine Learning in Additive Manufacturing: Application to Fused Filament Fabrication. *JOM* **2020**, *72*, 4695–4705. [CrossRef]
11. Aggarwal, C.C. *Neural Networks and Deep Learning: A Textbook*; Springer International Publishing: Cham, Switzerland, 2018; ISBN 9783319944630.
12. McKay, M.D.; Beckman, R.J.; Conover, W.J. A Comparison of Three Methods for Selecting Values of Input Variables in the Analysis of Output from a Computer Code. *Technometrics* **1979**, *21*, 239. [CrossRef]
13. Ye, K.Q. Orthogonal Column Latin Hypercubes and Their Application in Computer Experiments. *J. Am. Stat. Assoc.* **1998**, *93*, 1430. [CrossRef]
14. Aggarwal, C.C. Training Deep Neural Networks. In *Neural Networks and Deep Learning*; Aggarwal, C.C., Ed.; Springer International Publishing: Cham, Switzerland, 2018; pp. 105–167. ISBN 978-3-319-94462-3.
15. Izquierdo, J.; Crespo Márquez, A.; Uribetxebarria, J. Dynamic artificial neural network-based reliability considering operational context of assets. *Reliab. Eng. Syst. Saf.* **2019**, *188*, 483–493. [CrossRef]
16. Oparaji, U.; Sheu, R.-J.; Bankhead, M.; Austin, J.; Patelli, E. Robust artificial neural network for reliability and sensitivity analyses of complex non-linear systems. *Neural Netw.* **2017**, *96*, 80–90. [CrossRef] [PubMed]
17. Mehrer, J.; Spoerer, C.J.; Kriegeskorte, N.; Kietzmann, T.C. Individual differences among deep neural network models. *Nat. Commun.* **2020**, *11*, 5725. [CrossRef]
18. Kramer, M.A. Nonlinear principal component analysis using autoassociative neural networks. *AIChE J.* **1991**, *37*, 233–243. [CrossRef]
19. Hinton, G.E.; Salakhutdinov, R.R. Reducing the dimensionality of data with neural networks. *Science* **2006**, *313*, 504–507. [CrossRef]
20. Lipton, Z.C.; Berkowitz, J.; Elkan, C. A Critical Review of Recurrent Neural Networks for Sequence Learning. 2015. Available online: http://arxiv.org/pdf/1506.00019v4 (accessed on 30 March 2022).
21. Hochreiter, S.; Schmidhuber, J. Long short-term memory. *Neural Comput.* **1997**, *9*, 1735–1780. [CrossRef]
22. Zhao, X.; Shirvan, K.; Salko, R.K.; Guo, F. On the prediction of critical heat flux using a physics-informed machine learning-aided framework. *Appl. Therm. Eng.* **2020**, *164*, 114540. [CrossRef]
23. Daw, A.; Karpatne, A.; Watkins, W.; Read, J.; Kumar, V. Physics-Guided Neural Networks (PGNN): An Application in Lake Temperature Modeling. 2017. Available online: http://arxiv.org/pdf/1710.11431v3 (accessed on 30 March 2022).
24. Devicharan, R.; Garg, R. Optimization of the Print Quality by Controlling the Process Parameters on 3D Printing Machine. In *3D Printing and Additive Manufacturing Technologies*; Kumar, L.J., Pandey, P.M., Wimpenny, D.I., Eds.; Springer: Singapore, 2019; pp. 187–194. ISBN 978-981-13-0304-3.
25. Richter, M. SOS-Druckfehler Übersicht. Available online: https://einfach3ddruck.de/sos-druckfehler-schnell-und-effektiv-beheben/ (accessed on 3 March 2021).
26. Durão, L.F.C.S.; Barkoczy, R.; Zancul, E.; Lee Ho, L.; Bonnard, R. Optimizing additive manufacturing parameters for the fused deposition modeling technology using a design of experiments. *Prog. Addit. Manuf.* **2019**, *4*, 291–313. [CrossRef]
27. Spoerk, M.; Gonzalez-Gutierrez, J.; Lichal, C.; Cajner, H.; Berger, G.R.; Schuschnigg, S.; Cardon, L.; Holzer, C. Optimisation of the Adhesion of Polypropylene-Based Materials during Extrusion-Based Additive Manufacturing. *Polymers* **2018**, *10*, 490. [CrossRef]
28. Kujawa, M. The influence of first layer parameters on adhesion between the 3D printer's glass bed and ABS. In *Interdyscyplinarność Badań Naukowych 2017*; Oficyna Wydawnicza Politechniki Wrocławskiej: Wrocław, Poland, 2017.

29. Kapusuzoglu, B.; Mahadevan, S. Information fusion and machine learning for sensitivity analysis using physics knowledge and experimental data. *Reliab. Eng. Syst. Saf.* **2021**, *214*, 107712. [CrossRef]
30. Wenzel, S.; Slomski-Vetter, E.; Melz, T. Optimizing system reliability in additive manufacturing using PIML. *Int. J. Lightweight Mater. Manuf.* **2022**, *3*, 284–297. [CrossRef]

MDPI

Article

Reliability-Based Robust Design Optimization for Maximizing the Output Torque of Brushless Direct Current (BLDC) Motors Considering Manufacturing Uncertainty

Kyunghun Jeon [1], Donghyeon Yoo [1], Jongjin Park [1], Ki-Deok Lee [2], Jeong-Jong Lee [2] and Chang-Wan Kim [3,*]

[1] Graduate School of Mechanical Design & Production Engineering, Konkuk University, 120 Neungdong-ro, Gwangjin-gu, Seoul 05029, Korea
[2] Intelligent Mechatronics Research Center, Korea Electronics Technology Institute, Seongnam-si 13509, Gyeonggi-do, Korea
[3] School of Mechanical Engineering, Konkuk University, 120 Neungdong-ro, Gwangjin-gu, Seoul 05029, Korea
* Correspondence: goodant@konkuk.ac.kr

Abstract: In recent years, the deterministic design optimization method has been widely used to improve the output performance of brushless direct current (BLDC) motors. However, it does not contribute to reducing the failure rate and performance variation of products because it cannot determine the manufacturing uncertainty. In this study, we proposed reliability-based robust design optimization to improve the output torque of a BLDC motor while reducing the failure rate and performance variation. We calculated the output torque and vibration response of the BLDC motor using the electromagnetic–structural coupled analysis. We selected the tooth thickness, slot opening width, slot radius, slot depth, tooth width, magnet thickness, and magnet length as the design variables related to the shape of the stator and rotor that affect the output torque. We considered the distribution of design variables with manufacturing tolerances. We performed a reliability analysis of the BLDC motor considering the distribution of design variables with manufacturing tolerances. Using the reliability analysis results, we performed reliability-based robust design optimization (RBRDO) to maximize the output torque; consequently, the output torque increased by 8.8% compared to the initial BLDC motor, the standard deviation in output performance decreased by 46.9% with improved robustness, and the failure rate decreased by 99.2% with enhanced reliability. The proposed reliability-based robust design optimization is considered to be useful in the actual product design field because it can evaluate both the reliability and robustness of the product and improve its performance in the design stage.

Keywords: brushless direct current motor; manufacturing uncertainty; reliability-based robust design optimization; output torque; torque ripple; vibration analysis

Citation: Jeon, K.; Yoo, D.; Park, J.; Lee, K.-D.; Lee, J.-J.; Kim, C.-W. Reliability-Based Robust Design Optimization for Maximizing the Output Torque of Brushless Direct Current (BLDC) Motors Considering Manufacturing Uncertainty. *Machines* **2022**, *10*, 797. https://doi.org/10.3390/machines10090797

Academic Editors: Sven Matthiesen and Thomas Gwosch

Received: 18 August 2022
Accepted: 9 September 2022
Published: 10 September 2022

1. Introduction

Electric motors are power-generating devices used in various industries, such as the automobile, home appliance, and plant industries. With the recent strengthening of environmental regulations worldwide, the application field of electric motors has expanded, and accordingly, designs with high energy density (through high output and miniaturization) and weight reduction have been developed. However, an increase in energy density causes an increase in torque ripple, which is one of the primary causes of electric motor vibration. Therefore, it is necessary to study the design of electric motors with high energy density while considering torque ripple.

Many studies have been conducted using the finite element analysis (FEA) method to analyze the vibration characteristics of electric motors and to reduce vibrations. R. Islam et al. analyzed the torque waveform and cogging torque according to the permanent magnet shape of a permanent-magnet synchronous motor (PMSM) using a two-dimensional

(2-D) electromagnetic (EM) FEA. They applied a step-skew type motor to reduce cogging torque [1]. C. Studer et al. explained the principle of cogging torque generation in electric motors using 2-D EM FEA and presented a method for reducing cogging torque using design variables, such as stator tooth and permanent magnet shape [2]. M. Dai et al. calculated the torque ripple of a permanent-magnet brushless direct current (BLDC) motor using 2-D EM FEA. They reduced the torque ripple by adjusting the skew angle [3]. J. Hong et al. analyzed the EM force using three-dimensional (3-D) EM FEA and proposed a design to reduce the vibration caused by EM force by changing the stator pole and yoke shape of a switched reluctance motor (SRM) [4].

With the improvement of analysis technology and computing power, several studies have been conducted to improve the performance of electric motors by applying various optimization methods. Previously, studies using the deterministic design optimization (DDO) method were conducted to reduce torque ripple. Choi et al. conducted a DDO study to minimize torque ripple using the air gap shape of an SRM as a design variable [5]. Kim et al. conducted a DDO study for maximizing the output torque using the permanent magnet volume of the spoke-type BLDC motor as a design variable [6]. Vasilija conducted a DDO study using the genetic algorithm with the rotor shape as a design variable to minimize the cogging torque of the PMSM [7]. Lee et al. conducted a multi-objective optimization study for maximizing output torque and minimizing torque ripple of SRM using the stator and rotor shapes as design variables [8,9]. Kuci et al. conducted a topology optimization study to minimize the torque ripple of the PMSM using the rotor shape [10]. Choi et al. conducted a DDO study to effectively obtain a sinusoidal distribution of the air gap flux density of IPM using the permanent magnet shape [11]. Consequently, it was possible to improve the output and vibration performance of the motor. However, since the DDO method is designed by setting the design variable to a single fixed value, it cannot consider the characteristics of fluctuations in product performance due to the uncertainty that occurs in the manufacturing process.

Mass production often results in performance variations due to uncertainties arising during the manufacturing process. Uncertainties are generally caused by production and assembly tolerances, material properties, and the environments of use [12]. Variations in performance due to uncertainty cause product failures. Considering the uncertainties that occur during mass production, the need for research on probabilistic design optimization (PDO) has emerged to achieve product performance and quality standards [13]. Depending on the purpose, PDO can be classified into robust design optimization (RDO) and reliability-based design optimization (RBDO) methods. RDO is a method used to minimize fluctuations in product performance and quality [14], and RBDO is a method used to increase product reliability at a given probability level [15]. In recent years, many studies have been conducted to design a motor using the above two methods.

Kim et al. performed an RDO study to improve the robustness of the cogging torque of BLDC motors and to improve the vibration performance [16]. Lee et al. conducted an RDO study using the stator and rotor shapes and rotor eccentricity as design variables to reduce the back electromotive force of the interior permanent-magnet synchronous motor (IPMSM) [17]. Lee et al. performed the design experiments to determine the design factors affecting the cogging torque of a surface-mounted permanent-magnet synchronous motor (SPMSM). They conducted an RDO study to reduce the cogging torque in consideration of the uncertainty that occurs during the assembly process of the stator [18]. Kim et al. conducted an RDO study to reduce the cogging torque of SPMSM by setting the uncertainty through FEA and experiments [19]. Consequently, it was possible to design a motor that simultaneously improves motor performance and minimizes quality fluctuations. Ziyan et al. performed RBDO to reduce the cogging torque considering the magnetic flux density of the stator [20]. Mun et al. performed RBDO to reduce the cogging torque by considering the performance variation due to manufacturing tolerance and operating temperature [21].

Reliability-based robust design optimization (RBRDO), which integrates RDO and RBDO, is the only way to consider both product quality and reliability in the design stage. Jang et al. performed RBRDO considering the manufacturing uncertainty of SPMSM. They proposed a design that minimizes the back electromotive force and increases the reliability of cogging torque [12]. Kim et al. conducted an RBRDO study of BLDC motors and reduced the cogging torque and the failure rate of output torque [22]. Hao et al. calculated the uncertainty using Monte Carlo simulation and performed RBRDO to minimize the mass of solid rocket motors [23]. Jang et al. performed RBRDO of IPMSM considering manufacturing and material uncertainty. They proposed a design to reduce torque ripple and the failure rate of output torque, and they explained the superiority of RBRDO by comparing the results of RBRDO, DDO, and RBDO [24]. The above studies tried to reduce torque ripple or cogging torque through EM design changes to improve vibration performance. However, the change in vibration response after design optimization was not analyzed.

In this study, considering the uncertainty caused by the manufacturing tolerances of BLDC motors, we proposed RBRDO as a multi-objective optimization that simultaneously maximizes the output torque and minimizes its standard deviation. We performed the EM–structural coupled analysis to calculate the output torque, torque ripple, and vibration response of the BLDC motor. To perform RBRDO, we selected the tooth thickness, slot opening (SO) width, slot radius, slot depth, tooth width, magnet thickness, and magnet length as the design variables related to the shape of the stator and rotor that affect the output torque of the motor. We performed the reliability analysis of the BLDC motor considering the distribution of the design variables. We used the rate of change compared to the initial value of the torque ripple and the area of the permanent magnet as constraints. We verified the superiority of RBRDO by comparing the results of RBRDO with those of DDO and RDO. Further, we performed a vibration analysis to analyze the change in vibration response in RBRDO compared to the initial BLDC motor.

2. Electromagnetic–Structural Coupled Analysis

We used the EM–structural coupled analysis to calculate the vibration response of a BLDC motor. We calculated the EM force acting on the stator tooth through the EM FEA. By applying the calculated EM force as the load condition of the structural finite element model, we calculated the acceleration response of the motor due to the EM force. In this study, we neglected the tangential EM force. The tangential EM force is generally negligible because its effect on the motor vibration is less than the radial EM force [25].

In this study, we used a BLDC motor with a rated output of 1.5 kW, 4 poles, and 24 slots, as shown in Figure 1. Table 1 summarizes the specifications of the BLDC motor.

Figure 1. BLDC motor.

Table 1. Specifications of the BLDC motor.

Specifications	Quantity
Type	BLDC
Number of poles	4
Number of slots	24
Rated power	1.5 kW
Rated torque	7.17 N·m
Rated speed	2000 rpm

The EM forces generated in the air gap cause the EM vibration of an electric motor. Using the principle of virtual work, we can calculate the EM force as follows:

$$\frac{\partial}{\partial s}\int_{\Omega}\int_{0}^{H} B\cdot dH d\Omega = F_s, \tag{1}$$

where B is the magnetic flux density, H is the magnetic field, s is the x, y, and z axes in the Cartesian coordinate system, and Ω is the domain where the nodal force F_s is applied. Applying Equation (1) to the mesh element e gives the following equation:

$$\int_{e}\left(-B^T\cdot J^{-1}\cdot\frac{\partial J}{\partial s}\cdot H + \int_0^H B\cdot dH\left|J^{-1}\right|\frac{\partial |J|}{\partial s}\right)dV = F_s, \tag{2}$$

where J is the Jacobian matrix of e, and V is the total volume. The Jacobian matrix is determined according to the element type. In the linear case, the integral of B can be simplified as follows:

$$\int_0^H B\cdot dH = \int_0^H \mu H\cdot dH = \frac{\mu}{2}|H|^2, \tag{3}$$

where μ is the magnetic relative permeability. Using (1), (2), and (3), the local force F_s^i applied to a given node i can be formulated as follows:

$$\sum_{\forall e}\int_{e}\left(-B^T\cdot J^{-1}\cdot\frac{\partial J}{\partial s}\cdot H + \frac{\mu}{2}|H|^2\left|J^{-1}\right|\frac{\partial |J|}{\partial s}\right)dV = F_s^i. \tag{4}$$

Equation (4) can be expressed in the matrix form as follows:

$$-[\mathbf{B}]^{\mathrm{T}}[\mathbf{J}]^{-1}[\mathbf{H}]\nabla \mathrm{J} + \frac{\mu}{2}[\mathbf{H}]^2[\mathbf{J}]^{-1}\nabla \mathrm{J} = \{\mathrm{F_s}\}. \tag{5}$$

We performed an EM analysis using the 2-D cross-sectional model shown in Figure 2 to calculate the output torque, torque ripple, and EM force acting on the stator tooth of the BLDC motor. A three-phase alternating current at 2000 rpm rated operation was applied as the analysis condition.

As a result of the EM analysis, the output torque and torque ripple were calculated as 7.16 N·m and 3.46 N·m, respectively, as shown in Figure 3. Compared with the performance specification of the BLDC motor (7.17 N·m), the relative error is 0.14%.

The radial EM force acting on the stator tooth was calculated as shown in Figure 4. Vibration analysis was performed by inputting the calculated radial EM force into the stator structure.

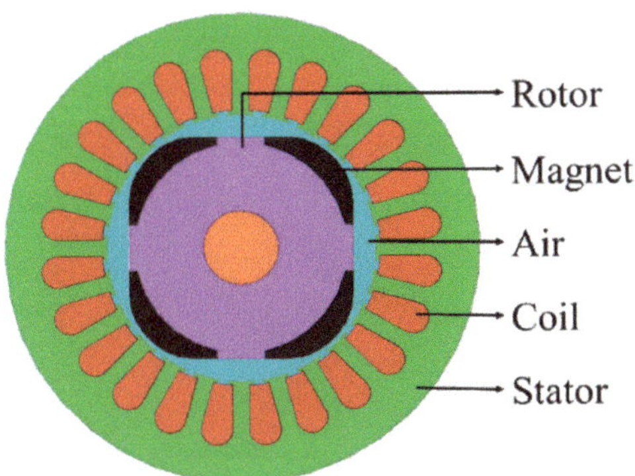

Figure 2. 2-D cross-section of the BLDC motor.

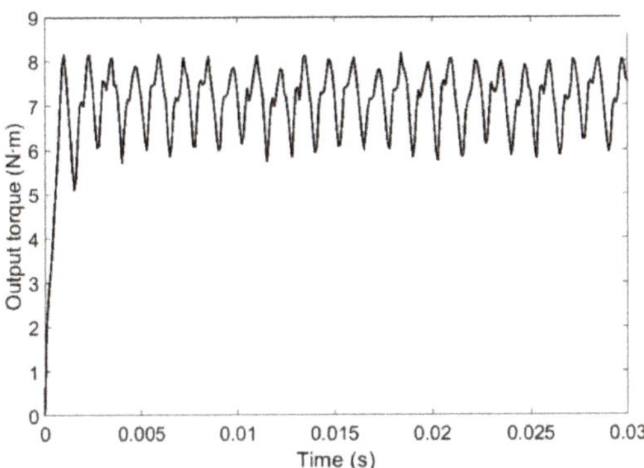

Figure 3. Output torque of the BLDC motor.

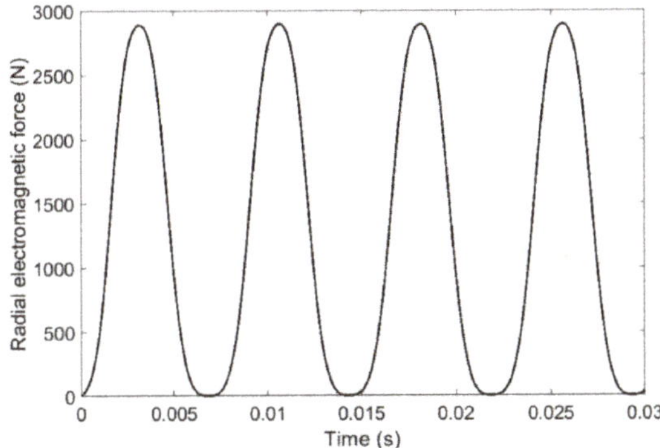

Figure 4. Radial EM force acting on a single stator tooth.

The mechanical properties of BLDC motors can be described using the following equation of motion:

$$[\mathbf{M}]\{\ddot{x}\} + [\mathbf{C}]\{\dot{x}\} + [\mathbf{K}]\{x\} = \{\mathrm{F}(t)\}, \quad (6)$$

where **[M]**, **[C]**, and **[K]** represent a mass matrix, damping matrix, and stiffness matrix, respectively. $\{x\}$ is the displacement vector and $\{\mathrm{F}(t)\}$ is the applied load vector.

The harmonic components of the EM force calculated through the EM analysis were applied to the stator tooth of the 3-D finite element model. Transient analysis using the EM–

structural coupled analysis method was performed to calculate the acceleration response in the finite element model of the BLDC motor shown in Figure 5. Fast Fourier transform was performed to obtain the normalized frequency response curve shown in Figure 6.

Figure 5. Finite element model of the BLDC motor.

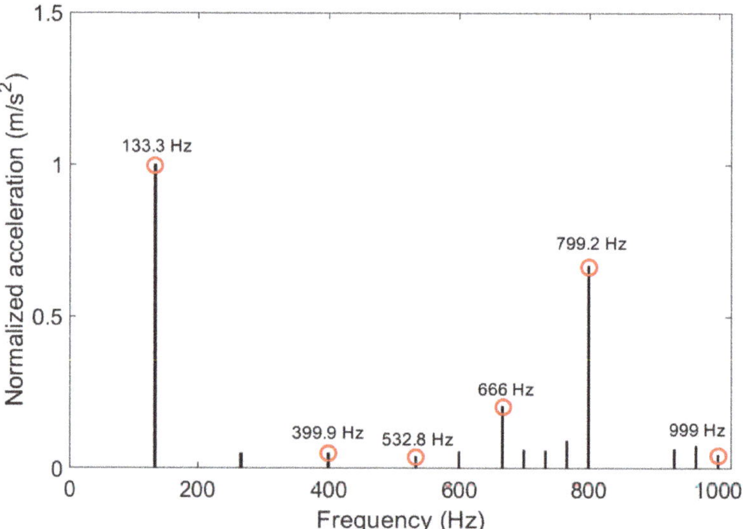

Figure 6. Normalized acceleration and peak frequencies of the finite element BLDC motor model.

The vibration measurement data in Figure 7 were used to verify the EM–structural coupled analysis results [26]. Table 2 summarizes the relative errors of the peak frequencies of the EM–structural coupled analysis and the vibration measurement results. The maximum relative error of the analysis and experiment for the frequency at which the peak acceleration value occurs in the range of 0 to 1000 Hz is less than 4%. We verified the vibration prediction accuracy of the analysis model. The reason for the difference of 133.3 Hz in the magnitude of the analysis and measurement results is that a sinusoidal current without harmonic components was applied to the input current. We performed a reliability analysis and RBRDO of the BLDC motor using the verified analysis model.

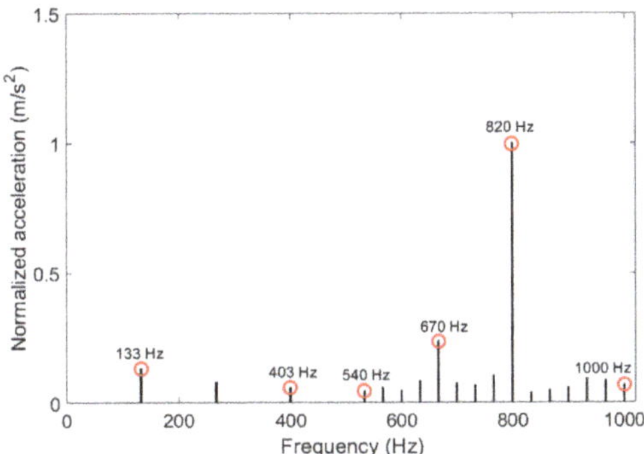

Figure 7. Normalized acceleration and peak frequencies of the measured point.

Table 2. Comparison of the peak frequencies of measured and simulated results.

Measured Peak Frequency (Hz)	Simulated Peak Frequency (Hz)	Relative Error (%)
133	133.3	0.23%
276	266.6	−3.48%
403	399.9	−0.77%
540	532.8	−1.33%
612	599.4	−2.06%
670	666	−0.60%
705	699.3	−0.81%
724	732.6	1.19%
780	765.9	−1.81%
820	799.2	−2.54%
943	932.4	−1.12%
986	965.7	−2.06%
1000	999	−0.10%

3. Probabilistic Design Optimization

The design optimization method is divided into DDO and PDO depending on whether the uncertainty of the design variables is considered. DDO is a traditional design optimization method in which the design variables are assumed to have a fixed value without considering uncertainty.

The DDO problem can be formulated as follows:

$$\begin{gathered} \min f(\mathbf{d}), \\ \textit{subject to } g_i(\mathbf{d}) < 0,\ i = 1,\ 2,\ \cdots, \end{gathered} \tag{7}$$

where f is the objective function for the design variable $\mathbf{d}$, and g is the constraint.

Because DDO does not consider the uncertainty of design variables, the product performance may vary, thereby not satisfying the requirements of the designer or causing product failures. Therefore, a PDO method that considers the uncertainty of design variables was used. PDO can be divided into RBDO and RDO depending on whether the design purpose is to satisfy the reliability or minimize the performance fluctuation.

RBDO is used to design a product that satisfies the reliability as per the requirement of the designer by quantitatively defining uncertainty. Therefore, reliability analysis is necessary to accurately predict the failure probability of a constraint condition. Reliability is a design condition in product design that refers to satisfying the required performance. The design variable is defined as a probability variable X to consider reliability quantitatively. A performance function representing the required performance is defined as g to evaluate the reliability of a product. As shown in Figure 8, the performance function is expressed as a probability density function, and at the performance function equals 0, it is divided into a feasible and an infeasible region that satisfies the constraint condition. The probability of product damage is obtained by calculating the area where g > 0 in the probability density function curve shown in Figure 8. Reliability increases as the probability of damage decreases [27].

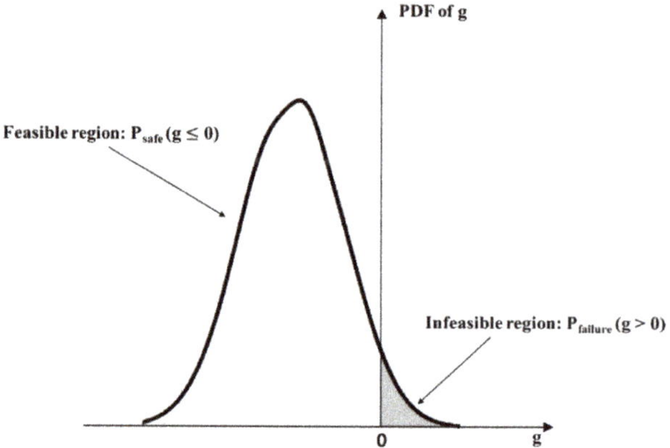

Figure 8. Probability density function of a performance function.

The RBDO problem can be formulated as follows:

$$\min h(\mathbf{d}),$$

$$subject\ to\ P_F(g_i(\mathbf{X}) > 0) \geq P_{t,\,i},\ i = 1,\ 2,\ \cdots, \tag{8}$$

where h is the objective function, g_i is the i-th constraint function, P_F is the failure rate under infeasible conditions, and $P_{t,\,i}$ is the i-th target value to guarantee the reliability of g_i.

Robust design is a method that additionally considers robustness in the existing design method and was first developed by Taguchi Genichi. The Taguchi method uses the S/N ratio and orthogonal array as evaluation indices for robustness to minimize the performance fluctuations due to noise factors [28,29]. The Taguchi method maximizes the S/N ratio to reduce the variation of the quality loss function and further uses an adjustment parameter such that the average of the quality loss function reaches the target value. Through this process, we can discover a design that reduces the fluctuation of the quality loss function and simultaneously satisfies the target performance.

However, since robust design uses an orthogonal arrangement table, it is difficult to consider a wide design range, and the design variables can be defined only in a discrete space. In addition, the general design requires many constraint conditions. The Taguchi method is inefficient in dealing with these constraints. RDO, a mathematically well-developed design optimization method, is used to solve the above problems [30].

RDO is used to find a design in which the product performance is insensitive to uncertainties, such as the variations of design variables. In this method, the mean of the quality loss function is optimized, and the variance is minimized while satisfying the design constraints. The RDO problem can be formulated as follows:

$$\begin{gathered} \min f(\mu_h,\ \sigma_h^2),\ h(\mathbf{X};\mathbf{d}), \\ subject\ to\ g_i(\mathbf{d}) \leq 0,\ i = 1,\ 2,\ \cdots, \end{gathered} \tag{9}$$

where f is the quality loss function consisting of mean μ_h and variance σ_h^2 for h. Since f is expressed as the sum of the mean and variance multiplied by a specific weight, RDO is a multi-objective optimization. $\mathbf{d}$ is a vector of design variables defined as $\mathbf{d} = \mu(\mathbf{X})$, where

μ represents the mean of the design probability variable X. The purpose of RDO is to find the design that is most insensitive to changes in probability variables within the effective design domain $g_i \leq 0$ [28].

RBRDO is a method that integrates the above two design optimization methods to minimize the performance fluctuation characteristics and simultaneously satisfy the reliability level required by the designer. In the problem formulation of RBRDO, the loss function of the product quality is minimized according to the probabilistic constraint shown in the following equation:

$$\begin{gathered} \min f\left(\mu_h, \sigma_h^2\right), h(\mathbf{X}; \mathbf{d}), \\ \textit{subject to } P_F(g_i(\mathbf{X}) > 0) \geq P_{t,\,i},\ i = 1, 2, \cdots. \end{gathered} \tag{10}$$

RDO and RBDO are integrated into one numerical model to solve Equation (10). Although such an integrated optimization problem requires a high computational cost, it has an increasing need because it enables a robust and reliable product design against changes in design variables.

4. Reliability-Based Robust Design Optimization of the BLDC Motor

4.1. Reliability Analysis of the BLDC Motor

The stator and rotor of the BLDC motor are generally manufactured through the stamping process. There are uncertainties in the manufacturing process due to various factors, such as manufacturing tolerances and assembly environment. These uncertainties can eventually lead to variations in motor performance. The performance variation of the motor may not satisfy the requirements of the designer and may increase the failure rate. Therefore, we require a design considering the uncertainty caused by the manufacturing tolerance of the BLDC motor.

The design variables that affect the motor's output performance are defined as shown in Figure 9. Table 3 summarizes the distribution of design variables caused by the uncertainty of manufacturing tolerances during the stamping process. In this study, we referred to the stamping uncertainty of the metal plate proposed in [31]. Baron et al. recommended a standard deviation of 0.06 mm for the stamping process. We performed a reliability analysis to calculate the performance variation of the BLDC motor. Table 4 summarizes the reliability analysis results. Figure 10 shows the probability distributions for the output torque and torque ripple of BLDC motors.

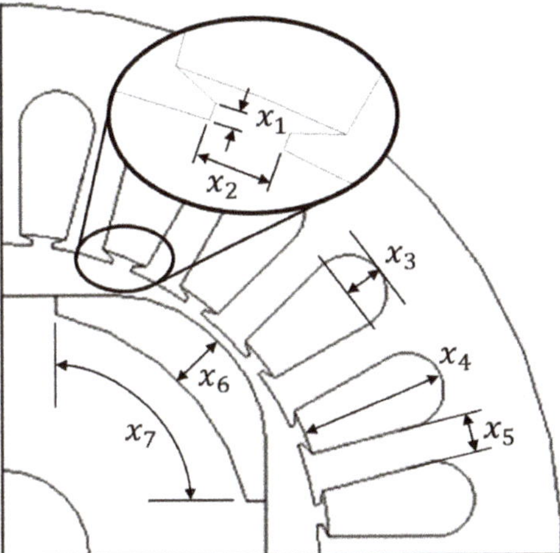

Figure 9. Design variables of the BLDC motor.

Table 3. Probabilistic distribution of the design variables of the BLDC motor.

Parameters	Unit	Mean	Standard Deviation	Distribution
Tooth thickness x_1	mm	0.5	0.06	Normal
SO width x_2	mm	2.18	0.06	Normal
Slot radius x_3	mm	4.05	0.06	Normal
Slot depth x_4	mm	15.86	0.06	Normal
Tooth width x_5	mm	4.19	0.06	Normal
Magnet thickness x_6	mm	6	0.06	Normal
Magnet length x_7	mm	37.53	0.06	Normal

Table 4. Reliability analysis of the BLDC motor.

Output torque	Mean (N·m)	7.16
	Standard deviation	0.49
Torque ripple	Mean (N·m)	3.46
	Probability of failure (%)	36.81

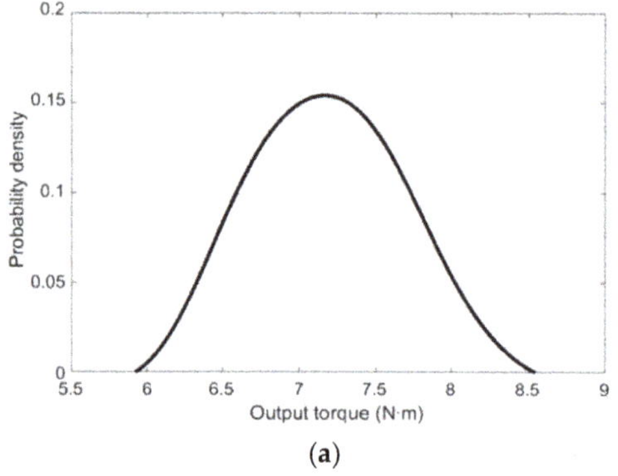

(**a**)

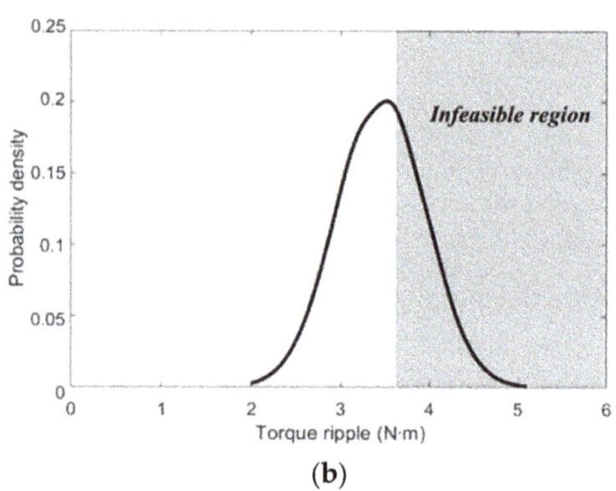

(**b**)

Figure 10. Probability distributions of (**a**) output torque and (**b**) torque ripple of the BLDC motor.

4.2. Design Optimization for Maximizing the Output Torque of the BLDC Motor

We applied various optimization methods to improve the output torque of the BLDC motor. We performed a DDO that does not consider the distribution of design variables due to the manufacturing uncertainty of BLDC motors. However, DDO has a limit in reducing the performance variation when considering the distribution of each variable. An RDO, which can minimize performance variation, was performed to overcome the limitations of DDO. Finally, we performed RBRDO to reduce performance variation and failure rate simultaneously. Table 5 summarizes the design variables and spaces of BLDC motors.

The purpose of the DDO of the BLDC motor is to maximize the output torque. The constraint conditions of DDO are torque ripple and magnet area, and the change is limited to +5% of the initial value. Equation (11) shows the formulation of the DDO of the BLDC motor.

$$\begin{aligned} &Find\ x_i(i = 1, \cdots, 7), \\ &Maximize\ \ T_{out}(x_i), \\ &Subject\ to\ g_1(x_i) \leq 1.05\ T_{ripp,initial}, \\ &\qquad g_2(x_i) \leq 1.05\ A_{initial}. \end{aligned} \tag{11}$$

x_i represents the seven design variables that determine the shape of the motor, T_{out} is a function of output torque, and g_1 and g_2 represent the functions of torque ripple and magnet area, respectively.

Table 5. Design variables and their bounds.

Design Variables	Unit	Design Spaces
Tooth thickness x_1	mm	$0.2 < x_1 < 0.98$
SO width x_2	mm	$1.96 < x_2 < 2.39$
Slot radius x_3	mm	$3.65 < x_3 < 4.45$
Slot depth x_4	mm	$11.19 < x_4 < 20.56$
Tooth width x_5	mm	$3.69 < x_5 < 4.68$
Magnet thickness x_6	mm	$3.45 < x_6 < 8.35$
Magnet length x_7	mm	$35.53 < x_7 < 39.53$

RDO of the BLDC motor is a multi-objective optimization that can minimize performance variation while maximizing performance by considering the distribution of design variables due to manufacturing uncertainty. The first and second objective functions are the maximization and minimization of the mean value of the output torque, respectively. The constraint conditions are the same as the formulation of the DDO. Equation (12) shows the RDO problem formulation.

$$\begin{aligned} &Find\ x_i(i = 1, \cdots, 7), \\ &Maximize\ \ F = \left(w_1 \tfrac{\mu_f}{\mu_{f0}} + w_2 \tfrac{\sigma_{f0}}{\sigma_f}\right), \\ &Subject\ to\ g_1(x_i) \leq 1.05\ T_{ripp,initial}, \\ &\qquad g_2(x_i) \leq 1.05\ A_{initial}. \end{aligned} \tag{12}$$

F, μ_f, and σ_f represent the mean and standard deviation, mean value, and standard deviation of the output torque, respectively. μ_{f0} and σ_{f0} are initial values of the mean and standard deviation of the output torque, respectively. w_1 and w_2 are weight factors of 0.5 and 1, respectively.

Equation (13) is the formula for the problem of RBRDO, which increases the robustness of the performance and reduces the failure rate by considering the manufacturing uncertainty of the BLDC motor. Figure 11 shows the RBRDO flowchart. The objective function is the same as the formulation of the RDO. The constraint condition is the failure rate; 99% reliability is achieved within the +5% change rate compared to the initial torque ripple. In this study, we used the system reliability optimization method for the RBRDO of the BLDC motor. This method is suitable when there are various constraints [32].

$$\begin{aligned} &Find\ x_i(i = 1, \cdots, 7), \\ &Maximize\ \ F = \left(w_1 \tfrac{\mu_f}{\mu_{f0}} + w_2 \tfrac{\sigma_{f0}}{\sigma_f}\right), \\ &Subject\ to\ Pr\left(g_1(x_i) \leq 1.05\ T_{ripp,initial}\right) \geq 0.99, \\ &\qquad g_2(x_i) \leq 1.05\ A_{initial}. \end{aligned} \tag{13}$$

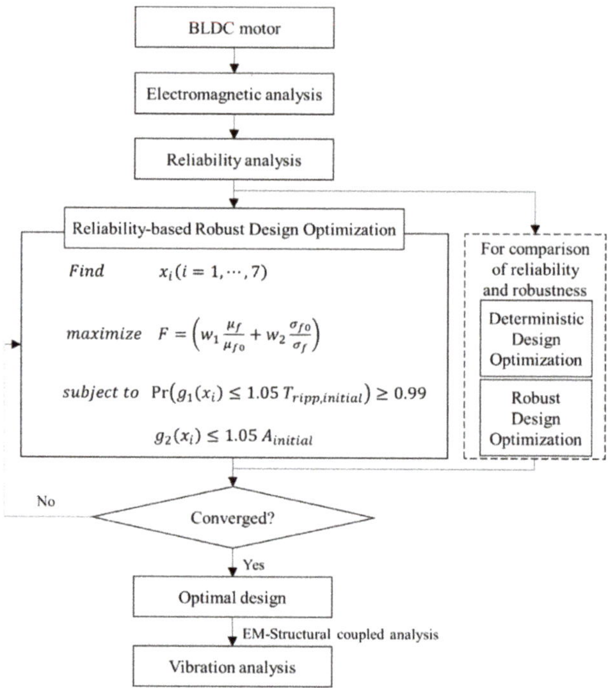

Figure 11. Flowchart of the RBRDO method.

5. Results and Discussion

5.1. RBRDO Results

Table 6 shows the changes in design variables and performance before and after RBRDO. Figure 12 shows the superiority of RBRDO by comparing the probability distribution results of the output torque and torque ripple of the initial BLDC motor and RBRDO. Consequently, the mean value of the output torque increased by 9% compared to the initial design to 7.79 N·m. The standard deviation of the output torque decreased by 45.8% compared to the initial design to 0.26, resulting in a robust design. The mean value of torque ripple decreased by 51.7% compared to the initial design to 1.67 N·m. The failure rate decreased by 99.2% compared to the initial design to 0.28%. The magnet area was 151.1 mm^2, which satisfied the constraint condition.

Table 6. Comparison of the design variables and performances in the initial design and RBRDO.

Design Variables and Performance		Initial Design	RBRDO	Rate of Change (%)
Tooth thickness (mm)		0.5	0.46	−8.0
SO width (mm)		2.18	2.37	+8.7
Slot radius (mm)		4.05	3.78	−6.7
Slot depth (mm)		15.86	11.2	−29.5
Tooth width (mm)		4.19	3.70	−11.9
Magnet thickness (mm)		6	5	−16.7
Magnet length (mm)		37.53	39.52	+5.3
Output torque	Mean (N·m)	7.16	7.79	+8.8
	Standard deviation	0.49	0.26	−46.9
Torque ripple	Mean (N·m)	3.46	1.67	−51.7
	Probability of failure (%)	36.8	0.28	−99.2
Magnet area (mm^2)		152.5	151.1	−0.92

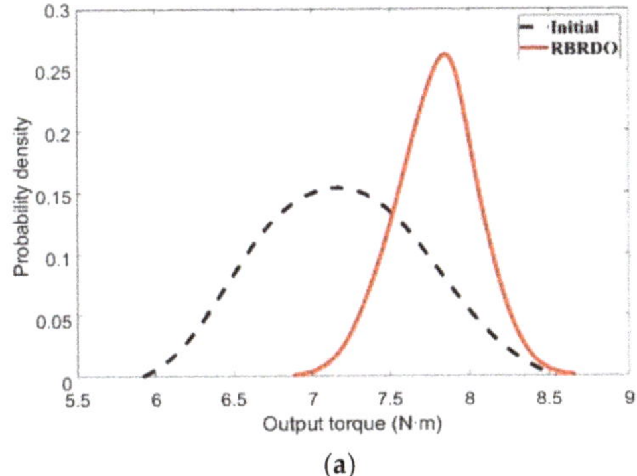

(a)

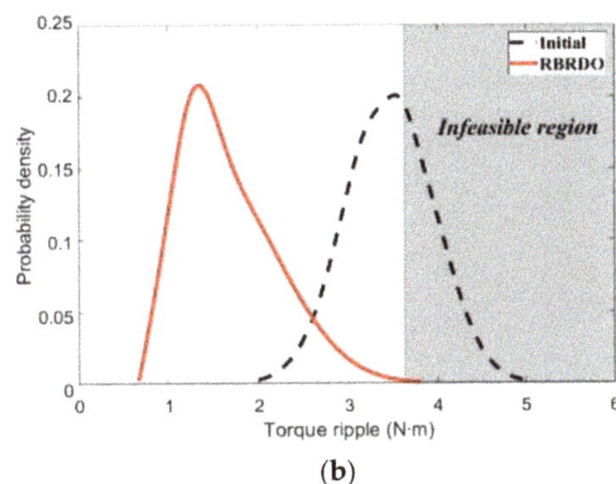

(b)

Figure 12. Comparison of probability distribution results of (**a**) output torque and (**b**) torque ripple in the initial design and RBRDO.

5.2. Comparison of Results of RBRDO with Those of DDO and RDO

In this study, we performed RBRDO to maximize the output torque and reduce the performance variation and failure rate simultaneously, considering the manufacturing uncertainty of the BLDC motor. We compared the design optimization results of RBRDO with those of the DDO and RDO methods to confirm that RBRDO provides better reliability and robustness than DDO or RDO. Compared to DDO, RBRDO considers uncertainty. Thus, it minimizes the standard deviation of performance to improve robustness. In addition, RBRDO can reduce the failure rate by improving reliability.

Table 7 summarizes each optimization result's output torque, torque ripple, and failure rate. Figure 13 shows the probability distributions of output torque and torque ripple as the result of reliability analysis for each optimization method. In the case of DDO, the average output torque was 8.49 N·m, and the optimal solution with the highest value was derived. However, considering the distribution of design variables, the standard deviation of the output torque was 0.54, which cannot reduce the performance variation that occurs during mass production. The torque ripple failure rate was 33.0%. In the case of RDO, the mean value of the output torque was 8.29 N·m, and the standard deviation was reduced by 50% compared to DDO, resulting in a robust design for output performance. However, the failure rate of 10.1% occurred because the constraint condition for the failure rate was not considered. In the case of RBRDO, the mean value of output torque was 7.79 N·m, and the standard deviation was 0.26. Compared to the initial design, it was possible to increase the output torque and reduce the variation of output performance. In addition, the torque ripple was 1.67 N·m, and the failure rate was the lowest at 0.28%.

Table 7. Comparison of DDO, RDO, and RBRDO results.

		Optimization Method		
		DDO	**RDO**	**RBRDO**
Output torque (N·m)	Mean	8.49	8.28	7.79
	Standard deviation	0.54	0.24	0.26
Torque ripple (N·m)	Mean	3.35	2.63	1.67
	Probability of failure (%)	33.0	10.1	0.28
Magnet area (mm^2)		160.0	153.2	151.1

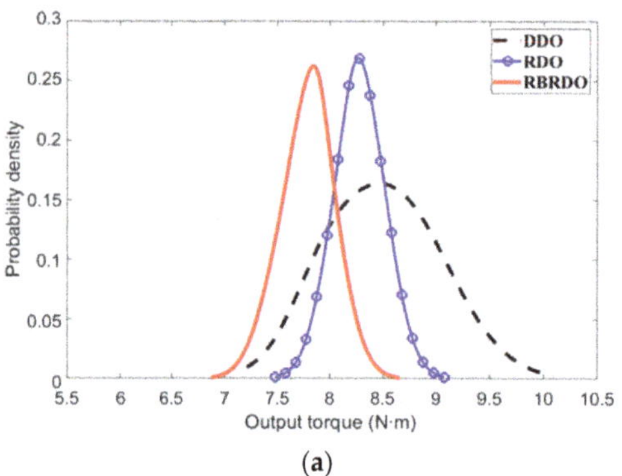

(a)

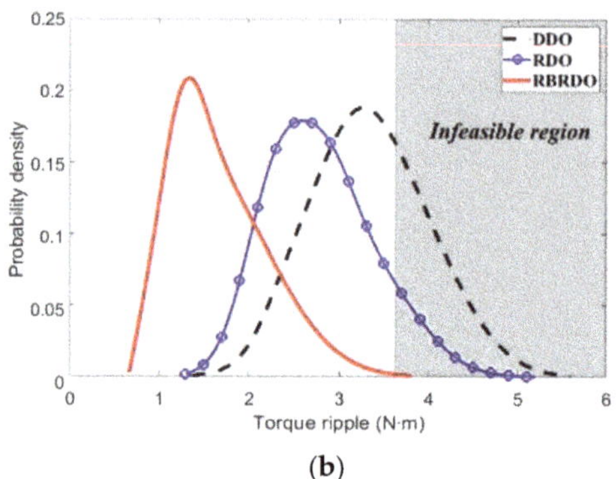

(b)

Figure 13. Comparison of DDO, RDO, and RBRDO results for (**a**) output torque and (**b**) torque ripple.

To compare the vibration responses of the initial BLDC motor and RBRDO model, EM–structural coupled analysis was performed. Figure 14 shows the frequency response graph of the vibration analysis results. It is a response to the EM force. The fundamental frequency of 133.3 Hz increased by 9% from 2.18 m/s^2 to 2.38 m/s^2 due to the energy density increase. However, the response at 400 Hz and the frequency component of torque ripple decreased (by 18.2%) from 0.11 m/s^2 to 0.09 m/s^2. We used the root mean square (RMS) method to compare the magnitude of the vibration response from 0 to 1000 Hz. The RMS value of the vibration response of the initial BLDC motor was 0.748. The RMS value of the vibration response of the RBRDO model increased by 2% to 0.763. However, since the output torque increased by 9% with RBRDO, we could confirm that a higher increase is possible in the output torque than the magnitude of the vibration response.

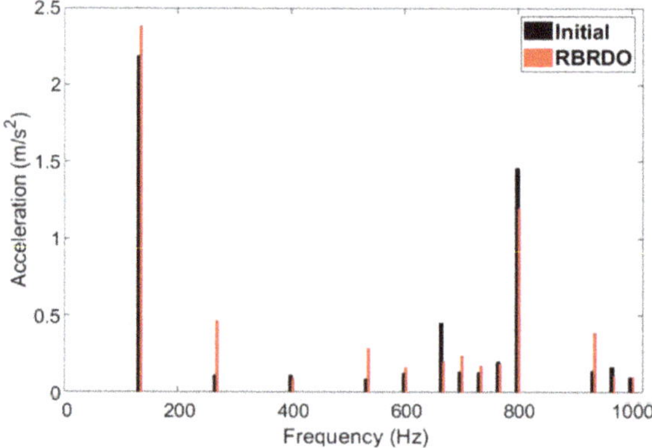

Figure 14. Comparison of vibration analysis results of initial BLDC motor and RBRDO.

6. Conclusions

In this paper, we proposed RBRDO to improve the output torque of the BLDC motor and simultaneously reduce the failure rate and output torque standard deviation caused by uncertainty due to manufacturing tolerance. Using the EM–structural coupled analysis, we calculated the output torque and torque ripple, when the BLDC motor rotated at the rated speed, and analyzed the vibration characteristics. We selected the tooth thickness, SO width, slot radius, slot depth, tooth width, magnet thickness, and magnet length as the design variables related to the shape of the stator and rotor, which affected the motor's output performance, to perform RBRDO. We performed a reliability analysis of the BLDC motor using the output torque and torque ripple calculated through EM analysis. We defined RBRDO as a multi-objective function to maximize the output torque and minimize the standard deviation of the output torque. We set the constraint condition to have a reliability of 99% within a rate of change of +5% compared to the initial torque ripple.

As a result of RBRDO, the optimal design of the mean value of output torque was 7.79 N·m, which increased by 8.8% compared to the initial BLDC motor, and the standard deviation decreased (by 46.9%) to 0.26. The mean value of torque ripple was reduced by 51.7% compared to the initial design to 1.67 N·m, and the failure rate decreased by 99.2%. We performed DDO and RDO to confirm that RBRDO showed better robustness and reliability, and all results were compared through reliability analysis considering the distribution of design variables. Comparing the results of RBRDO and DDO, we confirmed that the standard deviation of the output torque, which indicates the performance variation of the motor, was reduced by 51.9% through the improvement of robustness. Using RBRDO, the failure rate decreased by 97.2% through improved reliability. Therefore, the RBRDO of the BLDC motor maximizes the output torque while reducing failure rate and performance variation caused by manufacturing uncertainty. In addition, comparing the vibration response of the initial BLDC motor and the RBRDO model revealed that when the output torque increased by 9%, the vibration response increased by 2%; this confirmed that RBRDO could increase the energy density without affecting the vibration response.

The proposed procedure proved to be an effective design method for improving the output performance of a motor while considering both reliability and robustness. However, additional research should be conducted on the multi-objective reliability-based robust design optimization to reduce vibration while improving the output performance of the motor. The multi-objective reliability-based robust design optimization increases the computational cost exponentially. As a future work, the authors intend to develop a new solution for a multi-purpose, reliability-based robust design for improving the output performance of the motor and reducing vibration while considering the computational cost.

Author Contributions: Conceptualization, K.J. and C.-W.K.; Methodology, K.J. and C.-W.K.; Software, K.J.; Validation, K.J. and C.-W.K.; Formal Analysis, K.J. and D.Y.; Investigation, K.J.; Resources, K.J.; Data Curation, K.J.; Writing—Original Draft Preparation, K.J.; Writing, Review, and Editing, K.J., J.P. and C.-W.K.; Visualization, K.J.; Supervision, C.-W.K., J.-J.L. and K.-D.L.; Project Administration, C.-W.K.; Funding Acquisition, C.-W.K. All authors have read and agreed to the published version of the manuscript.

Funding: This paper was supported by Konkuk University Researcher Fund in 2021 and the Technology Innovation Program (20012518) funded By the Ministry of Trade, Industry and Energy (MOTIE, Korea).

Institutional Review Board Statement: Not applicable.

Informed Consent Statement: Not applicable.

Data Availability Statement: Not applicable.

Conflicts of Interest: The authors declare no conflict of interest.

References

1. Islam, R.; Husain, I.; Fardoun, A.; McLaughlin, K. Permanent magnet synchronous motor magnet designs with skewing for torque ripple and cogging torque reduction. In Proceedings of the 2007 IEEE Industry Applications Annual Meeting, New Orleans, LA, USA, 23–27 September 2007; pp. 1552–1559.
2. Studer, C.; Keyhani, A.; Sebastian, T.; Murthy, S.K. Study of cogging torque in permanent magnet machines. In Proceedings of the IAS'97. Conference Record of the 1997 IEEE Industry Applications Conference Thirty-Second IAS Annual Meeting, New Orleans, LA, USA, 5–9 October 1997; pp. 42–49.
3. Dai, M.; Keyhani, A.; Sebastian, T. Torque ripple analysis of a PM brushless DC motor using finite element method. *IEEE Trans. Energy Convers.* **2004**, *19*, 40–45. [CrossRef]
4. Hong, J.P.; Ha, K.H.; Lee, J. Stator pole and yoke design for vibration reduction of switched reluctance motor. *IEEE Trans. Magn.* **2002**, *38*, 929–932. [CrossRef]
5. Choi, Y.K.; Yoon, H.S.; Koh, C.S. Pole-shape optimization of a switched-reluctance motor for torque ripple reduction. *IEEE Trans. Magn.* **2007**, *43*, 1797–1800. [CrossRef]

6. Kim, H.W.; Kim, K.T.; Jo, Y.S.; Hur, J. Optimization methods of torque density for developing the neodymium free SPOKE-type BLDC motor. *IEEE Trans. Magn.* **2013**, *49*, 2173–2176. [CrossRef]
7. Sarac, V. Performance optimization of permanent magnet synchronous motor by cogging torque reduction. *J. Electr. Eng.* **2019**, *70*, 218–226. [CrossRef]
8. Lee, C.; Lee, J.; Jang, I.G. Shape optimization-based design investigation of the switched reluctance motors regarding the target torque and current limitation. *Struct. Multidiscip. Optim.* **2021**, *64*, 859–870. [CrossRef]
9. Lee, C.; Jang, I.G. Topology optimization of multiple-barrier synchronous reluctance motors with initial random hollow circles. *Struct. Multidiscip. Optim.* **2021**, *64*, 2213–2224. [CrossRef]
10. Kuci, E.; Henrotte, F.; Duysinx, P.; Geuzaine, C. Combination of topology optimization and Lie derivative-based shape optimization for electro-mechanical design. *Struct. Multidiscip. Optim.* **2019**, *59*, 1723–1731. [CrossRef]
11. Choi, J.S.; Izui, K.; Nishiwaki, S.; Kawamoto, A.; Nomura, T. Rotor pole design of IPM motors for a sinusoidal air-gap flux density distribution. *Struct. Multidiscip. Optim.* **2012**, *46*, 445–455. [CrossRef]
12. Jang, J.; Cho, S.G.; Lee, S.J.; Kim, K.S.; Kim, J.M.; Hong, J.P.; Lee, T.H. Reliability-based robust design optimization with kernel density estimation for electric power steering motor considering manufacturing uncertainties. *IEEE Trans. Magn.* **2015**, *51*, 8001904. [CrossRef]
13. Kim, N.K.; Kim, D.H.; Kim, D.W.; Kim, H.G.; Lowther, D.A.; Sykulski, J.K. Robust optimization utilizing the second-order design sensitivity information. *IEEE Trans. Magn.* **2010**, *46*, 3117–3120. [CrossRef]
14. Park, G.J.; Lee, T.H.; Lee, K.H.; Hwang, K.H. Robust design: An overview. *AIAA J.* **2006**, *44*, 181–191. [CrossRef]
15. Youn, B.D.; Choi, K.K.; Du, L. Enriched Performance Measure Approach for Reliability-Based Design Optimization. *AIAA J.* **2005**, *43*, 874–884. [CrossRef]
16. Kim, D.W.; Choi, N.S.; Lee, C.U.; Kim, D.H. Assessment of statistical moments of a performance function for robust design of electromagnetic devices. *IEEE Trans. Magn.* **2015**, *51*, 7205104. [CrossRef]
17. Lee, S.J.; Kim, K.S.; Cho, S.G.; Jang, J.; Lee, T.; Hong, J.P. Taguchi robust design of back electromotive force considering the manufacturing tolerances in IPMSM. In Proceedings of the 2012 Sixth International Conference on Electromagnetic Field Problems and Applications, Dalian, China, 19–21 June 2012; pp. 1–4.
18. Lee, S.G.; Kim, S.; Park, J.C.; Park, M.R.; Lee, T.H.; Lim, M.S. Robust Design Optimization of SPMSM for Robotic Actuator Considering Assembly Imperfection of Segmented Stator Core. *IEEE Trans. Energy Convers.* **2020**, *35*, 2076–2085. [CrossRef]
19. Kim, S.; Lee, S.G.; Kim, J.M.; Lee, T.H.; Lim, M.S. Robust design optimization of surface-mounted permanent magnet synchronous motor using uncertainty characterization by bootstrap method. *IEEE Trans. Energy Convers.* **2020**, *35*, 2056–2065. [CrossRef]
20. Ren, Z.; Ma, J.; Qi, Y.; Zhang, D.; Koh, C.S. Managing Uncertainties of Permanent Magnet Synchronous Machine by Adaptive Kriging Assisted Weight Index Monte Carlo Simulation Method. *IEEE Trans. Energy Convers.* **2020**, *35*, 2162–2169. [CrossRef]
21. Mun, J.; Lim, J.; Kwak, Y.; Kang, B.; Choi, K.K.; Kim, D.H. Reliability-based design optimization of a permanent magnet motor under manufacturing tolerance and temperature fluctuation. *IEEE Trans. Magn.* **2021**, *57*, 8203304. [CrossRef]
22. Kim, D.W.; Kang, B.; Choi, K.K.; Kim, D.H. A comparative study on probabilistic optimization methods for electromagnetic design. *IEEE Trans. Magn.* **2015**, *52*, 7201304. [CrossRef]
23. Hao, Z.; Haowen, L.; Pengcheng, W.; Guobiao, C.; Feng, H. Uncertainty analysis and design optimization of solid rocket motors with finocyl grain. *Struct. Multidiscip. Optim.* **2020**, *62*, 3521–3537. [CrossRef]
24. Jang, G.U.; Kim, C.W.; Bae, D.; Cho, Y.; Lee, J.J.; Cho, S. Reliability-based robust design optimization for torque ripple reduction considering manufacturing uncertainty of interior permanent magnet synchronous motor. *J. Mech. Sci. Technol.* **2020**, *34*, 1249–1256. [CrossRef]
25. Gieras, J.F.; Wang, C.; Lai, J.C. *Noise of Polyphase Electric Motors*; CRC Press: Boca Raton, FL, USA, 2018.
26. Cho, S.; Hwang, J.; Kim, C.W. A study on vibration characteristics of brushless dc motor by electromagnetic-structural coupled analysis using entire finite element model. *IEEE Trans. Energy Convers.* **2018**, *33*, 1712–1718. [CrossRef]
27. Kang, B.; Choi, K.K.; Kim, D.H. An efficient serial-loop strategy for reliability-based robust optimization of electromagnetic design problems. *IEEE Trans. Magn.* **2017**, *54*, 7000904. [CrossRef]
28. Taguchi, G. *Introduction to Quality Engineering: Designing Quality into Products and Processes*; Asian Productivity Organization: Tokyo, Japan, 1986.
29. Taguchi, G. *System of Experimental Design; Engineering Methods to Optimize Quality and Minimize Costs*; UNIPUB/Kraus International Publications: White Plains, NY, USA, 1987.
30. Park, G.J. Design of experiments. In *Analytic Methods for Design Practice*; Springer: London, UK, 2007; pp. 309–391.
31. Baron, J.; Hammett, P.; Smith, D. *Stamping Process Variation: An Analysis of Stamping Process Capability and Implications for Design, Die Tryout and Process Control*; Technical Report Prepared for the Auto Steel Partnership Program: Southfield, MI, USA, 1999.
32. Tillman, F.A.; Hwang, C.L.; Kuo, W. Optimization techniques for system reliability with Redundancy—A review. *IEEE Trans. Reliab.* **1977**, *26*, 148–155. [CrossRef]

MDPI

Article

New Control Strategy for Heating Portable Fuel Cell Power Systems for Energy-Efficient and Reliable Operation

Sebastian Zimprich, Diego Dávila-Portals, Sven Matthiesen * and Thomas Gwosch

Institute of Product Engineering, Karlsruhe Institute of Technology (KIT), 76131 Karlsruhe, Germany
* Correspondence: sven.matthiesen@kit.edu; Tel.: +49-721-60847156

Abstract: Using hydrogen fuel cells for power systems, temperature conditions are important for efficient and reliable operations, especially in low-temperature environments. A heating system with an electrical energy buffer is therefore required for reliable operation. There is a research gap in finding an appropriate control strategy regarding energy efficiency and reliable operations for different environmental conditions. This paper investigates heating strategies for the subfreezing start of a fuel cell for portable applications at an early development stage to enable frontloading in product engineering. The strategies were investigated by simulation and experiment. A prototype for such a system was built and tested for subfreezing start-ups and non-subfreezing start-ups. This was done by heating the fuel cell system with different control strategies to test their efficiency. It was found that operating strategies to heat up the fuel cell system can ensure a more reliable and energy-efficient operation. The heating strategy needs to be adjusted according to the ambient conditions, as this influences the required heating energy, efficiency, and reliable operation of the system. A differentiation in the control strategy between subfreezing and non-subfreezing temperatures is recommended due to reliability reasons.

Keywords: control strategies; environmental conditions; fuel cell; heating strategies; hydrogen; portable device; power system reliability; testing

Citation: Zimprich, S.; Dávila-Portals, D.; Matthiesen, S.; Gwosch, T. New Control Strategy for Heating Portable Fuel Cell Power Systems for Energy-Efficient and Reliable Operation. *Machines* **2022**, *10*, 1159. https://doi.org/10.3390/machines10121159

Academic Editor: Kim Tiow Ooi

Received: 27 October 2022
Accepted: 1 December 2022
Published: 3 December 2022

1. Introduction

Hydrogen is a high-potential alternative fuel source, which can help reduce the emission of greenhouse gases into the atmosphere. Using renewable energies, this element can be obtained through the electrolysis of water. Hydrogen fuel cells (FC) can rely on different principles and are found in various sizes; however, all of them generate electrical energy via an electrochemical reaction [1]. Proton exchange membrane fuel cells (PEMFCs) have been developed for a wide span of power outputs, ranging from micro PEMFCs with 100 mW to applications requiring several kW [2–5].

Unlike in the automotive field, smaller portable applications currently on focus on the market-leading lithium-ion batteries and do not take fuel cell systems into account [6]. Considering the energy density of batteries nowadays (usually between 50 and 200 kWh/kg), the added weight from the additional battery units poses a problem for most portable applications [7]. A comparison of the energy and power density for different types of energy storage systems was shown by Julien et al. [7]. Therefore, the search for lighter energy carriers could significantly impact the technology behind battery-powered machines. A weight reduction in the energy carriers could be realized by replacing traditional batteries with PEMFCs.

A proton exchange membrane fuel cell (PEMFC) requires a constant supply of fuel (i.e., hydrogen). A suitable option for this purpose is a metal hydride tank, which stores hydrogen at a low-pressure level compared to pressure tanks, in a dissolved state into metal particles. These may release or absorb the hydrogen gas depending on the pressure inside the tank. With respect to its discharge behavior, the tank presents a constant pressure for a wide range of hydrogen supply. Hence, a metal hydride tank shows an optimal behavior for the supply of hydrogen to smaller and portable PEMFCs, such as the tanks lower pressure level due to safety reasons.

Currently, there are mechanisms that demonstrate better energy and power density values when compared to lithium-ion batteries, which are the industry standard for most applications. This indicates that lithium-ion batteries could be substituted by alternative technologies, such as fuel cell systems. Especially for tasks with increased energy demands due to longer utilization periods, PEMFCs would be advantageous [8]. The key characteristics highlighting PEMFCs' superior functionality for portable applications are their lack of reliance on the power grid, their reliability, and their increased energy density when compared to batteries. PEMFC systems as energy carriers can have an energy density of about 850 Wh/kg. The benefits of PEMFC systems, when compared to classical battery systems, increase with an increasing operation time [7–9].

A portable fuel cells performance is challenged when operating at low (<0 °C) and increased ambient temperatures (>40 °C). Current and power drop with lower temperatures, which leads to a decrease in efficiency [10–13]. With every succeeding subfreezing start-up, the current density can drop further due to permanent damage in the membrane caused by ice formations [1,12]. By electrically heating up a PEMFC stack, it will achieve higher temperatures significantly faster when compared to passive heating caused by its own heat losses [14]. Therefore, the subfreezing start-up of PEMFCs sets a barrier towards its further commercialization [15]. This characteristic trait is found to negatively affect many portable devices. Potential portable applications can be power generators, drone applications, or power tools. Unfortunately, in everyday environments, subfreezing temperatures are fairly common and at times unavoidable. As a result, it is vital to look for appropriate solutions allowing devices powered by a PEMFC to function under subfreezing temperatures.

Due to the weight and size constraints of portable machinery, the available space to construct a solution is quite limited. Different strategies have been proposed to enable the subfreezing start-up of a PEMFC [16–18]. There are two common methods used for the subfreezing start-up of a PEMFC. The first method includes purging the device prior to shutting it down and humidifying it before start-up. This extracts the water inside the membrane, which results from the operation of the PEMFC. The second method heats up the device before start-up in order to reach temperatures above the freezing point of water [12–15,19,20]. However, the first method, which purges and humidifies the device, would require an additional subsystem. As this would require an adequate amount of space and add weight to the system, it would undermine the portable characteristics of the device. In addition, at subfreezing temperatures, this alone cannot reliably prevent the formation of ice, as there is no way to determine whether all water has been removed from the cell.

Many other patents and invention disclosures have solved the subfreezing start-up problem by circulating a previously heated liquid around the FC. This, however, is not suitable for portable applications, as this contradicts the weight and size restrictions [21–25]. By reversing the polarity of the cell, a reverse current flow through the PEMFC generates heat. This may cause degradation within the cell after several uses [15,26,27]. Therefore, this solution is not ideal for the subfreezing start-up of a PEMFC despite its simplicity as well as its low weight and volume.

Inserting a higher amount of hydrogen into the anode causes an exothermic reaction that heats the cathode [28]. Another possibility is to keep the stack temperature in a specified range by using electrical heaters, in turn preventing ice formations [29,30]. Keeping the stack temperature in a specified range over a long period of time would require a great deal of energy, which is not suitable for portable applications. More simplistic solutions have been suggested where the stack has electrical heaters that raise the temperature only prior to the actual start-up [31,32]. Given that this system's functionality adds minimal weight and volume to the device, it may be a suitable solution to heat the small PEMFCs during a subfreezing start-up. In addition, the hydrogen supply is highly dependent on the temperature and is critical for the PEMFCs operation. Therefore, the thermal management strategy for the hydrogen tank has to be considered in portable applications [9].

Due to weight and size constraints of portable machinery, only a limited amount of energy can be used for the heating. The used heating strategy needs to provide enough heating power to achieve this heating process for the PEMFC. As a result, the energy storage has to provide enough energy for the strategy. The control strategy has to take into account the strict weight and size constraints that a portable device imposes. Therefore, the influence of ambient temperature on the heating strategy is especially relevant in portable applications. The ambient conditions influence the required energy, reliability, as well as efficiency of the PEMFC system; however, its effects under portable constraints are currently unknown.

A tradeoff between the heating strategies and portable constraints has to be considered to ensure a reliable operation. The goal of this study is therefore to investigate and validate heating strategies for environments in which portable FC are subjected to subfreezing temperatures.

2. Materials and Methods

In the following chapter, the experimental procedure is presented. This includes the control strategies and modeling, the simulations, the individual experiments, as well as the procedures used to collect the experimental data.

The prototype used for this study is a modified portable cordless screwdriver that is based on a Festool PDC 18/4. It was modified to function using a 20 W PEMFC stack and a small energy buffer. The energy buffer has a maximum power output of 900 W and a capacity of 43.3 kJ. This setup is used for all experiments conducted in this study as well as for the necessary parametrizations needed for the simulation models [33]. This prototype is presented in detail in a preliminary study [33].

2.1. Control Strategy and Study Design

Prior to start-up, when warming up the PEMFC at subfreezing ambient temperatures, potential ice formations in the cathode melts before the internal reaction begins. As a result, damage to the membrane can be avoided. This may also influence the power output and efficiency of the cell. Through the use of different sensors, temperatures and ice formations could be directly measured and/or detected within the cell. Unfortunately, due to the lack of space within the PEMFC membrane, these sensors cannot be included in the design. This leads to uncertainties, which cannot be solved by simulation or component tests. In order to investigate these uncertainties, testing the overall system is necessary.

Therefore, the ambient temperature (T_a) as the test case (TC) and the temperature at which the heater is turned off (T_{HO}) as the control strategy (CS) are investigated in this study. The ambient temperature T_a is varied in two test cases. Test case one (TC1) is a non-subfreezing start-up with T_a = 5 °C. Test case two (TC2) is a subfreezing start-up with T_a = −3 °C.

The heating is performed until a certain measured temperature in the PEMFC is reached. When the temperature is reached the first time, the heater is turned off. By varying the temperature at which the heater is turned off (T_{HO}), the effect of the control strategies on the power output, efficiency, and reliability can be investigated. The PEMFC in this study has a nominal temperature is 50 °C, and this temperature was used for control strategy one (CS1) as T_{HO}. Control strategy two (CS2) has a temperature T_{HO} just over the freezing point of water, with T_{HO} = 5 °C. This demonstrates whether the heating strategy is enough for a safe subfreezing start-up as well as how much energy is required for the different strategies to function properly. The four possible combinations of TC and CS are given in Table 1. The four combinations are tested in this study by experiment and simulation.

Table 1. Control strategy study design with the four tests as combination of CS and TC.

		Ambient Temperature T_a	
		TC1: T_a = 5 °C	**TC2: T_a = −3 °C**
Turn-off temperature T_{HO} in the control strategy	CS1: T_{HO} = 5 °C	Test 1/1	Test 1/2
	CS2: T_{HO} = 50 °C	Test 2/1	Test 2/2

The heaters controller, specifically for the hydrogen tank, may deliver the same output for all experiments carried out at temperatures 15 °C and 25 °C. This is justified by the steady hydrogen flow maintained throughout these tests. This temperature range was obtained from the study conducted by Kyoung et al. It allows the pressure output of the hydrogen tank to lie within a range that allows for normal operation of the PEMFC [9].

The control strategy is made up of three parallel paths. It starts when the device is turned on. On the left side of Figure 1, the state flow diagram for the hydrogen supply control strategy to the PEMFC membrane is shown. Here, the valve is only opened when the temperature of the FC is above 5 °C. When the temperature in the tests decreases and drops below 5 °C, the valve is closed to prevent permanent damage to the membrane. In the middle path, the control strategy for the FC heater is shown. The heating starts when the FC temperature is below T_{HO}. Once T_{HO} is reached for the first time, the start-up heating is completed, and the heater is turned off.

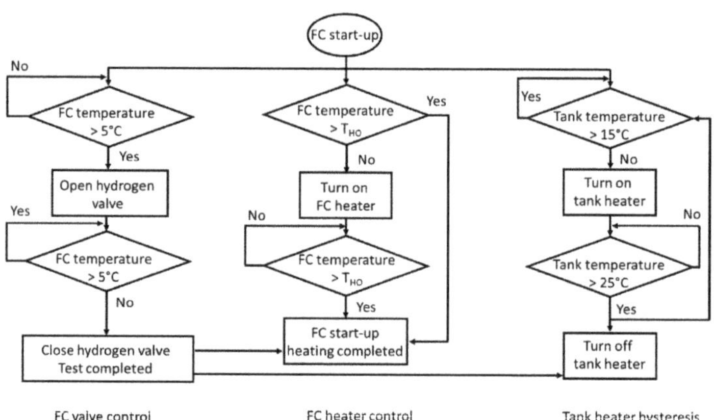

Figure 1. PEMFC's systems control strategy in this study.

The right path in Figure 1 shows the hysteresis controller for the metal hydride tank heater. The hysteresis controller keeps the tank temperature in a range between 15 °C and 25 °C. The tank heater, as well as the FC heater, are stopped by the left part if the FCs temperature drops below 5 °C, as the tests are completed there. The complete control strategy is shown in Figure 1.

2.2. Mathematical Simulation Model

The model design is based on the physical components of the cordless screwdriver. Figure 2 shows the structure of the proposed system, as well as the interaction between individual components. This design focuses on the system's thermal behavior as well as the necessary components needed to apply the heating control strategy. The grey-colored blocks were not considered in the model. Although they belong to the power supply subsystem, they were out of the scope of this study.

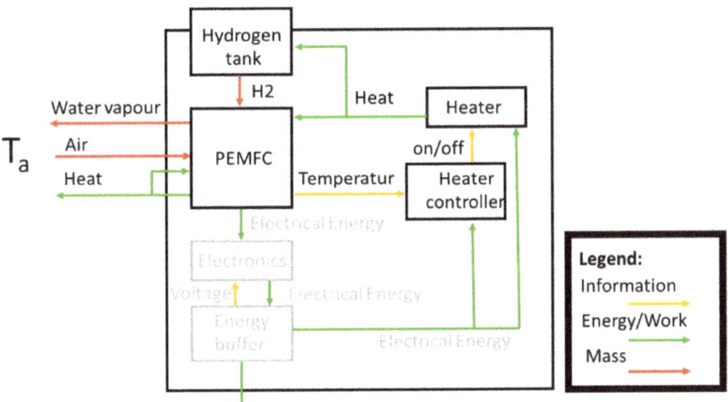

Figure 2. Functional structure of the fuel-cell powered screwdriver system.

The ambient temperature T_a is considered the environmental parameter in this model. It is assumed that similar parameters have little to no influence on the system's behavior. The thermal model used in this study takes two subsystems into account: the PEMFC and the hydrogen tank. Figure 3 shows the thermal circuit for these subsystems. Here, the thermal masses are represented by ellipses, while heat transfers are represented by rectangles. The heaters are represented as triangles. The PEMFC and the hydrogen tank are thermally isolated from each other. The environment is assumed to be an ideal temperature source with constant temperature T_a.

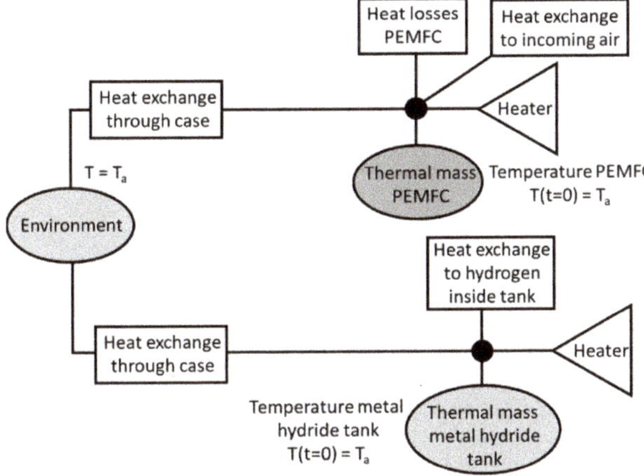

Figure 3. Thermal model of the PEMFC system.

Both the PEMFC and hydrogen storage are modeled as thermal masses with the starting temperature T_a. The thermal mass of the metal hydride tank is measured at 275 J/K. The thermal mass of the PEMFC is modelled with 81 J/K. Heat conduction, convection, and radiation are considered as well as heat exchange with fluids. The effect of the air between the hydrogen tank and the case is negligible. The PEMFCs heater has a heating power of 70 W, while the metal hydride tanks heater has a heating power of 30 W.

$$\dot{Q}_{A\rightarrow B} = \lambda_{A\rightarrow B} * (T_B(t) - T_A(t)) = \sum \dot{Q}_{cond} + \sum \dot{Q}_{conv} + \sum \dot{Q}_{rad} \quad (1)$$

In our model, the heat exchanged between two points is described as the product of the heat conductivity $\lambda_{A \to B}$ and the difference in temperature T between them. As shown in Equation (1), both variables were in this case time-dependent. Alternatively, this can also be expressed as the summation of all heat exchange mechanisms applied to the different points of the heat transfer chain. All three heat exchange mechanisms were considered for the five heat transfer paths, represented as rectangles in Figure 3. For the different heat transfer mechanisms, temperature-dependent coefficients were calculated. These were assumed to be constant throughout the simulation. The single coefficients were not validated by experiment, as this would require a complex experiment for each coefficient [34]. The balance of absorbed/released energy by the reaction inside the PEMFC was calculated using the gravimetric flows $\dot{m}$ for both the incoming reactants and the outgoing products. The corresponding specific enthalpy values were calculated h_x, using the gas's temperatures.

An electrical model of the PEMFC was implemented since its electrical output has the most significant influence on the power supply subsystem. It was modeled as a Thevenin's equivalent circuit with a diode, as proposed by Njoya et al. [35]. The PEMFC's losses as well as the heating power of the heaters were calculated from its electrical model and the heaters, respectively. Equation (2) displays the mathematical definition of its output voltage V. This depends on the open circuit voltage V_{OC}, the number of cells in the stack N, the Tafel slope A of the voltage-current curve, the exchange current i_O, and its current i_{FC} respectively, as well as the response time T_d. Its open-circuit voltage V_{OC} was calculated, as shown in Equations (3), as the product of the voltage constant at nominal operation K_C and the Nernst voltage E_n. The exchange current i_O, shown in Equation (4), is a function of the Boltzmann's constant k, the partial pressures of hydrogen p_{H2} and oxygen p_{O2}, the ideal gas constant R, the Planck's constant h, the activation energy barrier ΔG, and the temperature of operation T. Finally, Equation (5) shows that the Tafel slope A is a function of the ideal gas constant R, the temperature T, and the charge transfer coefficient α.

$$V = V_{OC} - N * A * ln\left(\frac{i_{fc}}{i_0}\right) * \frac{1}{s * T_d/3 + 1} \tag{2}$$

$$V_{OC} = K_C * E_n \tag{3}$$

$$i_0 = \frac{2 * 96485\left[\frac{A * s}{mol}\right] * k * (p_{H_2} + p_{O_2})}{R * h} * e^{-\frac{\Delta G}{R * T}} \tag{4}$$

$$A = \frac{R * T}{2 * \alpha * 96485\left[\frac{A * s}{mol}\right]} \tag{5}$$

The used 20 W PEMFC stack at nominal operation delivers 7.8 V and 2.6 A. The nominal operating temperature range lies at 55 °C and the nominal ambient temperature range between 5 °C and 30 °C. This of course reinforces the argument for using heating strategies for a subfreezing start-up. The nominal hydrogen pressure lies between 0.45 and 0.55 bar depending on the storage temperature. This shows that the temperature of the hydrogen storage needs to be managed as well since it has an important impact on its pressure. It is worth noting that this PEMFC's nominal efficiency is 40%.

The PEMFC had a power output depending on its temperature and electrical load. This demonstrates the importance of implementing heating strategies in order to generate the highest power output from the cell [36]. The implemented model for the metal hydride hydrogen storage for the adsorption pressure of the metal hydride is based on Kyoung et al. [9]. Equation (6) shows the outgoing flow from the tank Q_{out} in relation to the opening cross-section A_{out}, the difference of pressure between the tank and the environment Δp, and the density of the gas ρ.

$$Q_{out} = A_{out} * \sqrt{2 * \frac{\Delta p}{\rho}} \quad (6)$$

The flow between devices was transported through plastic tubes and was assumed to be adiabatic. The amount of heat absorbed by the incoming gases from the PEMFC's membrane was negligible for determining its temperature.

Table 2 shows the list of variables that were considered in the above-mentioned simulation. These variables were considered in the experiments as well.

Table 2. Observed variables in this study.

Controlled Variables	Measured Dependent Variables
Initial temperature of the PEMFC, i.e., ambient temperature	Average mass flow of hydrogen
Start and shutdown temperature of the PEMFC's heater	PEMFC's temperature
Start and shutdown temperature of the hydrogen storage's heater	Hydrogen storage's temperature
On/off states of heaters for the PEMFC and the hydrogen storage	PEMFC's voltage and current
On/off state of hydrogen valve	Voltage source's current

The simulation model considered all relevant subsystems for the investigated thermal behavior. It also included some simplifications along the entire system that left out variables that had a negligible impact on the studied behavior. The model was implemented into Matlab Simulink. The goal of these simulations was to virtually test the required heating energy for the heaters of the PEMFC and hydrogen tank. The PEMFC's temperature was evaluated in these simulations. These could then be related back to heating strategies aiding the subfreezing start-up of a PEMFC-powered portable device. This is achieved by using the XiL-approach (X-in-the-Lopp, Software-in-the-Loop SiL, Hardware-in-the-Loop HiL). This method is carried out by taking a component of a sub-system (HiL) or the algorithm of [37] the heating strategy (SiL) and repeatedly testing its influence on the entire system, while slight changes are made with every iteration.

The simulation was carried out with the four tests listed in Table 1. After that, experiments were carried out to take the PEMFC's efficiency and therefore the reliability of the system into account. The experimental setup is further explained in the following sub-section.

2.3. Test Bench for Experimental Investigation

Figure 4a shows the experimental setup. The prototype was held in place inside an enclosure where the ambient temperature T_a was set to the desired value. The refrigeration unit Huber Unistat 425 was used to cool the air inside [38]. The enclosure was made of 3 mm Plexiglas. The heat exchanger and ventilator were placed on top of the experiments enclosure. A near constant temperature within the enclosure replicated the ambient conditions typically found in real environmental conditions. The necessary sensors and controllers for the tests were attached to the prototype, as shown in Figure 4b.

Figure 5 shows the scheme of the experimental design. The sensors given in Table 3, heaters, and valve were attached to the prototype and connected to their respective control modules. This had a parallel bus interface that enabled the connection to the processor module. The processor module was connected via an Ethernet cable to the computer.

(a) (b)

Figure 4. (**a**) Experimental setup with cooling system; (**b**) prototype used with PEMFC heaters.

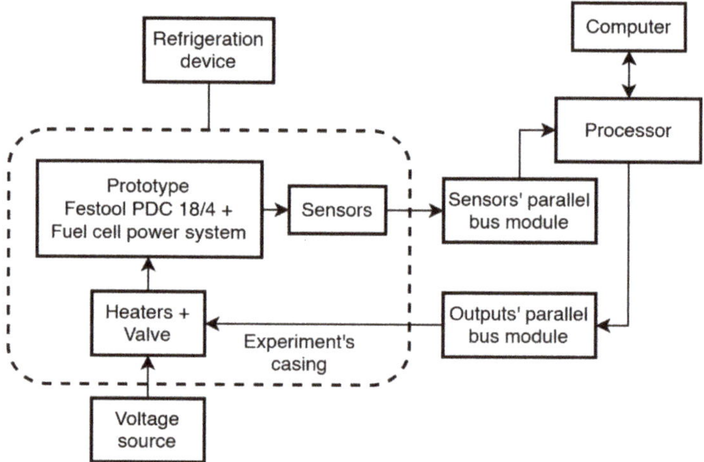

Figure 5. Scheme of the experimental design.

Table 3. Sensors used and variables measured in the experiments.

Sensor/Measuring Device	Variable
Scale	Average Hydrogen consumption
Thermocouple type J	PEMFC's temperature
Thermocouple type J	Hydrogen storage's temperature
Voltage Transducer type LV 25-P	PEMFC's voltage
Current Transducer type CASR	PEMFC's current
Voltage Transducer type LV 25-P	Battery's voltage
Current Transducer type CASR	Battery's current

A ADwin-Pro II was used as a processor for control and data acquisition [39]. The control algorithm was coded, compiled, and monitored using Matlab Simulink. In order to heat up the PEMFC and the hydrogen storage, heating foils were attached to them. Due to the limited space on the PEMFC and the hydrogen tank, multiple 10 W polyimide heating films were used for this study. The required heating energy for the control strategies is unknown. Therefore the heating foils were powered by a supply module EA PS9040-20T. It was set to provide a constant direct current voltage of 18 V, which represented the nominal voltage of the battery that was originally used in this prototype. The PEMFC's output was connected to a resistor with a resistance of 3 Ω, which simulated the ideal load for its nominal working point (i.e., 7.8 V@2.6 A). PEMFC's voltage and current are measured to rate the system's power output. The mean efficiency is calculated by dividing the total power output of the PEMFC by the chemical energy of the hydrogen used.

The required start-up time and heating energy is evaluated for simulation and experiment in the four tests. The start-up is completed when the fuel cell heater and the metal hydride tank heater are both shut off. At this point, the desired temperature T_{HO} is reached, and the hydride tank is in the temperature window between 15 °C and 25 °C. The tank temperature can drop below 15 °C afterward as hydrogen is released to the PEMFC. Therefore, further heating energy is used to keep the metal hydride tank in the desired temperature window between 15 °C and 25 °C. Since the start-up is already completed, this energy is therefore not included in the evaluation of the start-up energy. By comparing the required heating time and energy between experiment and simulation, the validity of the simulation is evaluated. For this, the relative error between simulation and experiment is calculated for the start-up time and heating energy.

3. Results

The following sections presents the results obtained for the conducted simulations and experimental investigations. Simulations were conducted to investigate the effect the control strategy and test cases have on the thermal behavior and required heating energy. Taking the reliable operation and efficiency into account, the experimental investigations were carried out to validate the simulations.

3.1. Simulation Results

The simulations of the PEMFC power supply system were performed, using the proposed heating strategies, for the four different test cases. The ambient temperature (i.e., the initial temperature) T_a of the system and the temperature at which the heating strategies shut down the heaters T_{HO} were varied in these simulations. Figure 6 displays the necessary heating energy over time for the different simulations that were conducted.

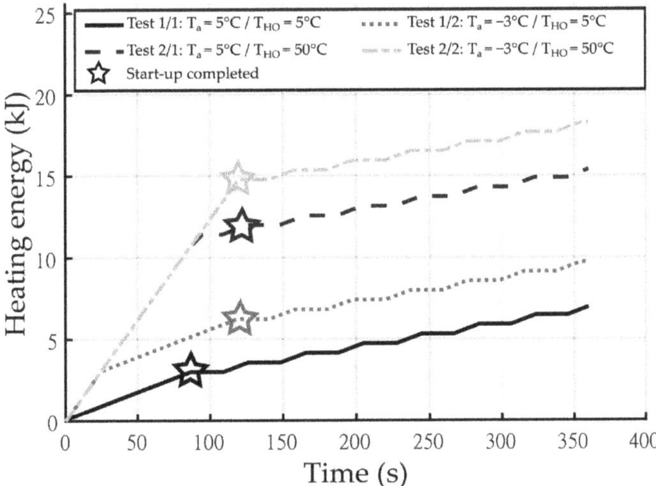

Figure 6. Heating energy required for the heaters of the PEMFC and hydrogen tank.

The tests showed a steep slope when both heaters (of the PEMFC and the hydrogen tank) were turned on. A smaller slope was observed when only the heater for the hydrogen tank was heating and no slope when both heaters were turned off.

The simulations with a lower ambient temperature T_a required more heating energy compared to those with a higher ambient temperature. This can be seen in Figure 6, when looking at the steeper slope for test 1/2, test 2/2, and test 2/1 at the beginning of the simulation compared to test 1/1. The steeper slope in the three tests is due to the fact that the fuel cell heater was switched on. In the other tests, only the tank heater was turned on at the start.

After some time, the slope decreased when the fuel cell heater was turned off, showing the hysteresis of the hydrogen tank's heater for the selected temperature range. More heating energy was required when the heating strategy shut down the heater of the PEMFC at higher temperatures T_{HO}. In the simulation, T_{HO} had a larger effect on the required heating energy than T_a.

3.2. Experimental Results

With the selected heating strategies and the different tests carried out (see Table 1), the conducted experiments demonstrated how control strategies in a portable system behave with respect to its components. Figure 7a shows the development of the PEMFC's average temperature over time. Test 1/1, starting at T_a = 5 °C, showed only a slow temperature rise during its operation, as T_{HO} was also set to 5 °C, and therefore, no energy was used to heat the PEMFC. However, energy was used to heat the metal hydride tank to ensure a steady hydrogen flow. The lower the starting temperature, the longer it took to reach higher temperatures. Changing the ambient temperature T_a to −3 °C in test 1/2 and shutting off the heater at T_{HO} = 5 °C showed a drop in the PEMFC's temperature. A temperature above 5 °C at a subfreezing ambient temperature could not be maintained without external heating, as the temperature dropped after turning off the heater. Test 2/1 and test 2/2 had a tendency to reach a stable operating temperature after shutting down the PEMFC's heaters. The longest time that the heating strategies took to reach their shutdown temperature was 270 s, which occurred in test 2/2. After the temperature T_{HO} is reached, the PEMFC heater shuts off. If the temperature drops, afterward, the heating is not restarted by the control strategy; instead, as in the study, the start-up behavior is investigated. However, the control strategy prevents a PEMFC operation below 5°C to prevent permanent damage to the membrane.

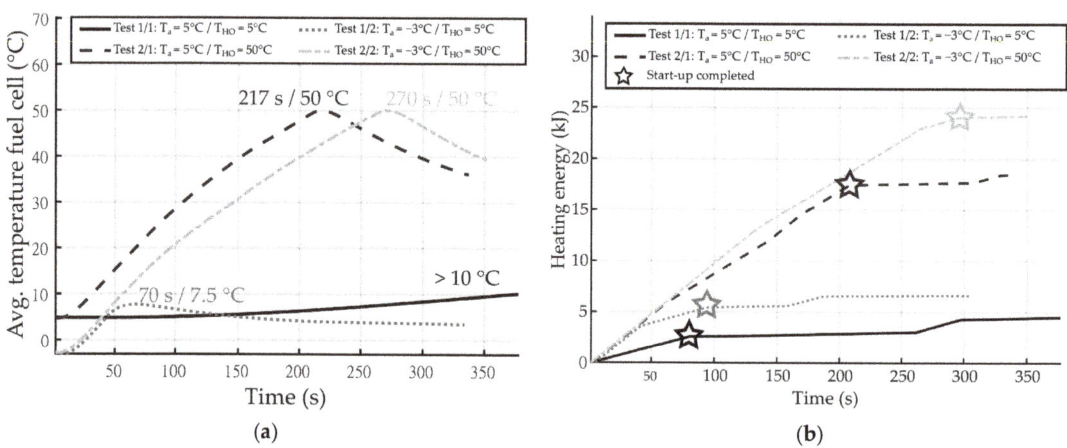

Figure 7. (**a**) PEMFC's average temperature development over time for different heating strategies; (**b**) heating energy of the PEMFC and hydrogen tank required in the four tests.

Figure 7b displays the heating energy required for the different tests. The control strategies that had higher shut-off temperatures T_{HO} for the heater, required more energy compared to those where the shut-off temperature was lower. The same applied for the experiments where the ambient temperature T_a was lower in comparison with those where it was higher. After reaching the temperature T_{HO}, when the PEMFC's heater shuts off, the energy curve's slope decreased, as only the hydrogen tank was heated to stay in a temperature range from 15 °C to 25 °C. In total, the maximum energy required throughout the tests was in test 2/2, with approximately 25 kJ. It is worth noting that test 2/2 had the lowest ambient temperature and highest shut-off temperature.

Table 4 compares the required start-up time and energy for the four tests in simulation and experiment. Test 1/1 had the smallest relative model error between simulation and experiment for the start-up time (8.0%) The smallest relative model error between simulation and experiment for the start-up heating energy (14.6%) was derived from test 1/2. The largest relative model error for the start-up time was shown in test 2/1 (−87.7%). The largest relative model error for the start-up energy was shown in test 2/2 (−38.6%). The negative relative model error indicates a larger value in the experiment compared to the simulation.

Table 4. Required start-up heating time and energy in simulation and experiment.

Test	Simulation Start-Up Time and Energy	Experiment Start-Up Time and Energy	Relative Model Error
1/1	86.8 s/3.0 kJ	80.3 s/2.6 kJ	8.0%/16.5%
1/2	118.6 s/6.2 kJ	94.2 s/5.4 kJ	−54.4%/14.6%
2/1	94.7 s/11.4 kJ	207.9 s/17.4 kJ	−87.7%/−34.4%
2/2	119.7 s/14.8 kJ	296.7 s/24.1 kJ	−59.6%/−38.6%

Figure 8a shows the required heating power. Since the heater of the hydrogen tank consumed approx. 30 W, and the heater of the PEMFC consumed approx. 70 W, the required power for heating the PEMFC and the power for heating the hydrogen tank can be separated. For all tests, energy is used to heat the hydrogen tank after the PEMFC is fully heated up. In test 1/1, the energy is only used to heat the hydrogen tank.

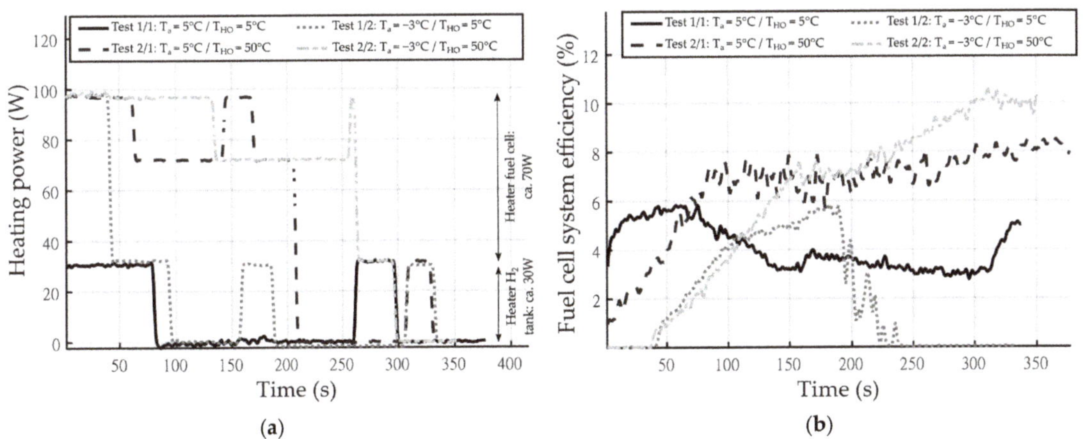

Figure 8. (**a**) Power consumption over time of the four for different tests; (**b**) PEMFC's efficiency over time for the four different tests.

Figure 8b shows the efficiency of the PEMFC system. Tests in which the system was heated up to T_{HO} = 50 °C reached a higher mean efficiency (test 2/1 = 6.55%, test 2/2 = 6.69%)

compared to the tests with heating up to T_{HO} = 5 °C (test 1/1 = 4.04%, test 1/2 = 3.39%). In test 1/1, the PEMFC's heaters were not turned on. Therefore, the curve corresponding to this test shows a lower efficiency for the system, with no significant rise of efficiency for the duration of the experiment. Test 1/2 reached a zero power output around 240 s. When compared to Figure 7a, it is seen that at this point, the PEMFC's temperature dropped below 5 °C. At this point, the PEMFC is switched off by the control strategy.

After conducting these four tests, further testing revealed a considerable drop in power output of the PEMFC for the same testing parameters.

4. Discussion

The following section discusses the proposed heating control strategies in order to reliably operate a portable FC system in subfreezing temperatures. First, the effect to the required energy is discussed. Then, the influence of the control strategies on efficiency and reliability is analyzed.

4.1. Required Energy and Thermal Behaviour

The variables measured in the experiments show the behavior that was expected from the simulative investigation. In addition, the experimental investigation allowed for the efficiency and reliability to be taken into account. The amount of power required by the heaters in all four tests was around 100 W (divided into approx. 30 W for the hydrogen tank's heater and 70 W for the PEMFC heater). This can be delivered by commercially available energy buffers, which are required in such a system as buffer storage between the PEMFC and the motor electronics. The ambient temperature T_a does influence the amount of time the heater is turned on. However, the temperature T_a does have an impact on the required energy for the heaters. More heating energy is saved if the system is set at higher temperatures, and the heating strategies shut down the heaters at lower temperatures. This behavior was also shown by Oszcipok et al. Here, the amount of energy required for heating the PEMFC is highly dependent on the ambient temperature T_a due to the heat transfer between the prototype and its surroundings [14].

When looking at the required heating energy for the tests, there are deviations between the simulations and the experiments. However, the qualitative progression between simulation and experiment matches all four tests. Both investigations showed that the energy consumption is higher when both heaters are turned on. This behavior can be seen in Figures 6 and 7b, where the three different slopes for the curve can be observed. The steepest slope corresponds to the case where both heaters were turned on. Next, the second steepest slope corresponds to the case where only the hydrogen tank's heater was on. Finally, the flattest the three curves (ca. 0%) corresponds to the test where both heaters were turned off.

Test 1/1 showed the smallest deviation between simulation and experiment, calculated with the relative model error, for the start-up time. This test had the smallest start-up time with a positive relative model error. The relative model error of the required start-up heating energy is smallest in test 1/2. The relative model error for the start-up time is negative in test 1/1 and larger in its absolute value compared to test 1/1. Test 2/2 had the longest start-up time and showed the largest absolute relative model error for the required heating energy. Both model errors were negative in this test. Therefore, in the four tests, the absolute relative model error for the heating energy increased with an increasing start-up time. The relative model error turned with increasing starting time from positive to negative. This indicates that the heating losses were underestimated in the simulation. The heat coefficients and heat flows were assumed too low. The error adds up in the transmitted energy with increasing time. This leads to an underestimation of the required heating energy in the simulation. With an increasing start-up time, this leads to an increasing negative relative model error. The underestimated heating losses does not have a large effect in test 1/1 due to the small temperature difference between T_a and T_{HO}, therefore resulting in a small start-up time. With the larger temperature difference in the three other tests, the effect of this error increases and leads to a negative relative model error. As the relative model error for the heating energy in test 1/1 is quite small at 16.5%, it can be assumed that the thermal masses were modelled in the right range. The experimental determination of heat transfer coefficients would reduce the error in the simulations and make the results more valid.

With the simulations, an energy buffer can be selected to carry out the heating strategies at an early development stage to enable frontloading. Furthermore, with the experiments, it can be determined if this energy buffer is sufficient for the device to function accordingly. The results shown in Figure 7a indicate that for the worst-case scenario (i.e., test 2/2: subfreezing start-up and heating strategy up to the PEMFC's nominal operating temperature), the amount of energy required from the energy buffer is ca. 24 kJ. This amount of energy can be provided by the energy buffer storage used in the prototype, which has a capacity of 43.3 kJ.

The ambient temperature T_a has an impact on the required heating energy in the control strategy. The temperature at which the heaters are turned off, T_{HO}, which is directly implemented in the control strategy, had a higher impact on the required energy, as seen in Figure 7b, than the ambient temperature T_a. This shows that portable fuel cell power systems are in principle capable of starting in subfreezing conditions. This also shows that the control strategy can have a higher impact on the energy requirements, and therefore on the design of the electrical system, compared to that of the ambient temperature.

The simulations carried out applied a handful of simplifications to the model. Notably, simplifications in the heat coefficient caused deviations in the results and are the main limitation for the simulations. Nevertheless, the tendencies presented by the simulation and tests are qualitative similar. As a result, this simulation model can be considered to assess how the system would behave in practice. With these simulations, it was possible to approximate how the PEMFC system would behave under different ambient temperatures and under different parameters for the heating strategies.

4.2. *Influence of the Control Strategies on Efficiency and Reliability*

As Datta et al. showed, the voltage and power output is lower for lower temperatures of the PEMFC. They also showed that the PEMFC's voltage is at 50% of its rated voltage when it is operated at 10 °C [19]. This phenomenon can be seen when looking at the efficiency of the fuel cell, which is highly dependent on its power output. The tests confirm this, as the results show that using a larger amount of heating energy during a cold start-up leads to the higher efficiency of the fuel cell system. However, since more energy is needed for heating, this reduces the overall efficiency in addition. For an optimal control strategy, the heating energy and the efficiency must be taken into account, which are in conflict with each other.

This is seen in Figure 8b, as test 2/1 and test 2/2 reached a higher efficiency but also required a higher amount of energy. This is indicated in Figure 7a. Another aspect is the tendency of the fuel cell's temperature to stabilize itself after the heater has been turned off. This shows that after reaching a sufficiently high temperature, the fuel cell's operation can produce enough heat to keep the device running even under a subfreezing ambient temperature T_a. This phenomenon was also shown by Oszcipok et al. [14]. This shows that using less heating energy for the fuel cell system could be sufficient for a reliable cold start-up. However, this is only the case if the ambient temperature T_a is not too low, as this does not necessarily lead to better results. When comparing the temperature development over time between test 1/1 and test 1/2, it is seen that for ambient temperatures T_a above 0 °C, the device is capable of heating itself. With this, it avoids damaging its membrane, and as a result, the heating strategies are not required to assure the reliability of the system under these conditions.

If the ambient temperature T_a is too low or lies below 0 °C, it cannot heat itself up fast enough to avoid damage to the membrane. This is due to the fact its temperature would drop below the freezing point of water. If it is too cold, and the fuel cell is not sufficiently heated, the heat exchange with the environment can cause the heating strategy to shut down the fuel cell to prevent permanent damages in the membrane. This happens due to the temperature dropping to values too close to or below the freezing point of water (see Figure 7a). Therefore, the goal of reducing the required heating energy is in conflict with optimizing the PEMFC's output.

The phenomenon shown by Cho et al. and by Datta et al. demonstrated how the fuel cell suffers an irreversible performance decay during a subfreezing start-up. On the other hand, Chiang et al. showed that the power output of a fuel cell is reduced for lower temperatures. Both of these phenomena could be reduced and/or totally avoided [12,13,19]. The results of this study could also help patents such as those proposed by Thompson et al. or Jang et al. by allowing these to be applied to a wider range of portable applications [40,41].

If the temperature of the fuel cell drops far enough, this would cause the water inside the cathode to begin to freeze. In turn, this would lead to an irreversible performance decay of the cell. Therefore, the heating strategies are vital for the reliable operation of the fuel cell under ambient temperatures T_a below 0 °C. For subfreezing temperatures, a more reliable control strategy is required, as supplying too little heating energy may cause permanent damage to the fuel cell. For ambient temperatures T_a above 0 °C, a more experimental control strategy to save heating energy can be implemented. For these reasons, the appropriate selection of a heating strategy for the cold start-up of a fuel cell system can significantly influence its efficiency and reliability. Due to reliability reasons, a differentiation in the control strategy between subfreezing and non-subfreezing through the heater shut off temperature is recommended.

5. Conclusions

The influence of the heating strategies during a cold start-up on a portable PEMFC system were investigated experimentally and through simulations. The necessary energy supply required to establish the efficiency and reliability of the fuel cell system were investigated.

To start, for a fuel cell system at an ambient temperature T_a of −3 °C, an energy buffer would have to deliver 25 kJ of energy to the heaters. Notably, this value can be obtained from traditional batteries. This energy buffer is strongly dependent on the control strategy, which is specified by the temperature at which the heater is turned off. It should be noted that the energy buffer impacts the efficiency of the fuel cell system. The ambient temperature, however, has a smaller impact on the required energy than the temperature at which the heater is turned off. The ambient temperature T_a impacts how fast the fuel cell will cool down after the heaters are shut off. The results also revealed that heating the fuel cell above 5 °C leads to a higher power output and efficiency of the fuel cell system.

When a higher efficiency of the fuel cell system is required, a greater energy supply is needed for the heating strategies. Using less heating energy (i.e., heating the PEMFC to a lower temperature) can save energy, but it may impact the reliable operation of the PEMFC system. It is proposed to adjust the control strategy regarding the ambient temperature whether it is subfreezing or non-subfreezing. For subfreezing ambient temperatures, a more reliable control strategy is required. This is to ensure its reliability, as subfreezing temperatures can cause permanent damage to the fuel cell membrane.

These findings help in dimensioning the energy buffer. The heat transfer coefficients are an uncertainty in the simulative design of such a system at an early development stage and should be validated with extra experiments. A miscalculation of the heat transfer coefficients can harm the reliability although a subfreezing start-up would be possible. Sufficient energy must be considered for the portable device to perform its function. The fuel cell loads the energy buffer in operation. Enough charge must be left in the energy buffer to heat up the fuel cell system the next time it is used. Therefore, the fuel cell can only switch off when this critical amount of energy is available in the energy buffer.

The findings with respect to the heating strategies help to improve reliability and efficiency during operation of a portable fuel cell system. Subfreezing temperatures were identified as a critical factor of the control strategy. Therefore, the distinction between subfreezing and non-subfreezing ambient temperatures has to be considered in the control strategy to ensure its reliability.

Author Contributions: Conceptualization, S.Z. and D.D.-P.; data curation, S.Z. and D.D.-P.; formal analysis, D.D.-P.; investigation, S.Z. and D.D.-P.; project administration, S.M.; supervision, S.M.; validation, S.Z.; visualization, S.Z. and D.D.-P.; writing—original draft, S.Z. and D.D.-P.; writing—review and editing, S.Z., D.D.-P., T.G. and S.M. All authors have read and agreed to the published version of the manuscript.

Funding: This research received no external funding.

Institutional Review Board Statement: Ethical review and approval were waived for this study due to the fact that it was not a clinical study and did not bear the characteristics of a medical experiment. Only measurement data were recorded non-invasively during an ordinary routine work procedure of one subject.

Informed Consent Statement: Not applicable.

Data Availability Statement: The data that support the findings of this study are available from the corresponding author upon reasonable request.

Conflicts of Interest: The authors declare no conflict of interest.

References

1. Dicks, A.L.; Rand, D.A.J. *Fuel Cell Systems Explained*; Wiley: London, UK, 2018; ISBN 9781118613528.
2. Fraunhofer-Institut für Keramische Technologien und Systeme IKTS. LTCC-PEMFC. Available online: https://www.ikts.fraunhofer.de/content/dam/ikts/abteilungen/elektronik_mikrosystemtechnik/Hybride_Mikrosysteme/IKTS_Ltcc-pemfc.pdf (accessed on 1 March 2022).
3. Ballard Power Systems, Inc. FCveloCity-HD: Fuel Cell Power Module for Heavy Duty Motive Applications. Available online: https://www.ballard.com/docs/default-source/spec-sheets/fcvelocity-hd.pdf?sfvrsn=2debc380_4 (accessed on 23 August 2021).
4. Kaya, K.; Hames, Y. (Eds.) A study on fuel cell electric unmanned aerial vehicle. In Proceedings of the 4th International Conference on Power Electronics and their Applications, Elazig, Turkey, 25–27 September 2019.
5. Krcum, M.; Gudelj, A.; Juric, Z. (Eds.) Fuel cells for marine application. In Proceedings of the 46th International Symposium Electronics in Marine, Zadar, Croatia, 18 June 2004.
6. Weydanz, W. (Ed.) *Power Tools: Batteries*; Elsevier: Amsterdam, The Netherlands, 2009; ISBN 978-0-444-52093-7.
7. Julien, C.; Mauger, A.; Vijh, A.; Zaghib, K. *Lithium Batteries*; Springer International Publishing: Cham, Switzerland, 2016; ISBN 978-3-319-19107-2.
8. Wilberforce, T.; Alaswad, A.; Palumbo, A.; Dassisti, M.; Olabi, A.G. Advances in stationary and portable fuel cell applications. *Int. J. Hydrogen Energy* **2016**, *41*, 16509–16522. [CrossRef]
9. Kyoung, S.; Ferekh, S.; Gwak, G.; Jo, A.; Ju, H. Three-dimensional modeling and simulation of hydrogen desorption in metal hydride hydrogen storage vessels. *Int. J. Hydrogen Energy* **2015**, *40*, 14322–14330. [CrossRef]
10. Al-Baghdadi, M.A.R.S.; Al-Janabi, H.A.K.S. Numerical analysis of a proton exchange membrane fuel cell. Part 1: Model development. *Proc. Inst. Mech. Eng. Part A J. Power Energy* **2007**, *221*, 917–929. [CrossRef]
11. Kazim, A.; Lund, P. Basic parametric study of a proton exchange membrane fuel cell. *Proc. Inst. Mech. Eng. Part A J. Power Energy* **2006**, *220*, 847–853. [CrossRef]
12. Cho, E.; Ko, J.-J.; Ha, H.Y.; Hong, S.-A.; Lee, K.-Y.; Lim, T.-W.; Oh, I.-H. Effects of Water Removal on the Performance Degradation of PEMFCs Repetitively Brought to <0 °C. *J. Electrochem. Soc.* **2004**, *151*, A661. [CrossRef]
13. Chiang, M.-S.; Chu, H.-S. Effects of Temperature and Humidification Levels on the Performance of a Proton Exchange Membrane Fuel Cell. *Proc. Inst. Mech. Eng. Part A J. Power Energy* **2006**, *220*, 435–448. [CrossRef]
14. Oszcipok, M.; Zedda, M.; Hesselmann, J.; Huppmann, M.; Wodrich, M.; Junghardt, M.; Hebling, C. Portable proton exchange membrane fuel-cell systems for outdoor applications. *J. Power Sources* **2006**, *157*, 666–673. [CrossRef]
15. Amamou, A.A.; Kelouwani, S.; Boulon, L.; Agbossou, K. A Comprehensive Review of Solutions and Strategies for Cold Start of Automotive Proton Exchange Membrane Fuel Cells. *IEEE Access* **2016**, *4*, 4989–5002. [CrossRef]
16. Fuller, T.F.; Wheeler, D.J. Start up of Cold Fuel Cell. Patent 17733198, 22 October 1998.
17. Fuss, R.L.; Thompson, E.L. Control System and Method for Starting a Frozen Fuel Cell. U.S. Patent 22237702, 16 August 2002.
18. Min, K.Y.; Jun, K.J.; Uk, K.S.; Hyun, L.J.; Woo, L.N.; Jae, S.I.; Suk, S.W. Purge System for Fuel Cell with Improved Cold Start Performance. U.S. Patent 87231310, 31 August 2010.
19. Datta, B.K.; Velayutham, G.; Goud, A.P. Fuel cell power source for a cold region. *J. Power Sources* **2002**, *106*, 370–376. [CrossRef]
20. Pistono, A.O.; Rice, C.A. Subzero water distribution in proton exchange membrane fuel cells: Effects of preconditioning method. *Int. J. Hydrogen Energy* **2019**, *44*, 22098–22109. [CrossRef]

21. Hobmeyr, R.T.J.; Wexel, D.M. Cold Start Pre-Heater for a Fuel Cell System. U.S. Patent 7368196B2, 3 February 2004.
22. Clingerman, B.J.; Kirklin, M.C.; Rainville, J.D. Cold Start Compressor Control and Mechanization in a Fuel Cell System. U.S. Patent 68490607, 12 March 2007.
23. Limbeck, U. Method to Cold-Start Fuel Cell System at Sub-Zero Temperatures. U.S. Patent 58864505, 4 February 2005.
24. Lin, B. Thermal Control of Fuel Cell for Improved Cold Start. U.S. Patent 2006024378, 21 June 2006.
25. Limbeck, U. Cold Start Facility for Fuel Cells at below Zero Temperatures Has Heated Fluid Circulated through Cell. Patent 102004005935, 6 February 2004.
26. Travassos, M.A.; Lopes, V.V.; Silva, R.A.; Novais, A.Q.; Rangel, C.M. Assessing cell polarity reversal degradation phenomena in PEM fuel cells by electrochemical impedance spectroscopy. *Int. J. Hydrogen Energy* **2013**, *38*, 7684–7696. [CrossRef]
27. Luo, L.; Huang, B.; Cheng, Z.; Jian, Q. Rapid degradation characteristics of an air-cooled PEMFC stack. *Int. J. Energy Res.* **2020**, *44*, 4784–4799. [CrossRef]
28. Plant, L.B.; Rock, J.A. Method of Cold Start-Up of a PEM Fuel Cell. Patent EP00124165, 7 November 2000.
29. Nobuo, F.; Kimihide, H.; Kenji, K.; Tadaichi, M.; Hiroaki, M.; Shinya, S.; Naohiro, Y. Fuel Cell System and Method of Starting the Frozen Fuel Cell System. U.S. Patent 20060141309A1, 20 November.
30. Iwasaki, Y.; Wakabayashi, K. Freeze Protected Fuel Cell System. Patent 03014169, 24 June 2003.
31. Scott, D. FCPM Freeze Start Heater. U.S. Patent 86471604, 9 June 2004.
32. Docter, A.; Frank, G.; Konrad, G.; Lamm, A.; Mueller, J.T. Fuel Cell and Method of Cold Start-up of Such a Fuel Cell. Patent 03004709, 4 March 2003.
33. Zimprich, S.; Matthiesen, S.; Gwosch, T. Hydrogen as Energy Supply for Mobile Applications Using the Example of a Power Tool. *Konstruktion* **2022**, 65–69. [CrossRef]
34. Barroso, J.; Renau, J.; Lozano, A.; Miralles, J.; Martín, J.; Sánchez, F.; Barreras, F. Experimental determination of the heat transfer coefficient for the optimal design of the cooling system of a PEM fuel cell placed inside the fuselage of an UAV. *Appl. Therm. Eng.* **2015**, *89*, 1–10. [CrossRef]
35. Njoya, S.M.; Tremblay, O.; Dessaint, L.-A. A generic fuel cell model for the simulation of fuel cell vehicles. In Proceedings of the 2009 IEEE Vehicle Power and Propulsion Conference (VPPC), Dearborn, MI, USA, 7–10 September 2009; pp. 1722–1729.
36. Horizon Fuel Cell Technologies. H-20 Fuel Cell Stack: User Manual. Available online: https://2fca7bd6-9a97-4678-bc9f-28d3354f6ea2.filesusr.com/ugd/047f54_17ecd01808c4ffa1b478d67ed1e45470.pdf (accessed on 19 August 2021).
37. Schroder, T.; Gwosch, T.; Matthiesen, S. Comparison of Parameterization Methods for Real-Time Battery Simulation Used in Mechatronic Powertrain Test Benches. *IEEE Access* **2020**, *8*, 87517–87528. [CrossRef]
38. Huber Kältemaschinenbau AG. Huber Unistat 425 with Pilot ONE. Available online: https://www.huber-online.com/en/product_datasheet.aspx?no=1050.0010.01 (accessed on 1 March 2022).
39. Jäger Computergesteuerte Messtechnik GmbH. ADwin-Pro II: System and Hardware Description. 2018. Available online: https://smt.at/wp-content/uploads/adwin-pro-ii-system-and-hardware-description.pdf (accessed on 1 March 2022).
40. Fuss, R.L.; Thompson, E.L. Combustion-Thawed Fuel Cell. Patent 11212905, 22 April 2005.
41. Sang, J.K.; Han, K.J. End Cell Heater for Fuel Cell. U.S. Patent 20210007185, 14 July 2017.

Article

Model-Based Control Design of an EHA Position Control Based on Multicriteria Optimization

Matthias Dörr *, Felix Leitenberger, Kai Wolter, Sven Matthiesen and Thomas Gwosch

Karlsruhe Institute of Technology (KIT), 76131 Karlsruhe, Germany
* Correspondence: matthias.doerr@kit.edu; Tel.: +49-721-608-47041

Abstract: For the control of dynamic systems such as an Electro-Hydraulic Actuator (EHA), there is a need to optimize the control based on simulations, since a prototype or a physical system is usually not available during system design. In consequence, no system identification can be performed. Therefore, it is unclear how well a simulation model of an EHA can be used for multicriteria optimization of the position control due to the uncertain model quality. To evaluate the suitability for control optimization, the EHA is modeled and parameterized as a grey-box model using existing parameters independent of test bench experiments. A method for multi-objective optimization of a controller is used to optimize the position control of the EHA. Finally, the step responses are compared with the test bench. The evaluation of the step responses for different loads and control parameters shows similar behavior between the simulation model and the physical system on the test bench, although the essential phenomena could not be reproduced. This means that the model quality achieved by modeling is suitable as an indication for the optimization of the control by simulation without a physical system.

Keywords: mechatronic system modeling; electro-hydraulic actuator; control system; control optimization; mechanical system; test bench; position control

Citation: Dörr, M.; Leitenberger, F.; Wolter, K.; Matthiesen, S.; Gwosch, T. Model-Based Control Design of an EHA Position Control Based on Multicriteria Optimization. *Machines* **2022**, *10*, 1190. https://doi.org/10.3390/machines10121190

Academic Editor: Christoph M. Hackl

Received: 27 October 2022
Accepted: 6 December 2022
Published: 8 December 2022

1. Introduction

For fast linear movements under high loads, highly integrated Electro-Hydraulic Actuators (EHA) are often used in aerospace and industrial applications as an alternative to conventional hydraulic actuators. An EHA is a self-contained actuator that operates by electrical power. EHAs consist of at least a hydraulic cylinder, a hydraulic line system, a hydraulic pump, a motor, and power electronics.

Implementing a closed-loop position control for an EHA is difficult due to nonlinearities [1] and uncertainties such as friction or the dynamics of the piston pump [2]. Therefore, a primary research focus is the development and validation of different control approaches. For this purpose, classical Proportional–Integral–Derivative (PID) control, control including fuzzy logic [3–5], observer-based controls [6–11], adaptive trajectory controls [12,13], sliding mode control [14–16], or hybrids of the named approaches [17] are used to solve this problem. For application, the parameterization method is also of great importance in addition to the investigation of control schemes.

There are different methods for adjusting controls: metaheuristics, analytical methods, and multi-objective optimization, some of which rely on machine learning. For metaheuristics and other manual experience-based "trial-and-error" approaches, as often performed in industry, a high effort is required, especially if multiple operating points, such as different loads, need to be tested. Therefore, these methods are primarily used when a physical EHA, but not a suitable model of an EHA, is available for testing.

The mathematical modeling in terms of poles and zeros for the use of analytical methods is difficult for the EHA, since nonlinearities, such as friction and uncertainties in the parameterization, can lead to significant deviations in the overall result. A solution to overcome this problem is the system identification of a physical EHA on a test bench. Izzuddin et al. [18] obtained a linear transfer function in discrete form from multi-sine, as well as continuous step input via Auto-Regressive Exogenous (ARX) system identification, and compared control algorithms on the obtained simulation and tests on the test bench. Similar approaches were used to obtain an ARX model from experimental data and to validate the same controller in simulation and on the test bench, which was used to gather the input for system identification [19–21]. Since a physical system must be available to perform the system identification, this approach is unsuitable for early design optimization in product development, where mature prototypes are unavailable.

In addition to the previously mentioned approaches, a frequently used approach is a model-based optimization of the control system. For example, Wonohadidjojo et al. [22] developed an analytical model of the electrohydraulic servo system for PID optimization with Particle Swarm Optimization (PSO), modeling friction and internal leakage. Shern et al. [23] used a PID optimization with improved PSO, which chose between optimization of settling time or overshoot for the analytical model. The PSO was improved by combining two fitness functions for overshoot and settling time with linear weight summation [24]. Other optimization algorithms used for PID optimization for an EHA were, for example Genetic Algorithms [25–27], the Nelder–Mead approach [28], a hybrid algorithm of PSO, and gravitational search algorithms [29] or beetle antennae search algorithms [30].

Performance validation is either based on simulations or simulations and bench testing separately, similar to the development of other control methods. Comparison of simulation results with test bench results is not conducted, except in cases where the simulation model is parameterized with test bench results, by system identification, usually with the ARX model. In early product development, there is usually no physical EHA that can be tested on a test bench. Thus, no data are available for system identification. The data available in the early stages of product development are generally limited to individual components, but not to the system behavior of the EHA.

For an early optimization of the controller in the early stages of product development, it is necessary to optimize the control with a simulation model that is independent of the measurement data from a physical EHA which was tested on a test bench beforehand. The problem is that the validity of simulation models of an EHA parameterized only by data of individual components is unclear, and thus control optimization cannot be performed.

Therefore, this paper investigates how to model an EHA with a grey-box model. This grey-box model is used to optimize the controller parameters with a PSO. Finally, the optimized controller parameters are set on an EHA on a test bench, allowing the comparison of the dynamic behavior of the EHA on the test bench with the dynamic behavior of the simulation model. The main contributions of this article are summarized as follows:

1. Modeling of an EHA with a comprehensive grey-box model independent of data from bench tests, with the publication of all relevant parameters.
2. Application of a method for multi-objective optimization of PID control for optimal control parameters of the EHA using a simulation model and two load cases.
3. Comparison of the system behavior between the simulation model and the test bench with the help of step responses using the optimized control parameters for both load cases.

2. Materials and Methods

2.1. Electro-Hydraulic Actuator and Simulation Model

In this study, an EHA (BAS.50/32.U04.201.200 MI.BA100.G2.P42, AHP Merkle GmbH, Gottenheim, Germany) was investigated. The EHA consists of a double-acting hydraulic cylinder, a supply line system, an internal gear pump, and a permanent magnet-excited synchronous machine. The change in volume flow is continuous and unaffected by valve influences. Therefore, this EHA is a pump-controlled hydraulic system. Compared to valve-controlled systems, they are more efficient because there is no pressure drop (Manring and Fales 2019). An external inverter is used as power electronics to control the EHA (MOVIAXIS MXA81A-016-503-00/XFE24A, SEW-Eurodrive GmbH, Bruchsal, Germany).

In general, the method for multi-objective optimization of the PID position control is independent of the modeling software and could have been performed with modeling software for multi-domain systems such as MATLAB Simscape or Simcenter Amesim. The simulation model in this paper was created with MATLAB 2020a using the extension Simulink (version 10.1). The fifth-order Dormand–Prince equation was used as a fixed step solver. At first, subsystems were created based on their function. Those can be described with the process elements of transformer, converter, source, and sink. Figure 1 shows all subsystems of the simulation model on the top level of the description. Power quantities are exchanged between these subsystems. Examples of power quantities are the current and the voltage U for the electrical power or the torque M and the speed n for the mechanical rotary power. The central element is the control unit (CU), which adjusts the motor's rotational velocity based on the desired position.

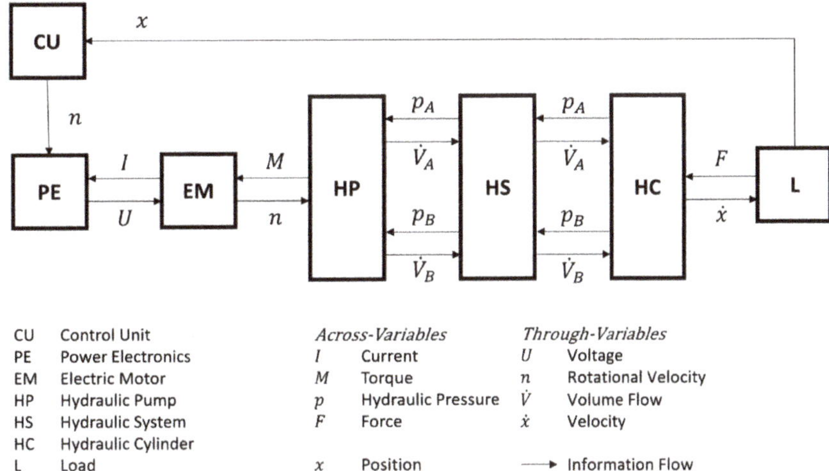

Figure 1. The system structure of the EHA is based on [31]. The central element is the control unit (CU), which is optimized within the scope of this paper.

In the following, the subsystems control unit (CU), power electronics (PE), electric motor (EM), hydraulic pump (HP), hydraulic system (HS), and hydraulic cylinder (HC) are briefly described. It is specified which physical effects are taken into account or are neglected. All relevant parameters are listed in Appendix A. The subsystem load (L) is described in Section 2.3. *Test bench for validation of the simulation model.*

2.1.1. Control Unit (CU)

The control unit contains the parts concerning the position control of the hydraulic cylinder. In this study, the type of control is a PID feedback control. The controller is implemented as a discrete parallel PID controller using the PID blockset of MATLAB Simulink library. The controller uses a sampling frequency of 2 kHz. The position signal is influenced by external effects, which are indicated by noise. This noise can be characterized by a system identification of the used position sensor in the EHA. Three relevant frequencies of white noise with different amplitudes can be identified, which are described in Table 1. These white noises are added to the position signal and filtered with a kHz low-pass filter.

Table 1. The white noise of the position sensor.

Frequency	Amplitude
15 Hz	0.01 mm
200 Hz	0.005 mm
400 Hz	0.003 mm

2.1.2. Power Electronics (PE)

The power electronics consist of a torque-based, field-oriented controller for the internal permanent magnet synchronous motor (PMSM) and an outer-loop speed controller. Therefore, the "Interior PM Controller" model from the powertrain blockset of the MATLAB Simulink library was used [32–36]. The overall inverter efficiency is constant. The current controller uses a sampling frequency of 16 kHz. The "Interior PM Controller" model is able to calculate the optimum current regulator gains based on the parameters of the electric motor. The chosen parameters are listed in Table A1 in the Appendix A.

An outer-loop speed controller was implemented, which is a PI feedback control based on the control unit of the inverter of SEW [37]. The acceleration pre-control gain is set to zero. The parameters are taken from the configuration on the test bench.

2.1.3. Electric Motor (EM)

The electric motor is modeled as a three-phase Interior Permanent Magnet Synchronous Motor (PMSM) with sinusoidal back electromotive force. For this purpose, the "Interior PMSM" model from the Powertrain blockset of the MATLAB Simulink library is used [38,39]. Torque is used as the mechanical input. Physical inertia, viscous damping, and static friction are considered in the model.

2.1.4. Hydraulic Pump (HP)

The hydraulic pump is an internal gear pump modeled as a gray box. The assignment between mechanical and hydraulic energy is performed with a black box model via a data sheet provided by the manufacturer. The speed and the pressure difference are used to determine the torque to be applied by the electric motor. The correlations between the parameters are integrated with a lookup table using the Akima spline as an interpolation and extrapolation method.

Due to the experimental determination of the data, this includes leakage and friction, but only for stationary conditions. Thus, the losses considered are independent of acceleration and pressure changes. The influence of rotational inertia is considered in the electric motor subsystem.

The volumetric flow rate is calculated using the volumetric displacement per rotation unit and the rotational speed. Compressibility is neglected so that incoming and outgoing volumetric flows are equal in magnitude. Since we only consider step responses in our study, we have also neglected the thermal aspect of the fluid system, which is otherwise very relevant. An example of the consideration of temperature and power dissipation is shown in [40]. Therefore, all fluid parameters are time-invariant and independent of pressure or temperature.

2.1.5. Hydraulic System (HS)

The hydraulic system describes all hydraulic parts except the hydraulic pump and cylinder. Therefore, the supply pipes between the hydraulic pump, cylinder, and reservoir are modeled. The hydraulic diagram of the hydraulic system is depicted in Figure 2.

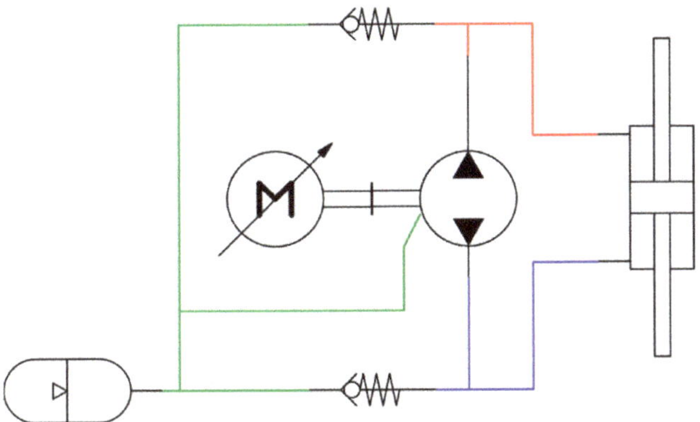

Figure 2. Hydraulic diagram of the EHA.

The behavior of the supply pipes is characterized by the pressure loss Δp due to local hydraulic resistances in the pipelines and their elbows. The pressure loss Δp is calculated with the Darcy–Weisbach equation (Equation (1)) and is dependent on the pressure loss coefficient ζ, which is determined differently for pipe friction and the pipe elbow (Equation (2)). All other parameters such as the density ρ_{oil}, viscosity μ_{oil}, etc. are considered to be time-invariant. The compressibility of the fluid is neglected. The pressure losses are therefore dependent on the volume flow $\dot{V}$.

$$\Delta p = \frac{\zeta * \rho_{oil}}{2} * \left(\frac{\dot{V}}{A}\right)^2 \tag{1}$$

$$\zeta = \zeta_{pipe} + \zeta_{elbow} \tag{2}$$

In the case of pipe friction, the pressure loss coefficient ζ_{pipe} (Equation (3)) is determined via the Darcy friction factor λ (Equation (4)), which depends on the Reynolds number Re (Equation (5)). For transient flow, the Hagen–Poiseuille equation is used. For turbulent flow in smooth conduits, the Blasius correlation is utilized. A Reynolds number greater than 100,000 is not expected.

$$\zeta_{pipe} = \lambda * \frac{L}{R} \tag{3}$$

$$\lambda = \begin{cases} \frac{64}{Re}, & Re < 2320\ (transient) \\ \frac{0.3164}{Re^{0.25}}, & 100,000 > Re \geq 2320\ (turbulent) \end{cases} \tag{4}$$

$$Re = \frac{2 * r * \left|\dot{V}\right|}{A * \mu_{oil}} \tag{5}$$

The pressure loss in the elbow ζ_{elbow} is characterized by the tabulated pressure loss coefficient provided in the Crane Technical Paper [41]. The hydraulic resistances can thus be transferred into characteristic diagrams and calculated separately for both supply pipes. The pipes are considered stiff and the compressibility of the hydraulic fluid is neglected. Therefore, the capacity and inductivity of the pipes are not considered. There is no leakage. The inertia of the hydraulic fluid is considered.

The reservoir must maintain the system pressure. Therefore, it contains a valve that opens when the pressure falls below the expected system pressure. Hydraulic fluid is flowing into the system and is therefore increasing the applied pressure.

2.1.6. Hydraulic Cylinder (HC)

The hydraulic cylinder is a double-rod linear actuator. The modeling is based on the description of Glöckler [42] and uses a force equilibrium. In this subsystem, due to the high pressures and large volumes, the compressibility of the hydraulic fluid is taken into account. The stiffness of the oil column in the cylinder depends on the bulk modulus of the hydraulic oil. The elasticity of the cylinder barrel is neglected.

The volumetric flow of the hydraulic fluid into the actuator is controlled by the output flow of the hydraulic system. The internal leakage is not considered, because it is equal to zero for new systems. The friction force F_{Fr} is modeled as Coulomb friction F_C for static states and viscous friction for dynamic states. The Coulomb friction is constant and the viscous friction is linear dependent on the velocity of the hydraulic cylinder, see Equation (6). All fluid parameters are time-invariant. They are therefore assumed as temperature-independent.

$$F_{Fr} = sgn(x) * (F_C + f_{viscous} * |x|) \tag{6}$$

All relevant parameters are listed in Appendix A.

2.2. *Method for Multi-Objective Optimization of PID Control*

A method for the optimization of a PID control proposed in [43] is used to optimize the PID parameters of the presented EHA using the simulation model. The method is based on a multi-step incremental optimization of the multi-dimensional problem. First, the method is briefly presented. Second, the parameters selected in this study are presented.

For this method, the following five criteria are used: Integrated Time Weighted Square Error (ITSE) criterion Q_{ITSE}, the Q_{t_r}, the settling time Q_{t_s}, the overshoot Q_{h_O} and the noise Q_{Noise}. The mathematical equation of the target functions is shown in Equations (7)–(11). For the EHA, the variables error e, actual position y, step height of the input variable U_s, and time of overshoot t_{h_O}, settled final position y_{final} and the speed of the motor as the output of the controller n are used.

$$Q_{ITSE} = \int_0^\infty t * e^2(t)\ dt \tag{7}$$

$$Q_{t_r} = t_1(y = 0.9 * U_s) - t_0(y = 0.1 * U_s) \tag{8}$$

$$Q_{t_s} = \mathrm{t}*\left(\left|\frac{\mathrm{y(t)} - U_s}{U_s}\right| < 0.1\right) - \mathrm{t}_{h_O} \tag{9}$$

$$Q_{h_O} = \frac{\max\left(\mathrm{y(t)} - y_{final}\right)}{y_{final}} \tag{10}$$

$$Q_{Noise} = \sum|\Delta n(t)| \tag{11}$$

The PID values are optimized using the five criteria presented with a cost function. The main criterion f_1 shift a percentage of the exponent with a certain weight g. The weight g is selected by the user. Furthermore, the exponential function of the main criterion forms a valley around the desired absolute value. The partial costs of the secondary criteria f_2, on the other hand, have continuously decreasing costs towards lower values. This forces the optimization algorithm to search for a solution on the multidimensional Pareto front, which lies around the desired value of the shifted main criterion.

$$f_1\left(x, x_{ref}, \alpha, g\right) = \mathrm{e}^{\alpha*\left|\frac{x - g*x_{ref}}{x_{ref}}\right|} \tag{12}$$

$$f_2\left(x, x_{ref}, \alpha\right) = \mathrm{e}^{\alpha*\frac{x - x_{ref}}{x_{ref}}} \tag{13}$$

Furthermore, criterion f_3 is introduced that represents stability criteria and nonlinear boundary conditions. This results in the overall cost function, which is represented by the following equation.

$$F(x_1,\ldots,x_n) = \prod_{i=1}^{n} f_1\left(x_i, x_{i,ref}, \alpha_i\right) * f_2\left(x_k, x_{k,ref}, \alpha, g\right) * f_3(s), \\ i \in \boldsymbol{A},\ k \in \boldsymbol{B} \\ \boldsymbol{A} = \{1,\ldots,n\},\ \boldsymbol{B} = \{1,\ldots,n\},\ \boldsymbol{A} \cap \boldsymbol{B} \tag{14}$$

The method uses an iterative and relative approach for optimization. Based on the created model, the first step is to determine the possible search space for the PID parameters. In this study, the method of Ziegler and Nichols is used to determine a first—mostly non-matching—parameter set of PID values. Subsequently, an initial optimization is performed with the PSO algorithm. The purpose of this step is to obtain a rough estimate which reflects the system-dependent behavior. The optimization uses only the ITSE criterion Q_{ITSE}. With the obtained PID values the first set of criteria can be determined. These initial values are used to calculate the above mentioned normalization. Based on the nature of a Pareto-optimal problem, the secondary criteria will be degraded during this optimization. If the optimization result does not correspond to the desired goal, another cycle can be started. Here, the main criterion and an associated improvement with the weight g are defined. The result of the previous optimization is used as a reference.

For this study, the objective of controlling the EHA is defined as follows: Reaching the set point as quickly as possible without considerable overshoot and with a smooth controller output. Therefore, three optimization loops are performed including the first optimization with the ITSE criterion. To achieve the control objective, the second optimization aims to minimize the control output by 20% (g = 0.8), and the third aims to reduce the overshoot by 20% (g = 0.8). The limits for PSO in the optimization loops refer to the previous parameters of Ziegler and Nichols, and the previous loop, respectively. The limits are 10% and 1000% for the proportional gain as well as 0.1% and 100,000% for the integral part and derivative part. Table 2 shows all setting values for the optimization in this study.

Table 2. Selected parameters for the application of the method for multi-objective optimization of PID control in this study.

Population size of PSO in all loops	125
Generations of PSO in all loops	10
Objective in system characterization	ITSE ($g = 1$)
Objective in first optimization loop	Rise time ($g = 0.8$)
Objective in second optimization loop	Overshoot ($g = 0.8$)
Range of first system characterization loop	$[0.1 * K_{ZN} \quad 10 * K_{ZN}]$
Range of second optimization loop	$[0.001 * K_{Loop1} \quad 1000 * K_{Loop1}]$
Range of third optimization loop	$[0.001 * K_{Loop2} \quad 1000 * K_{Loop2}]$

2.3. Test Bench for Validation of the Simulation Model

The focus of this study is on the position control of the EHA. To validate the simulation model, a test bench was used to measure and evaluate step responses. A step height of 5 mm is selected for this purpose, as 5% of the maximum stroke is a suitable test case [44].

The test bench is controlled by the measurement and control system Adwin-Pro2 (Jäger Computergesteuerte Meßtechnik GmbH, Lorsch, Germany). The EHA is controlled by an external inverter (MOVIAXIS MXA81A-016-503-00/XFE24A, SEW-Eurodrive GmbH, Bruchsal, Germany), which is coupled to the measurement and control system via an EtherCAT real-time network with a sampling frequency of 2 kHz. As described in the modeling, the control system of the EHA is a cascaded control system. The outer control loop of the EHA is closed with the help of a magnetostrictive linear position sensor (BTL0LP3, Balluff GmbH, Neuhausen auf den Fildern, Germany) via Adwin-Pro2. The position control is performed with a sampling frequency of 2 kHz. The other internal control loops (speed and current control) are implemented directly on the inverter. The speed control is performed with a sampling frequency of 4 kHz and the current control runs with a sampling frequency of 16 kHz. The control system, the systems involved, and the measurement equipment used are shown in Figure 3.

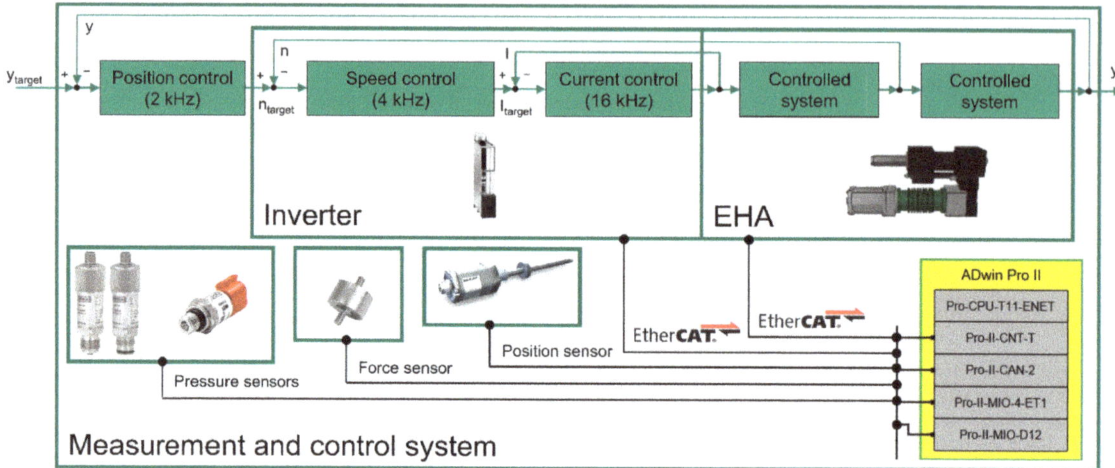

Figure 3. Representation of the control system, the systems involved, and the sensors used.

Two modules are used as loads. In the first load case, an inertial load realized with weights is applied to the EHA. A mass of 11.83 kg was chosen as the weight. The mass is shown in Figure 4a. In aerospace applications, the load increases due to the wind when the actuator extends the wing flap. Therefore, the actuator is moved against a compression spring for the second load case. A spring stiffness of 241.381 N/mm was selected for validation. The compression spring is shown in Figure 4b.

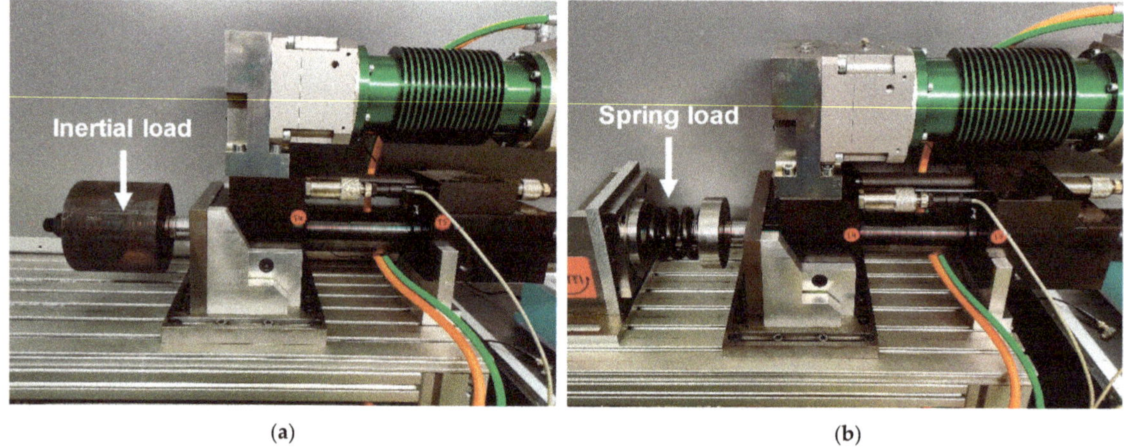

(a) (b)

Figure 4. Modules for applying different loads to the EHA: (**a**) Inertia of 11.83 kg Modell and (**b**) Compression spring with stiffness of 241.381 N/mm.

3. Results

In this section, the optimization results of the PID control based on the simulation model are presented. The simulation model is validated by comparing the various step responses of the simulation model with the step responses obtained on the test bench.

3.1. Results of the Method for Multi-Objective Optimization of PID Control for the Simulation Model

The results in this section were obtained using the simulation model of the EHA. According to the optimization method, initial parameters were first extracted for Ziegler and Nichols. As shown in Table 2, the search space for the PSO is determined based on these initial parameters. For the inertial load, the results of Ziegler and Nichols, the system characterization based on the ITSE criteria, and the optimization loops based on overshoot and rise time are shown in Table 3. The corresponding step responses are shown in Figure 5.

Table 3. PID parameters for Ziegler and Nichols, the system characterization, and the two optimization loops for the inertia of 11.8 kg.

Inertia Load of 11.8 kg	K_P	K_I	K_D
Ziegler and Nichols	13,800	345	0.0007246
System characterization (ITSE)	7827.556886	806.115288	0.724638
First optimization loop (Rise time)	9441.263070	0.806115	25.386430
Second optimization loop (Overshoot)	6167.990306	495.702891	0.025386

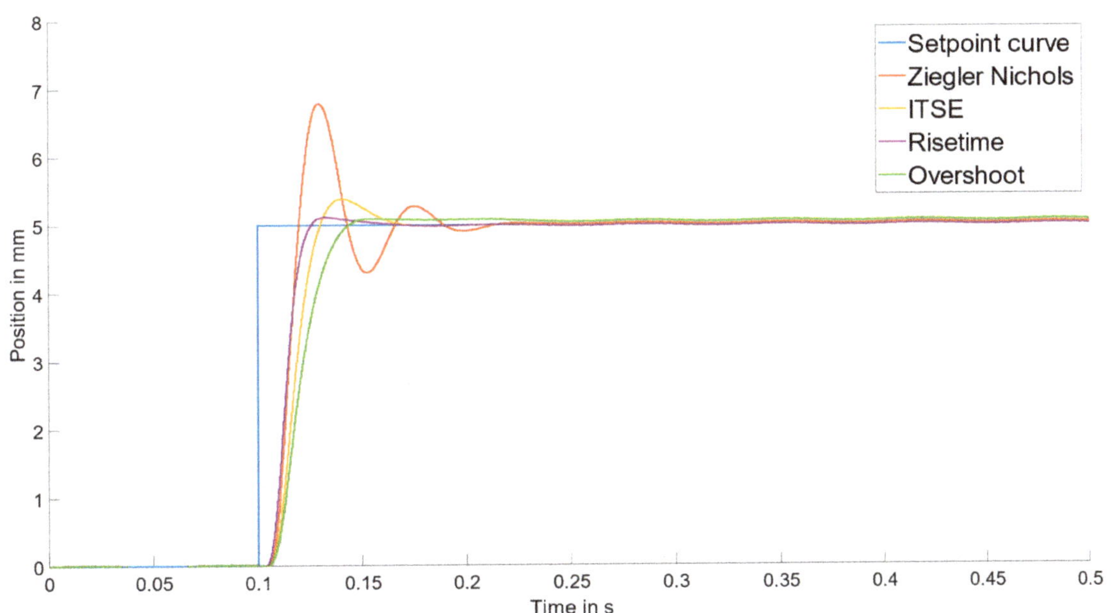

Figure 5. Step responses for the EHA simulation model for the PID parameters presented in Table 3 for the inertia of 11.8 kg.

For the spring load, the results of Ziegler and Nichols and the three optimization loops are shown in Table 4. The corresponding step responses are shown in Figure 6.

Table 4. PID parameters for Ziegler and Nichols, and the three optimization loops for the spring load of a spring stiffness of 241.381 N/mm.

Spring Stiffness 241.381 N/mm	K_P	K_I	K_D
Ziegler and Nichols	13,800	345	0.0007246
System characterization (ITSE)	7663.036665	19,810.42218	0.677059
First optimization loop (Rise time)	8844.394595	19.810422	24.370225
Second optimization loop (Noise)	6441.340341	16809.27533	0.015312

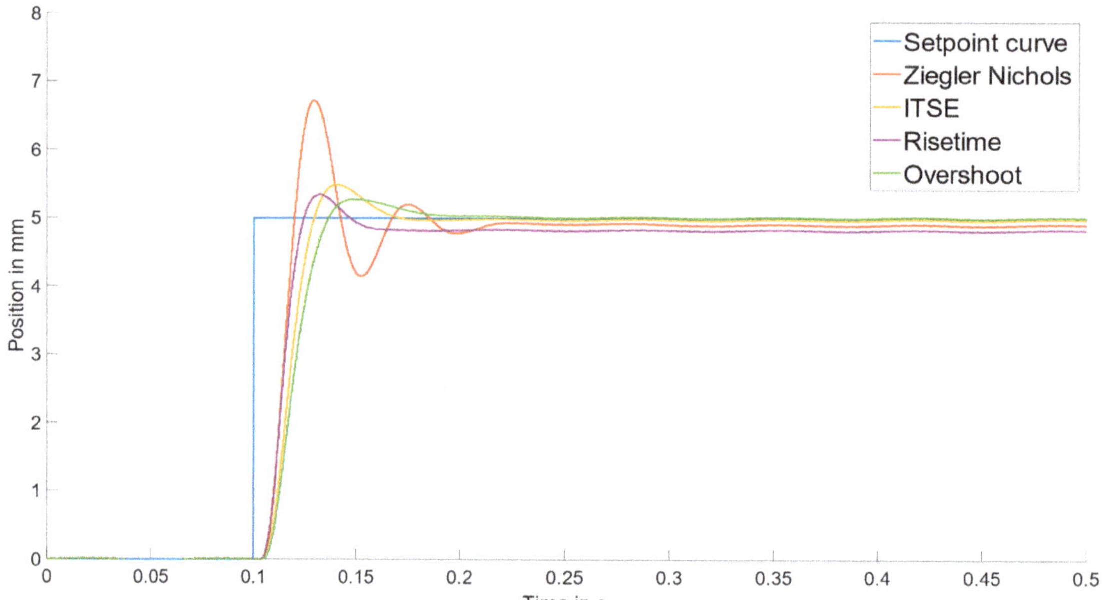

Figure 6. Step responses for the EHA simulation model for the PID parameters presented in Table 4 and the spring load of a spring stiffness of 241.381 N/mm.

3.2. Validation of the Simulation Model for the Electro-Hydraulic Actuator

To validate the simulation model, a comparison is made between the simulated step responses and those measured on the test bench. The step responses for the inertial load using the PID parameter set of Table 3 are shown in Figure 7. The step responses for the spring load for the PID parameters of Table 4 are shown in Figure 8.

The validation results from Figure 7 are described in the following. For Ziegler and Nichols, the step response shows approximately similar behavior with a significant deviation in overshoot and a deviation in rise time. For the system characterization, there are also deviations: While the slope is similar at the start, a sudden change of the gradient can be observed on the test bench at the amplitude of 4 mm. Furthermore, the overshoot is significantly lower for the test bench. For the first optimization loop, a similar behavior except for a smaller overshoot is obtained. For the second optimization loop, a similar behavior with only a tiny change of the gradient at 4 mm is observed.

The validation results from Figure 8 are described in the following. The validation results for the spring load also show significant differences for all four parameter sets. Partially, the results are similar to those for the inertial load in Figure 7. Therefore, the differences in the inertial load are also described.

For Ziegler and Nichols, the step response shows approximately similar behavior to the inertial load. For system characterization, the step response shows approximately similar behavior to the inertial load, except that the settling time of the test bench results is much slower. For the first optimization loop, the results differ significantly: the overshoot is much more significant in the simulation. For the first optimization loop, as shown in Section 3.1, the simulation results show a permanent deviation due to the small integral part. The permanent deviation is not observed on the results of the test bench. For the second optimization loop, in contrast to the inertial load, different behavior is shown. Although the overshoot is the same, it occurs later, and the settling time is significantly slower for the test bench results.

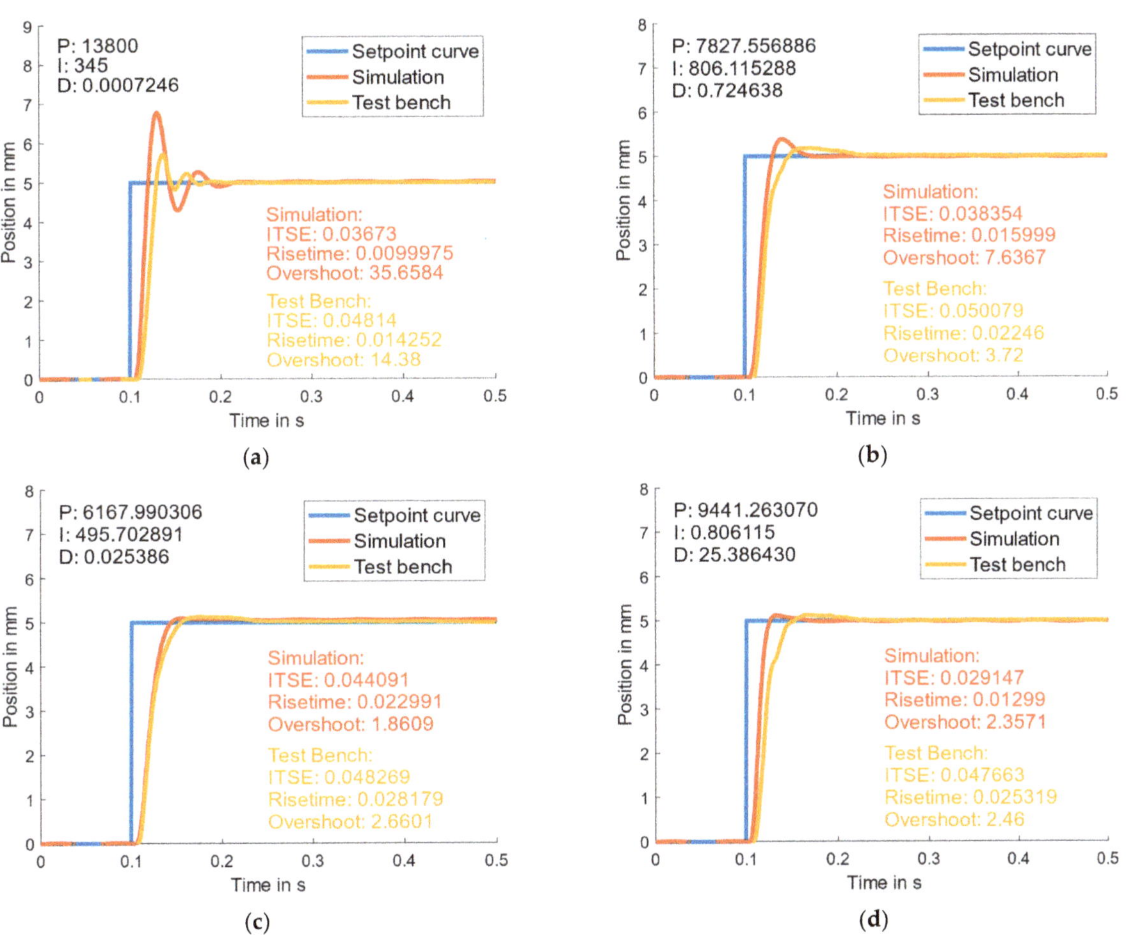

Figure 7. Comparison of the simulated and on the test bench measured step response of the EHA with the following PID parameters obtained with the method for multi-objective optimization of PID control for the inertial load: (**a**) Ziegler Nichols characterization [K_P = 13,800; K_I = 345; K_I = 0.0007246], (**b**) system characterization [K_P = 7827.556886; K_I = 806.115288; K_I = 0.724638], (**c**) first optimization loop [K_P = 9441.263070; K_I = 0.806115; K_I = 25.386430], and (**d**) second optimization loop [K_P = 6167.990306; K_I = 495.702891; K_I = 0.025386].

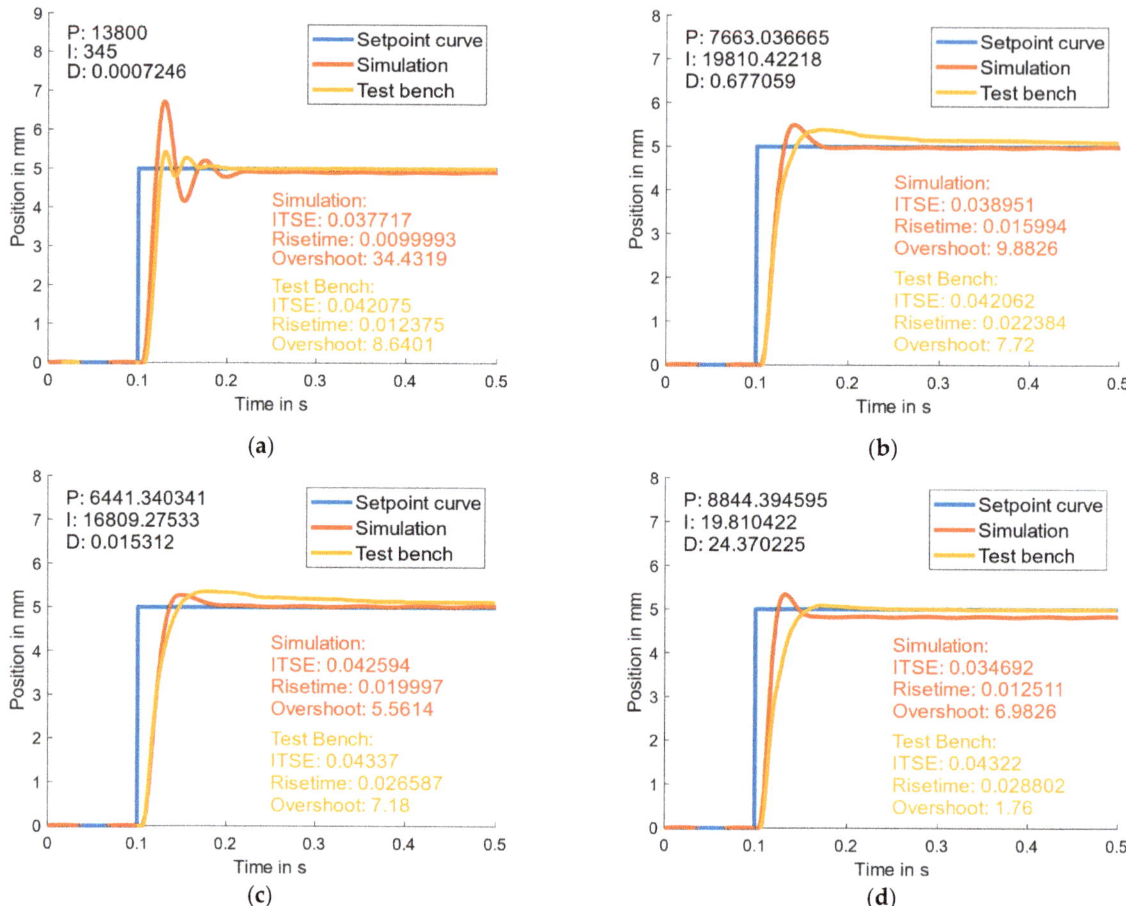

Figure 8. Comparison of the simulated and on the test bench measured step response of the EHA with the following PID parameters obtained with the method for multi-objective optimization of PID control for the spring load: (**a**) Ziegler Nichols characterization [K_P = 13,800; K_I = 345; K_D = 0.0007246], (**b**) system characterization [K_P = 7663.036665; K_I = 19,810.42218; K_D = 0.677059], (**c**) first optimization loop [K_P = 8844.394595; K_I = 19.810422; K_D = 24.370225], and (**d**) second optimization loop [K_P = 6441.340341; K_I = 16,809.27533; K_D = 0.015312].

4. Discussion

In this section, the results of the method for multi-objective optimization of PID control on the simulation model are discussed. Then, the validation of the simulation model using a test bench is discussed as the main result. Finally, limitations and further research directions are elaborated.

4.1. Discussion of the Results of the Method for Multi-Objective Optimization of PID Control

The results show the expected results for the inertial load in Figure 7 for the four parameter sets. While there is a strong overshoot in Ziegler and Nichols, as expected, the three optimizations show a local Pareto-optimum concerning the primary optimization criterion. From the PID values in Table 3, it can be seen that the proportional gain differs significantly. For the optimization of rise time, a high derivative part and a low integral part are found. For the optimization of ITSE and overshoot, a high integral part and a low derivative part are found.

The results for spring loading in Figure 8 also show the expected results in all cases. The results are similar to those for the inertia. For Ziegler and Nichols, almost similar behavior was obtained with the same PID values, but a permanent deviation occurs due to the constant force of the compressed spring. The PID values of Tables 3 and 4 differ mainly in the significantly higher integral part. As expected, the derivative part is most significant for the rise time with the smallest integral part.

The results show that the optimization method, which was designed and validated on rotating systems, also works for translational systems such as the EHA. For the validation of the simulation, the method must reach different local optima so that the validation can be performed with different PID values. Thus, the method has served its purpose and the local optima are suitable for validating the simulation. The use of both loads has proven to be suitable since, in this way, different integral parts and also different operating points can be validated.

The selection of the best PID values depends on the requirements of the overall system, for example, how fast the system should respond and how significant the overshoot should be. The method is thus a suitable way to optimize the PID control based on simple predefined objectives. However, a valid simulation model is a prerequisite. Therefore, the simulation model was validated in Section 3.2 by comparing these step responses with the step responses obtained on the test bench.

4.2. Validation of the Simulation Model with the Test Bench

The validation results for the four parameter sets of the inertial load show partially significant differences between the step responses obtained with the simulation model and on the test bench.

The deviation of the rise time is due to different behavior between the results of the simulation and test bench. While the increase in the position of the paths is similar at the beginning, a sudden change in the slope is observed on the test rig at an amplitude of 4 mm in most cases. Examples are the system characterization, first and second optimization loop for inertial load, where the change of the gradient is present in different magnitudes. For the second optimization loop, a similar behavior with only a tiny change of the gradient at 4 mm is observed. This can be explained by an oscillation between the outer loop of the position controller and the inner loop of the speed controller, which is shown in Figure 3. This oscillation is not adequately represented in the simulation model for the shown operating points shown. A second explanation is the influence of a limitation in the inverter, which was not represented in the simulation.

The deviation in overshoot is due to the damping of the system, which is difficult to simulate in absolute values. It seems that the hardware system has a higher damping coefficient than in the simulation. A second cause is probably the previously discussed change in the gradient.

Another phenomenon is the different response to a high integral part of the controller in the spring load. While the simulation quickly reaches the setpoint at a high integral part, a low integral part results in a permanent offset of the amplitude. The results of the test bench reach the setpoint very slowly at a high integral part of the controller. With a low integral part, the settling time is very slow. This could be explained by the control implemented in the inverter. It must be noted that modeling the control implemented in the inverter is a highly complex task, since not all information, such as the stored motor model, are available. Furthermore, it is unclear how the internal filters of the inverter are designed in detail.

Overall, by comparing the step responses of the simulation model with the test bench results for the inertia or spring load, the key phenomenon is the change in slope at an amplitude of 4 mm, which can be explained by the interaction of the cascaded control on the test rig, across almost all results.

The deviation between the results obtained with the simulation model and with the test bench shows that the simulation model of an EHA parameterized only on data of individual components is not completely valid for control optimization. Nevertheless, it indicates the behavior of the control system that can already be used in product development if no physical EHA is available. Thus, the modeling approach and the modeling itself provide added value to the application. If a physical EHA is available, performing a system identification, e.g., via Regressive Exogenous (ARX), as conducted by [19–21], can be performed to obtain a better fit between the results of the simulation model and the test bench.

4.3. Limitations and Further Research Directions

The results show a strong dependence of the internal control loop on the speed of the inverter. Therefore, modeling and validation of the internal control loop should be a basic prerequisite for the successful execution of a grey-box or white-box simulation. However, if the internal control of the simulation model is not matched to the test bench at a high level of detail, the probability of success is low. In this case, interactions in the cascaded control, as in our study, can influence the overall result. One possibility would be to use the method for optimizing the PID control to parameterize the speed control loop in the simulation model.

The quality of the grey-box simulations depends mainly on the amount and quality of the available data and information. In our case, we had many parameters available, as can be seen from Appendix A. The problem was mainly the lack of information about the cascaded control. Thus, the success of a similar approach depends on the available information.

In the grey-box simulation, many phenomena, such as sensor noise, were implemented. However, there are other phenomena, such as the capacity and inductivity of the pipes, that could have been improved with the appropriate information. To fully evaluate the system reliability of the EHA with the optimized position control used here, the thermal domain would also need to be considered, as temperature also has an impact on the dynamics of the system.

In our study, we investigated an EHA with two loads at a defined step response. Only some of the operating points that the EHA can perform were investigated. More operating points, and also other EHA, should be considered for transferring the results. This does not affect the core statement of this work, which is that the grey-box simulation is only suitable as an indication for control optimization. This can be assumed because the difficulties are present in all EHAs, especially due to the cascaded control. Instead, the amount and quality of information, as mentioned above, take an overriding role.

Given the already achieved convergence of results when using a grey-box model, there is still great potential and further research directions. Future work should investigate if it is possible to obtain better results with more information. To draw insights from the comparison, the representation of the parameters and the mapped phenomena or physical effects, as in this study, are necessary.

Another research direction is to use the simulation model of the EHA as a digital twin by connecting it directly to the EHA on the test bench. By continuously comparing the digital and physical EHA, the deviations can be identified and used to improve the simulation by parameterizing physical parameters, such as friction, that are difficult by generic white-box models. A high model quality can be achieved without relying on a black-box model, which can only be transferred to similar EHAs to a limited extent.

To fully evaluate the system reliability of the EHA with the position control optimized here, the thermal domain should also be considered as well, as temperature also has an impact on the dynamics of the system. For this purpose, thermal coupling systems must be used so that all relevant domains are also taken into account in the sense of the digital twin, to obtain a high degree of reliability concerning the findings obtained.

Author Contributions: Conceptualization, M.D.; methodology, F.L., K.W. and M.D.; software, F.L., K.W. and M.D.; validation, M.D.; formal analysis, F.L. and K.W.; investigation, M.D. and F.L.; data curation, M.D.; writing—original draft preparation, M.D., F.L. and K.W.; writing—review and editing, M.D., F.L., K.W., T.G. and S.M.; visualization, M.D. and F.L.; supervision, T.G. and S.M.; project administration, M.D.; funding acquisition, S.M. and T.G. All authors have read and agreed to the published version of the manuscript.

Funding: This research was funded by the Federal Ministry for Economic Affairs and Climate Action under the funding number 20Y1910E. The authors of this publication are responsible for its contents. The authors gratefully acknowledge the support.

Data Availability Statement: The data that support the findings of this study are available from the corresponding author upon reasonable request.

Conflicts of Interest: The authors declare no conflict of interest.

Abbreviations

The following abbreviations are used in this manuscript:

EHA	Electro-Hydraulic Actuators
PID	Proportional–Integral–Derivative
ARX	Auto-Regressive Exogenous
PSO	Particle Swarm Optimization
CU	Control Unit
PE	Power Electronics
EM	Electric Motor
PMSM	Permanent Magnet Synchronous Motor
HC	Hydraulic Pump
HP	Hydraulic System
HC	Hydraulic Cylinder
ITSE	Integrated Time Weighted Square Error

Appendix A

Table A1. Relevant parameters of the simulation model.

Parameter	Value	
Power Electronics (PE)		
Current controller proportional gain D-axis	12.6	V/A
Current controller proportional gain Q-axis	20.1	V/A
Current controller integral part	2320	V/A∗s
Current controller bandwidth	200	Hz
Speed controller frequency	2000	Hz
Speed controller proportional gain	1000	1/s
Speed controller integral part	100	$1/s^2$
Setpoint speed filter	0.63×10^{-3}	1/s
Actual speed filter	1.23×10^{-3}	1/s
Acceleration precontrol filter	1.25×10^{-3}	s
Electric Motor (EM)		
Nominal speed	40	1/s
Nominal torque	24.8	Nm
Maximal torque	90	Nm
Nominal power motor	6200	W
Reduced rotational inertia	1566	kg/mm^2
Number of rotor polepairs	5	
DC link voltage	560	V
Nominal current	13.8	A
Current limit	55.2	A
D-inductance (field)	10	mH

Table A1. *Cont.*

Parameter	Value	
Q-inductance (torque)	16	mH
Resistance (at 20°C)	1.5	Ω
Flux linkage established by magnets	0.214	Vs
Factor of induced voltage	0.137	V(RMS)/rpm
Limit speed for static friction	0.001	rad/s
Static friction torque	0.5	Nm
Viscous damping	0.005	Nm*s/rad
Hydraulic System (HS)		
Hydraulic fluid density	864	kg/m^3
Hydraulic fluid bulk modulus	1.448×10^9	N/m^2
System pressure	2	bar
Tube A length	160	mm
Tube B length	420	mm
Tube A inner radius	9.0	mm
Tube B inner radius	9.0	mm
Elbow A quantity	1	
Elbow B quantity	3	
Elbow bending radius	40	mm
Hydraulic Cylinder (HC)		
Piston radius	25	mm
Rod radius	16	mm
Dead volume A-Side	2.90×10^{-5}	m^3
Dead volume B-Side	2.90×10^{-5}	m^3
Moving mass of piston and rod	4.19	kg
Coulomb friction force	150	N
Viscous friction coefficient	200	Ns/m

References

1. Sohl, G.A.; Bobrow, J.E. Experiments and simulations on the nonlinear control of a hydraulic servosystem. *IEEE Trans. Control Syst. Technol.* **1999**, *7*, 238–247. [CrossRef] [PubMed]
2. Doan, N.; Yoon, J.I.; Ahn, K.K. Position control of Electro hydrostatic actuator (EHA) using a modified back stepping controller. *J. Korean Soc. Fluid Power Constr. Equip.* **2012**, *9*, 16–22. [CrossRef]
3. Kundu, S.; Bhattacharjee, R.; Chaudhuri, S. Evaluation of Fuzzy-Logic based Position Control Strategies for an Electrohydraulic Actuation System. In Proceedings of the 2021 International Conference on Advances in Electrical, Computing, Communication and Sustainable Technologies (ICAECT), Bhilai, India, 19–20 February 2021; pp. 1–7.
4. Bessa, W.M.; Dutra, M.S.; Kreuzer, E. Sliding Mode Control with Adaptive Fuzzy Dead-Zone Compensation of an Electro-hydraulic Servo-System. *J. Intell. Robot. Syst.* **2010**, *58*, 3–16. [CrossRef]
5. Guo, Q.; Shi, G.; Wang, D. Adaptive Composite Fuzzy Dynamic Surface Control for Electro-Hydraulic-System, with Variable-Supply-Pressure. *Asian J. Control* **2020**, *22*, 521–535. [CrossRef]
6. Guo, K.; Wei, J.; Fang, J.; Feng, R.; Wang, X. Position tracking control of electro-hydraulic single-rod actuator based on an extended disturbance observer. *Mechatronics* **2015**, *27*, 47–56. [CrossRef]
7. Nguyen, M.H.; Dao, H.V.; Ahn, K.K. Active Disturbance Rejection Control for Position Tracking of Electro-Hydraulic Servo Systems under Modeling Uncertainty and External Load. *Actuators* **2021**, *10*, 20. [CrossRef]
8. Won, D.; Kim, W.; Shin, D.; Chung, C.C. High-Gain Disturbance Observer-Based Backstepping Control With Output Tracking Error Constraint for Electro-Hydraulic Systems. *IEEE Trans. Control Syst. Technol.* **2015**, *23*, 787–795. [CrossRef]
9. Chao, Q.; Zhang, J.; Xu, B.; Huang, H.; Pan, M. A Review of High-Speed Electro-Hydrostatic Actuator Pumps in Aerospace Applications: Challenges and Solutions. *J. Mech. Des.* **2019**, *141*, 050801. [CrossRef]
10. Yao, J.; Wang, P.; Dong, Z.; Jiang, D.; Sha, T. A novel architecture of electro-hydrostatic actuator with digital distribution. *Chin. J. Aeronaut.* **2021**, *34*, 224–238. [CrossRef]
11. Yang, X.; Ge, Y.; Deng, W.; Yao, J. Observer-based motion axis control for hydraulic actuation systems. *Chin. J. Aeronaut.* **2022**, *in press*. [CrossRef]
12. Nguyen, M.T.; Doan, N.C.N.; Hyung, G.P.; Kyoung, K. Trajectory control of an electro hydraulic actuator using an iterative backstepping control scheme. *Mechatronics* **2015**, *29*, 96–102. [CrossRef]
13. Yang, X.; Deng, W.; Yao, J. Neural Adaptive Dynamic Surface Asymptotic Tracking Control of Hydraulic Manipu-lators With Guaranteed Transient Performance. *IEEE Trans. Neural Netw. Learn. Syst.* **2022**, *2022*, 3141463. [CrossRef]

14. Chen, G.; Liu, H.; Jia, P.; Qiu, G.; Yu, H.; Yan, G.; Ai, C.; Zhang, J. Position Output Adaptive Backstepping Control of Electro-Hydraulic Servo Closed-Pump Control System. *Processes* **2021**, *9*, 2209. [CrossRef]
15. Cerman, O.; Hušek, P. Adaptive fuzzy sliding mode control for electro-hydraulic servo mechanism. *Expert Syst. Appl.* **2012**, *39*, 10269–10277. [CrossRef]
16. Lin, Y.; Shi, Y.; Burton, R. Modeling and Robust Discrete-Time Sliding-Mode Control Design for a Fluid Power Electrohydraulic Actuator (EHA) System. *IEEE ASME Trans. Mechatron.* **2013**, *18*, 1–10. [CrossRef]
17. Feng, L.; Yan, H. Nonlinear Adaptive Robust Control of the Electro-Hydraulic Servo System. *Appl. Sci.* **2020**, *10*, 4494. [CrossRef]
18. Izzuddin, N.H.; Faudzi, A.M.; Johari, M.R.; Osman, K. System identification and predictive functional control for electro-hydraulic actuator system. In Proceedings of the 2015 IEEE International Symposium on Robotics and Intelligent Sensors (IRIS), Langkawi, Malaysia, 18–20 October 2015; pp. 138–143.
19. Liang, X.W.; Mohd Faudzi, A.A.; Ismail, Z.H. System Identification and Model Predictive Control using CVXGEN for Electro-Hydraulic Actuator. *Int. J. Integr. Eng.* **2019**, *11*, 04018. [CrossRef]
20. Ishak, N.; Yusof, N.M.; Azahar, W.N.A.W.; Adnan, R.; Tajudin, M. Model identifiction and controller design of a hydraulic cylinder based on pole placement. In Proceedings of the 2015 IEEE 11th International Colloquium on Signal Processing & Its Applications (CSPA), Kuala Lumpur, Malaysia, 6–8 March 2015; pp. 198–202, ISBN 978-1-4799-8249-3.
21. Rahmat, M.F.; Rozali, S.M.; Wahab, N.A.; Jusoff, K. Modeling and Controller Design of an Electro-Hydraulic Actuator System. *Am. J. Appl. Sci.* **2010**, *7*, 1100–1108. [CrossRef]
22. Wonohadidjojo, D.M.; Kothapalli, G.; Hassan, M.Y. Position Control of Electro-hydraulic Actuator System Using Fuzzy Logic Controller Optimized by Particle Swarm Optimization. *Int. J. Autom. Comput.* **2013**, *10*, 181–193. [CrossRef]
23. Shern, C.M.; Ghazali, R.; Horng, C.; Jaafar, H.I.; Chong, C.S.; Md Sam, Y. Performance Analysis of Position Tracking Control with PID Controller using an Improved Optimization Technique. *Int. J. Mech. Eng. Robot. Res.* **2019**, *8*, 401–405. [CrossRef]
24. Shern, C.M.; Ghazali, R.; Horng, C.S.; Soon, C.C.; Ghani, M.F.; Sam, Y.M.; Has, Z. The Effects of Weightage Values with Two Objective Functions in iPSO for Electro-Hydraulic Actuator System. *J. Adv. Res. Fluid Mech. Therm. Sci.* **2021**, *81*, 98–109. [CrossRef]
25. Bellad, K.; Hiremath, S.S.; Singaperumal, M.; Karunanidhi, S. Optimization of PID Parameters in Electro-Hydraulic Actuator System Using Genetic Algorithm. *Appl. Mech. Mater.* **2014**, *592–594*, 2229–2233. [CrossRef]
26. Elbayomy, K.M.; Zongxia, J.; Huaqing, Z. PID Controller Optimization by GA and Its Performances on the Electro-hydraulic Servo Control System. *Chin. J. Aeronaut.* **2008**, *21*, 378–384. [CrossRef]
27. Zaki Fadel, M.; Rabie, M.; Youssef, A. Optimization of Control Parameters Based on Genetic Algorithm Technique for Integrated Electrohydraulic Servo Actuator System. *Int. J. Mechatron. Autom.* **2020**, *6*, 24–37.
28. Tajjudin, M.; Ishak, N.; Ismail, H.; Rahiman, M.H.F.; Adnan, R. Optimized PID control using Nelder-Mead method for electro-hydraulic actuator systems. In Proceedings of the 2011 IEEE Control and System Graduate Research Colloquium, Shah Alam, Malaysia, 27–28 June 2011; pp. 90–93. [CrossRef]
29. Rahmat, M.F.; Marhainis Othman, S.; Rozali, S.M.; Has, Z. Optimization of Modified Sliding Mode Control for an Electro-Hydraulic Actuator System with Mismatched Disturbance. In Proceedings of the 2018 5th International Conference on Electrical Engineering, Computer Science and Informatics (EECSI), Malang, Indonesia, 16–18 October 2018; pp. 1–6.
30. Fan, Y.; Shao, J.; Sun, G. Optimized PID Controller Based on Beetle Antennae Search Algorithm for Electro-Hydraulic Position Servo Control System. *Sensors* **2019**, *19*, 2727. [CrossRef]
31. Leitenberger, F.; Gwosch, T.; Matthiesen, S. Architecture of the Digital Twin in Product Validation for the Application in Virtual-Physical Testing to Investigate System Reliability. In Proceedings of the 32nd Symposium Design for X. Design for X Symposium, Online, 27–28 September 2021. [CrossRef]
32. Torque-Based, Field-Oriented Controller for an Internal Permanent Magnet Synchronous Motor—Simulink. Available online: https://www.mathworks.com/help/releases/R2021a/autoblks/ref/interiorpmcontroller.html (accessed on 23 September 2022).
33. Morimoto, S.; Sanada, M.; Takeda, Y. Wide-speed operation of interior permanent magnet synchronous motors with high-performance current regulator. *IEEE Trans. Ind. Appl.* **1994**, *30*, 920–926. [CrossRef]
34. Li, M.; He, J.; Demerdash, N.A.O. A flux-weakening control approach for interior permanent magnet synchronous motors based on Z-source inverters. In Proceedings of the 2014 IEEE Transportation Electrification Conference and Expo (ITEC), Dearborn, MI, USA, 15–18 June 2014; pp. 1–6. [CrossRef]
35. Briz, F.; Degner, M.W.; Lorenz, R.D. Analysis and design of current regulators using complex vectors. *IEEE Trans. Ind. Appl.* **2000**, *36*, 817–825. [CrossRef]
36. Briz, F.; Diez, A.; Degner, M.W.; Lorenz, R.D. Current and flux regulation in field-weakening operation [of induction motors]. *IEEE Trans. Ind. Appl.* **2001**, *37*, 42–50. [CrossRef]
37. SEW Eurodrive. System Manual—MOVIAXIS®Multi-Axis Servo Inverter. Available online: https://download.sew-eurodrive.com/download/pdf/20062532.pdf (accessed on 23 September 2022).
38. Kundur, P.S. *Power System Stability and Control*; McGraw-Hill Education: New York, NY, USA, 1993.
39. Anderson, P.M. *Analysis of Faulted Power Systems*; IEEE Press: New York, NY, USA, 1995; ISBN 978-0-780-31145-9.
40. Qu, S.; Fassbender, D.; Vacca, A.; Busquets, E. A Cost-Effective Electro-Hydraulic Actuator Solution with Open Circuit Architecture. *TJFP* **2021**, *22*, 2224. [CrossRef]

41. Engineering Department. *Flow of Fluids Through Valves, Fittings and Pipe: Crane Technical Paper No. 410*; Crance Company: Stamford, CT, USA, 2009; ISBN 1-40052-712-0.
42. Glöckler, M. *Simulation Mechatronischer Systeme*; Springer Fachmedien: Wiesbaden, Germany, 2018; ISBN 978-3-658-20702-1.
43. Wolter, K. A Method for User-Friendly PID-Parameter Optimization for Highly Dynamic Component Test Benches. In Proceedings of the FISITA 2020 Web Congress, Online, 24 November 2020.
44. Frischemeier, S. Electrohydrostatic actuators for aircraft primary flight control—Types, modelling and evaluation. In Proceedings of the 5th Scandinavian International Conference on Fluid Power, SICFP'97, Linköping, Sweden, 28–30 May 1997. [CrossRef]

 *machines*

Article

Dynamic Analysis of a High-Contact-Ratio Spur Gear System with Localized Spalling and Experimental Validation

Zhenbang Cheng [1,2], Kang Huang [1,3], Yangshou Xiong [4,*] and Meng Sang [1,3]

[1] School of Mechanical Engineering, Hefei University of Technology, Hefei 230009, China; chzhbang@mail.hfut.edu.cn (Z.C.); hfhuang98@hfut.edu.cn (K.H.); sangmeng1995@mail.hfut.edu.cn (M.S.)
[2] School of Mechanical and Vehicle Engineering, West Anhui University, Luan 237012, China
[3] AnHui Key Laboratory of Digit Design and Manufacture, Hefei 230009, China
[4] AnHui Province Key Lab of Aerospace Structural Parts Forming Technology and Equipment, Hefei 230009, China
* Correspondence: xiongys@hfut.edu.cn

Abstract: The dynamic characteristics and tooth spalling fault features are studied for the high-contact-ratio spur gear bearing system. The bending torsional dynamic model is proposed in this study for the gear bearing system with an ellipsoid spalling fault. This model also considers time-varying meshing stiffness, tooth friction, fractal gear backlash, and comprehensive transmission error. The meshing stiffness of the system is evaluated using the potential energy method. The bifurcation diagram, time-domain waveform, Poincaré map, phase map, frequency spectrum, and related three-dimensional map are used as tools to analyze the system's dynamic response qualitatively. The results reveal that the system's motion with ellipsoid tooth spalling defect exhibits rich dynamic behavior. The response of the proposed dynamic model is consistent with experimental results in the frequency domain. Therefore, the developed dynamic model can predict the system's vibration behavior with localized spalling fault. Hence, it could also provide a theoretical foundation for future spall defect diagnosis of the gear transmission system.

Keywords: high-contact-ratio gear; tooth spalling; meshing stiffness; nonlinear dynamic behavior

Citation: Cheng, Z.; Huang, K.; Xiong, Y.; Sang, M. Dynamic Analysis of a High-Contact-Ratio Spur Gear System with Localized Spalling and Experimental Validation. *Machines* **2022**, *10*, 154. https://doi.org/10.3390/machines10020154

Academic Editors: Sven Matthiesen and Thomas Gwosch

Received: 12 January 2022
Accepted: 14 February 2022
Published: 18 February 2022

1. Introduction

The contact ratio is an important index to indicate the smoothness and the uniformity of the gear system's transmission load [1]. By increasing the addendum coefficient, enhancing the number of teeth, reducing the pressure angle, and adjusting the modification coefficient, an involute spur gear with a contact ratio greater than two can be obtained, called high-contact-ratio spur gear [2]. Compared with standard gears, high-contact-ratio spur gear has more pairs of teeth meshing at the same time. As a result, high-contact-ratio gear has the advantages of high load carrying capacity [3]. Thus, it is widely used in automobiles, power tools, and other fields. As a common failure form of gear, tooth spalling is frequently encountered under excessive load, which initially occurs at the local position of the tooth surface. Localized tooth spalling induces serious damage to the gear transmission system gradually. Three main aspects are considered while characterizing tooth spalling fault: meshing stiffness, dynamic characteristics, and experimental validation. The model of damaged gear pairs is analyzed to estimate the mesh stiffness. The mesh stiffness is imported into the gear dynamic model to evaluate the non-linear dynamic characteristics. The dynamic response obtained from the simulation is compared with the experimental results to verify the dynamic model.

Tooth spalling affects the tooth profile and results in a change of meshing stiffness. Several researchers have developed many methods to estimate the meshing stiffness of the gears. Wang [4] and Zhan [5] have developed the finite element methods. The finite element method can accurately simulate the contact state of the system, but it is time-consuming. The analytical method is very efficient in terms of computation [6]. For example, Sun [7] studied the mesh stiffness of spur gear pairs with tooth modifications based on the thin slice assumption. For the gear with rectangular tooth spalling, Chaari researched the mesh stiffness of spur-gear pair [8], Jiang [9] and Han [10] studied the helical gear, while Luo studied the planetary gear [11]. In references [12,13], the meshing stiffness of gears with rectangular spalls of various widths, lengths, and positions were compared. Saxena et al. [14] estimated gear mesh stiffness for rectangular, circular, and V-shaped spalling. It is assumed that the bottom of the tooth spalling is flat in the literature. In practice, the shape of the tooth spall usually has a curvilinear bottom with a gradually changing dent depth rather than a suddenly decreased tooth thickness. Compared to rectangular solid, the ellipsoid is close to the actual geometry of the tooth spalling. This geometric model makes the calculation of mesh stiffness more accurate. The mesh stiffness of the gear with ellipsoid tooth spall is analyzed in detail in this work.

The estimation of mesh stiffness lays a foundation for gear dynamics. Dynamic analysis is valuable for operating status monitoring and gear fault diagnosis [15,16]. Using statistical methods to evaluate spall severity was suitable under low velocity and low excitation [17]. Dadon et al. [18] researched the effect of different gear imperfections on fault detection. Ma and Chen [19] studied the differences in vibration signals of tooth crack and tooth spall. Modification coefficients were introduced to research the impact of the spalls on gear dynamics [20]. Chen et al. [21] compared the mesh characteristics of helical gears with spalling faults using analytical and finite-element methods. Based on the theoretical and experimental study, Huangfu et al. [22] investigated the meshing and dynamic characteristics of a spalled gear system. Luo et al. [23] demonstrated that tooth spall and sliding friction have an evident effect on kinetic performance. Shi et al. [24] discussed the dynamic characteristics of the gear with double-teeth spalling fault. In terms of global dynamics, Ma [25] employed the singularity theory to evaluate the bifurcation characteristics of the gear system with tooth spalling.

Two drawbacks in these previous models are found. Mesh stiffness of the high-contact-ratio gear with ellipsoid spalling defects is not calculated. Secondly, motion state analysis of the gear with localized spalling defects under excitation frequency and gear backlash is not considered. The above deficiencies are the main contributions of this study. Highlights of this paper are listed as follows.

(1) Modeling of the mesh stiffness of the high-contact-ratio gear system with localized ellipsoid spalling.
(2) Bifurcation characteristic of the high-contact-ratio gear system with localized ellipsoid tooth spalling fault is discussed.
(3) Experiments are carried out for vibration measurement to validate the proposed dynamic model.

The rest of the article is arranged as follows. Section 2 describes the modeling of the mesh stiffness through the precise tooth profile equation of the high-contact-ratio gear and the ellipsoid equation. Section 3 illustrates the proposed dynamic model of the gear-bearing system. Section 4 discusses the numerical results. Section 5 is about the vibration experiments. The conclusions are presented in Section 6.

2. Mesh Stiffness Computation

The alternating meshing process of two and three gear pairs causes the variation of mesh stiffness. It also plays the role of internal excitation of the gear system. It is of great significance to develop an analytical model to calculate the mesh stiffness.

2.1. Accurate Tooth Profile Equation

The tooth profile of high contact ratio gear comprises tooth tip arc line, involute tooth profile, and tooth root transition curve. The motion of a rack-type cutting tool is equivalent to the meshing of the rack and pinion. In the process of machining, the cutter's machining pitch line is always tangent to the gear's machining pitch circle [26]. The profile of the rack-type cutting tool is shown in Figure 1. As presented in Figure 2, during gear machining, the involute part of the tooth profile is cut directly by the straight part of the cutter (BC), and the fillet part of the cutter (AB) cuts the transition part. The transition part is the isometric curve of the extended involute. This extended involute is depicted by the center of the tool's rounded corner. An x–y coordinate system with the center of the gear as the origin is shown. The gear teeth are usually considered as a cantilever beam of a variable cross-section. The coordinates of any contact point: i, on the involute part are expressed as follows.

$$\begin{cases} x_i = r_b[(\alpha_i + \theta_b)\sin\alpha_i + \cos\alpha_i] \\ y_i = r_b[(\alpha_i + \theta_b)\cos\alpha_i - \sin\alpha_i] \end{cases} \tag{1}$$

r_b denotes the radius of the gear base circle. α_i is the working load angle acting on the contact point: i. θ_b is half of the tooth base arc angle (θ_b = $(\pi/2$ + $2X{\cdot}\tan\alpha_0)/N$ + $inv\alpha_0$). x is the displacement coefficient. N is the number of teeth. $inv(\cdot)$ represents the involute function of the pressure angle. Coordinates of point: j, on the transition curve are defined as follows.

$$\begin{cases} x_{\rm j} = r\cos\varphi - (a_1/\sin\gamma + r_\rho)\sin(\gamma - \varphi) \\ y_{\rm j} = r\sin\varphi - (a_1/\sin\gamma + r_\rho)\cos(\gamma - \varphi) \end{cases} \quad (\alpha_0 \le \gamma \le \pi/2) \tag{2}$$

r and r_ρ are the radii of the gear pitch and the top corner of the tool, respectively. a is the distance from the center of the top corner to the center line of the tool (a = $(h_a{}^*$ + $c^*)m$ − r_ρ, a_1 = a − xm, φ = $(a_1/\tan\gamma$ + $b)/r$, b = $\pi m/4$ + $h_a{}^*{\cdot}m{\cdot}\tan\alpha_0$ + $r_\rho{\cdot}\cos\alpha_0$).

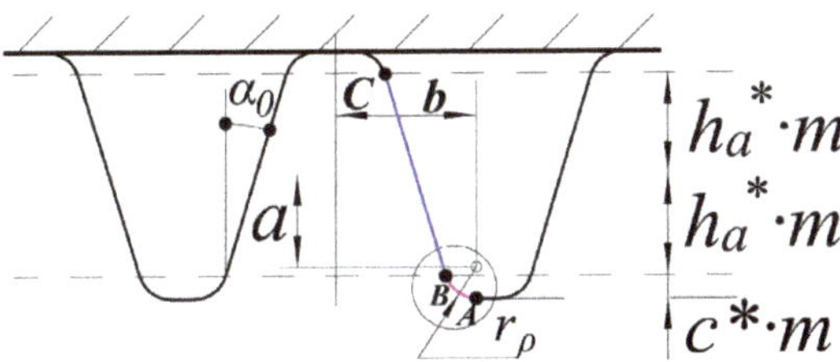

Figure 1. Rack-type cutting tool profile.

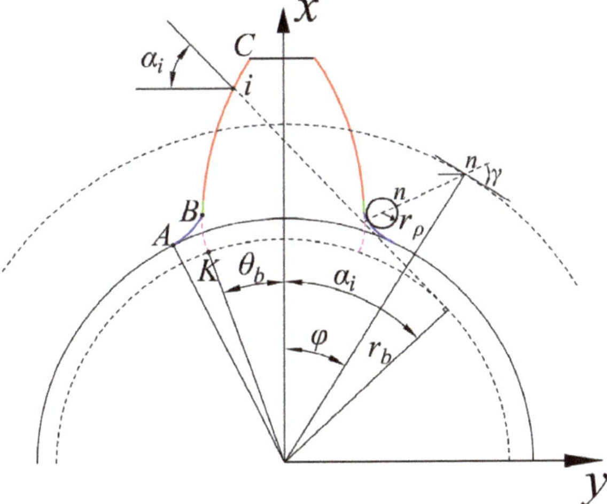

Figure 2. Tooth profile curve diagram.

2.2. Analytical Model of Meshing Stiffness

Many tooth surface spalls have progressively varying depth and a curved base surface in practice as shown in Figure 3. The ellipsoid tooth spall is formed by removing the intersection of the gear and the ellipsoid. As shown in Figure 4, the ellipsoid tooth spall's maximum width, length, and center depth are w_s, l_s, and h_s, respectively. The starting and ending position is denoted by x_{start} *and* x_{end}, respectively. θ_s is the angle between the tangent of tooth spall and involute curve. In this work, the position (x_{start}, θ_s) and the severity (w_s, l_s, h_s) are fixed.

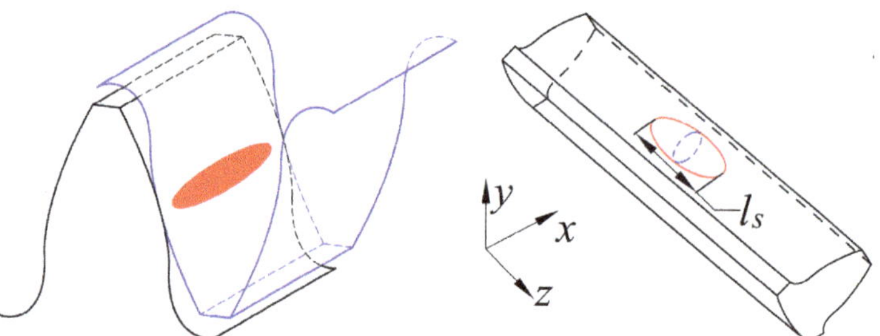

Figure 3. The ellipsoid tooth spall on the pinion.

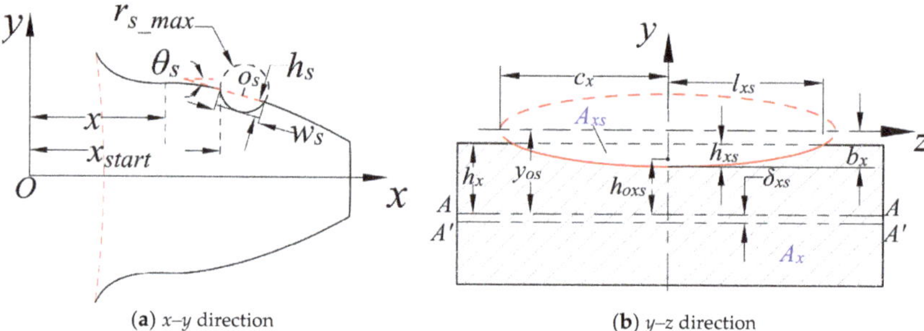

Figure 4. A cross-sectional view of the ellipsoid tooth spall.

Table 1 displays the comparison of key geometry parameters of the ellipsoid tooth spall for healthy and spall gear.

Table 1. The ellipsoid tooth spall data.

Case	w_s (mm)	l_s (mm)	h_s (mm)	θ_s (°)	x_{start} (mm)
Healthy gear	0	0	0	0	0
Spall gear	4	16	2	18	66

An ellipsoid's parametric function is defined as follows.

$$\frac{x^2}{a^2} + \frac{y^2}{b^2} + \frac{z^2}{c^2} = 1 \tag{3}$$

a, b, and c are the radii of the ellipsoid along x, y, and z axis. Equations (4) and (5) are derived assuming a and b as equal.

$$a = b = r_{s\_max} = \frac{(0.5w_s)^2 + h_s^2}{2h_s} \tag{4}$$

$$c = \frac{1}{2}\sqrt{\frac{r_{s\_max}^2 l_s^2}{2r_{s\_max}h_s - h_s^2}} \tag{5}$$

In the x–y plane, the coordinates of the ellipsoid center are derived as follows.

$$\begin{cases} x_{os} = x_{start} + r_{s\_max}\cos\left(\frac{\pi}{2} - \theta_s - \arcsin\left(\frac{0.5w_s}{r_{s\_max}}\right)\right) \\ y_{os} = -\tan(\theta_s)(x_{os} - x_{start}) + y(x_{start}) + \frac{r_{s\_max} - h_s}{\cos\theta_s}, \quad (h_s < r_{s\_max}) \end{cases} \tag{6}$$

Figure 4b shows the cross-section of the ellipsoid tooth spall at a distance x, The ellipse equation is as follows.

$$\frac{y_s{}^2}{b_x{}^2} + \frac{z_s{}^2}{c_x{}^2} = 1 \tag{7}$$

$$b_x = \sqrt{b^2(1 - \frac{(x_{os} - x_t)^2}{a^2})} \tag{8}$$

$$c_x = \sqrt{c^2(1 - \frac{(x_{os} - x_t)^2}{a^2})} \tag{9}$$

The length l_{xs} is calculated is expressed as follows.

$$l_{xs} = \sqrt{c_x^2\left(1 - (y_{os} - h_x)^2/b_x^2\right)} \tag{10}$$

As shown in Figure 4b, h_x is the half-height of the gear tooth cross-section at a distance x. The gear tooth contact length L_e is given as follows.

$$L_e = L - 2l_{xs} \tag{11}$$

L denotes the gear tooth width. The maximum depth of the ellipsoid tooth spall at a distance x is determined as follows.

$$h_{xs} = b_x - (y_{os} - h_x) \tag{12}$$

The corresponding area of the portion of the ellipse is deduced as follows.

$$A_{xs} = 2\int_{-b_x}^{-(y_{os}-h_x)} \sqrt{\left(1 - \frac{y_s{}^2}{b_x^2}\right)c_x^2 dy_s} \tag{13}$$

The cross-section area A_x of the ellipsoid tooth spall at a distance x is given below.

$$A_x = 2h_x L - A_{xs} \tag{14}$$

A_x causes a shift in the cross-neutral section's axis. The corresponding displacement δ_{xs} between these two central axes is calculated as follows.

$$\delta_{xs} = \frac{A_{xs}h_{oxs}}{A_x} \tag{15}$$

The area moment of inertia of the ellipse segments is expressed as follows.

$$I_{zs} = \int_{-b_x}^{-(y_{os}-h_x)} 2y_s{}^2\sqrt{\left(1 - \frac{y_s^2}{b_x^2}\right)c_x^2 dy_s} - A_{xs}(y_{os} - h_{oxs})^2, \quad (h_x \le y_{os}) \tag{16}$$

Under the tooth spall conditions, the area moment of inertia of the gear tooth cross-section with respect to the new axis is modified as follows.

$$I_z = \frac{2Lh_x^3}{3} + 2Lh_x\delta_{xs}^2 - \left(I_{zs} + A_{xs}(h_{oxs} + \delta_{xs})^2\right) \quad (17)$$

Figure 5 shows the gear profile. Segments AB and BC represent the transition curve and involute curve, respectively. P is the meshing point, and the corresponding pressure angle is α_p. F denotes the meshing force that is decomposed into F_x and F_y, respectively. Based on the potential energy method [27], comprehensive mesh stiffness $k(t)$ is deduced.

$$k(t) = \sum_{i=1}^{3} \frac{1}{\frac{1}{k_{hi}} + \frac{1}{k_{bi}^p} + \frac{1}{k_{si}^p} + \frac{1}{k_{ai}^p} + \frac{1}{k_{fi}^p} + \frac{1}{k_{bi}^g} + \frac{1}{k_{si}^g} + \frac{1}{k_{ai}^g} + \frac{1}{k_{fi}^g}} \quad (18)$$

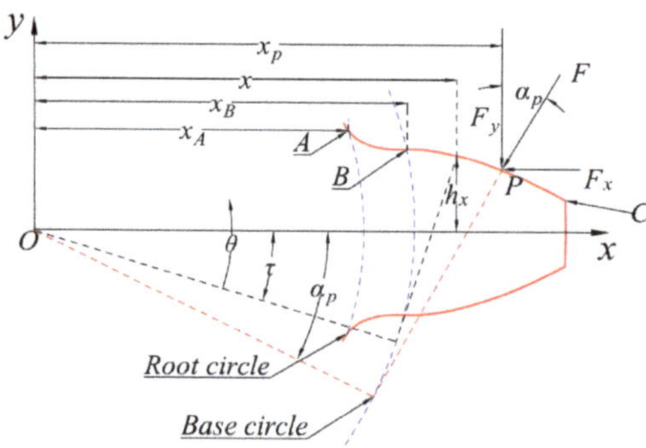

Figure 5. Gear profile.

Subscript i (i = 1, 2, 3) denotes the pair of the engaging gear tooth. Based on the tooth profile equations, each of the stiffness is expressed as follows.

$$\frac{1}{k_a} = \begin{cases} \int_{x_A}^{x_B} \frac{\sin^2\alpha_p}{EA_1}dx_1 + \int_{x_B}^{x_p} \frac{\sin^2\alpha_p}{EA_2}dx_2 & (x_B < x \le x_{start}) \\ \int_{x_A}^{x_B} \frac{\sin^2\alpha_p}{EA_1}dx_1 + \int_{x_B}^{x_s} \frac{\sin^2\alpha_p}{EA_2}dx_2 + \int_{x_s}^{x_p} \frac{\sin^2\alpha_p}{EA_3}dx_3 & (x_{start} < x \le x_{end}) \\ \int_{x_A}^{x_B} \frac{\sin^2\alpha_p}{EA_1}dx_1 + \int_{x_B}^{x_s} \frac{\sin^2\alpha_p}{EA_2}dx_2 + \int_{x_s}^{x_e} \frac{\sin^2\alpha_p}{EA_3}dx_3 + \int_{x_e}^{x_p} \frac{\sin^2\alpha_p}{EA_4}dx_4 & (x_{end} < x) \end{cases} \quad (19)$$

$$\frac{1}{k_b} = \begin{cases} \int_{x_A}^{x_B} \frac{M_1^2}{EI_1}dx_1 + \int_{x_B}^{x_p} \frac{M_2^2}{EI_2}dx_2 & (x_B < x \le x_{start}) \\ \int_{x_A}^{x_B} \frac{M_1^2}{EI_1}dx_1 + \int_{x_B}^{x_s} \frac{M_2^2}{EI_2}dx_2 + \int_{x_s}^{x_p} \frac{M_3^2}{EI_3}dx_3 & (x_{start} < x \le x_{end}) \\ \int_{x_A}^{x_B} \frac{M_1^2}{EI_1}dx_1 + \int_{x_B}^{x_s} \frac{M_2^2}{EI_2}dx_2 + \int_{x_s}^{x_e} \frac{M_3^2}{EI_3}dx_3 + \int_{x_e}^{x_p} \frac{M_4^2}{EI_4}dx_4 & (x_{end} < x) \end{cases} \quad (20)$$

$$\frac{1}{k_s} = \begin{cases} \int_{x_A}^{x_B} \frac{1.2\cos^2\alpha_p}{GA_1}dx_1 + \int_{x_B}^{x_p} \frac{1.2\cos^2\alpha_p}{GA_2}dx_2 & (x_B < x \le x_{start}) \\ \int_{x_A}^{x_B} \frac{1.2\cos^2\alpha_p}{GA_1}dx_1 + \int_{x_B}^{x_s} \frac{1.2\cos^2\alpha_p}{GA_2}dx_2 + \int_{x_s}^{x_p} \frac{1.2\cos^2\alpha_p}{GA_3}dx_3 & (x_{start} < x \le x_{end}) \\ \int_{x_A}^{x_B} \frac{1.2\cos^2\alpha_p}{GA_1}dx_1 + \int_{x_B}^{x_s} \frac{1.2\cos^2\alpha_p}{GA_2}dx_2 + \int_{x_s}^{x_e} \frac{1.2\cos^2\alpha_p}{GA_3}dx_3 + \int_{x_e}^{x_p} \frac{1.2\cos^2\alpha_p}{GA_4}dx_4 & (x_{end} < x) \end{cases} \quad (21)$$

$$\frac{1}{k_f} = \frac{\cos^2\alpha_P}{EL}\left\{L^*\left(\frac{u_f}{s_f}\right)^2 + M^*\left(\frac{u_f}{s_f}\right) + P^*\left(1 + Q^*\tan^2\alpha_P\right)\right\} \quad (22)$$

$$\frac{1}{k_h} = \frac{\pi E L_e}{4(1-v^2)} \tag{23}$$

E, G, and v stand for Young's modulus, shear modulus, and Poisson's ratio, respectively. A_1 and I_1 denote the cross-sectional area and corresponding inertia within the transition curve, respectively. A_2, A_4 and I_2, I_4 are the cross-sectional areas of the healthy part of the involute curve and the related area moments of inertia. For the spall part of the gear profile, A_3 and I_3 are evaluated by Equations (14) and (17), respectively. u_f, s_f, L^*, M^*, P,* and Q^* are given in reference [28].

Figure 6 shows the curve of mesh stiffness. The incidence of tooth spall reduces the mesh stiffness. A localized spall fault is considered. Hence, it occurs once per revolution. Thus, compared with the healthy gear, the reduction in mesh stiffness occurs mainly in the double-tooth and the triple-tooth zones.

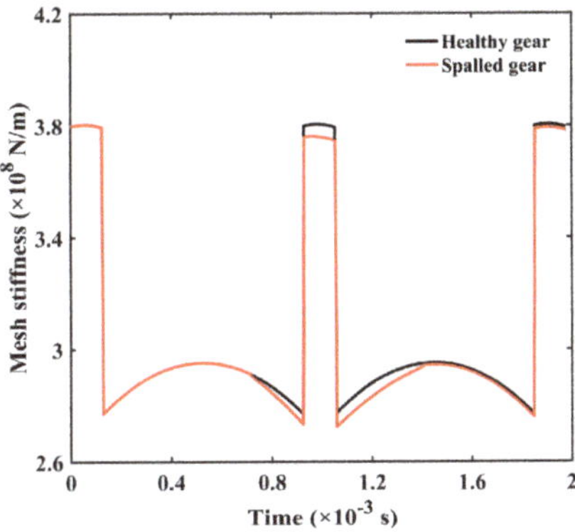

Figure 6. Time-varying mesh stiffness.

3. Dynamic Model of System

The support of the rolling bearing is assumed to be rigid. The gear-bearing translation-torsion dynamic lumped parameter model is established, as is shown in Figure 7, considering the influence of gear backlash, damping, comprehensive transmission error, friction force, and time-varying meshing stiffness.

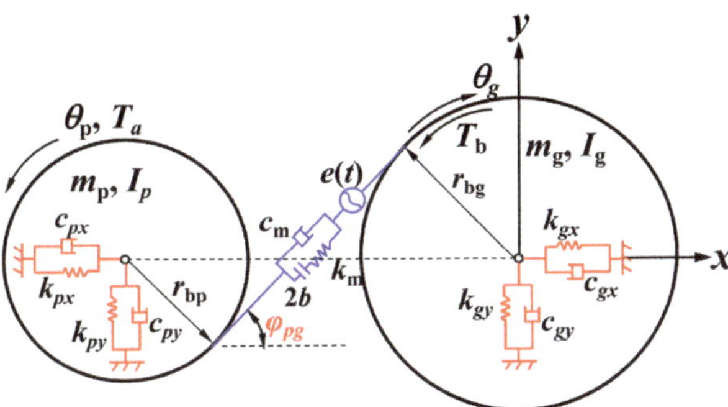

Figure 7. Gear dynamic model.

3.1. Gear-Bearing System

m, I, T, and θ represent the mass, moment of inertia, torque, and torsional angular displacement, respectively. Subscripts p and g indicate quantities associated with pinion and gear, respectively. r_b, k_x, and k_y denote the base circle radius, vertical radial support stiffness, and horizontal radial support stiffness. The time-varying meshing stiffness, meshing damping, and backlash of the gear pair are expressed by k_m, c_m, and $2b$, respectively. $e(t)$ is the comprehensive static error along the tangent direction of the gear base circle.

As shown in Figure 7, the dynamic model has six degrees of freedom, including two rotational degrees of freedom and four translational degrees of freedom along with the horizontal and vertical directions. The generalized coordinate array is expressed as follows.

$$q = \left[\theta_p, \theta_g, x_p, x_g, y_p, y_g\right]^T \tag{24}$$

3.2. Dynamic Meshing Force and Frictional Force

The relative displacement between pinion and gear along the line of action is expressed as follows.

$$\delta(t) = r_{bp}\theta_p - r_{bg}\theta_g + (x_p - x_g)\cos\varphi_{pg} + (y_p - y_g)\sin\varphi_{pg} - e(t) \tag{25}$$

The dynamic meshing force between gears consists of elastic meshing forces caused by time-varying stiffness and viscous meshing forces caused by meshing damping, denoted as follows.

$$F_d = c_m\dot{\delta}(t) + k_m f(\delta(t), b(t)) \tag{26}$$

The gap function formula can be calculated as follows.

$$f(\delta(t), b) = \begin{cases} \delta(t) - b(t) & (\delta(t) > b(t)) \\ 0 & (|\delta(t)| \le b(t)) \\ \delta(t) + b(t) & (\delta(t) < -b(t)) \end{cases} \tag{27}$$

$b(t)$ represents the gear backlash. Based on fractal theory, it is expressed as follows [29].

$$b(t) = b_0 - \frac{R_{a1}}{R_{ac}(D_1)}\sum_{k=0}^{+\infty} \lambda^{(D_1-2)k}\sin(\lambda^k t) - \frac{R_{a2}}{R_{ac}(D_2)}\sum_{k=0}^{+\infty} \lambda^{(D_2-2)k}\sin(\lambda^k t) \tag{28}$$

b_0, λ, R_{a1}/R_{a2}, and D_1/D_2 denote initial gear backlash, characteristic scale coefficient, actual surface roughness, and fractal dimension, respectively. $R_{ac}(D)$ is the function to get the corresponding R_a with a particular fractal dimension, which can be obtained from the reference [30]. The friction force between tooth surfaces during gear meshing is deduced as follows.

$$F_f = \eta\mu F_{\rm d} \tag{29}$$

The direction of friction is variable, and the direction coefficient depends on the following formula.

$$\eta = sign(\omega_{\rm p}KN_1 - \omega_g KN_2) \tag{30}$$

Friction force arms (Figure 8) are deduced as follows.

$$\begin{aligned} KN_1 &= \sqrt{(r_1 + r_2)^2 - (r_{b1} + r_{b2})^2} - \sqrt{r_{a2}{}^2 - r_{b2}{}^2} + r_{b1}\omega_1 t \\ KN_2 &= \sqrt{r_{a2}{}^2 - r_{b2}{}^2} - r_{b1}\omega_1 t \end{aligned} \tag{31}$$

The friction coefficient is related to relative sliding velocity, tooth surface roughness, contact pressure, lubrication situation, etc. Hence, it is particularly difficult to predict the friction coefficient value. The tooth friction model mainly includes the Coulomb friction model, Buckingham empirical formula, Benedict Kelly model, and the friction model based on Elastohydrodynamic lubrication (EHL) theory. Among these, the friction model proposed by Xu [31] reasonably agrees with the measured data, so it is usually adopted to predict the friction coefficient in this paper. It is based on non-Newtonian, thermal EHL theory, and multiple linear regression analysis. The friction coefficient calculation results are shown in Figure 9.

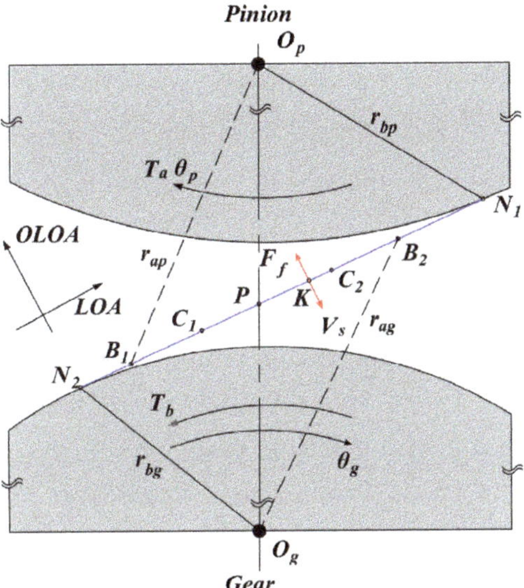

Figure 8. Geometrical relationship of gear pair.

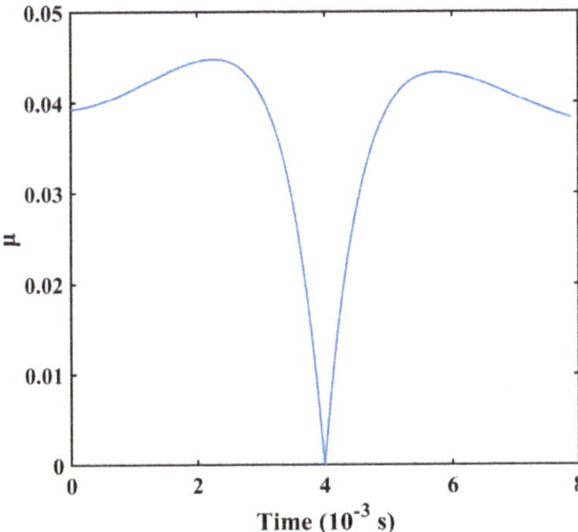

Figure 9. Time–varying friction coefficient.

The friction torque on the gear can be expressed as follows.

$$\begin{aligned} M_p &= \eta\mu F_d KN_1 \\ M_g &= \eta\mu F_d KN_2 \end{aligned} \tag{32}$$

3.3. *Differential Equations of Motion*

Based on the time-varying meshing stiffness, transmission error, the fractal backlash, the frictional force, and Newton's law of motion, the differential equation of vibration motion of the gear-bearing coupling system is deduced as follows.

$$\begin{cases} I_p\ddot{\theta}_p + F_d r_{bp} - M_p = T_a \\ I_g\ddot{\theta}_g - F_d r_{bg} + M_g = -T_b \\ m_p\ddot{x}_p + c_{px}\dot{x}_p + k_{px}x_p + F_d\cos\varphi_{pg} + F_f\sin\varphi_{pg} = 0 \\ m_g\ddot{x}_g + c_{gx}\dot{x}_g + k_{gx}x_g - F_d\cos\varphi_{pg} - F_f\sin\varphi_{pg} = 0 \\ m_p\ddot{y}_p + c_{py}\dot{y}_p + k_{py}y_p + F_d\sin\varphi_{pg} - F_f\cos\varphi_{pg} = 0 \\ m_g\ddot{y}_g + c_{gy}\dot{y}_g + k_{gy}y_g - F_d\sin\varphi_{pg} + F_f\cos\varphi_{pg} = 0 \end{cases} \tag{33}$$

Setting $x_1 = \theta_p$, $x_2 = x_1'$, $x_3 = \theta_g$, $x_4 = x_3'$, $x_5 = x_p$, $x_6 = x_5'$, $x_7 = x_g$, $x_8 = x_7'$, $x_9 = y_p$, $x_{10} = x_9'$, $x_{11} = y_g$, and $x_{12} = x_{11}'$, the above equation is transformed into the following form.

$$\begin{cases} x_1' = x_2 \\ x_2' = \frac{T_a}{I_p} + \frac{M_p}{I_p} - \frac{F_d r_{bp}}{I_p} \\ x_3' = x_4 \\ x_4' = \frac{-T_b}{I_g} - \frac{M_g}{I_g} + \frac{F_d r_{bg}}{I_g} \\ x_5' = x_6 \\ x_6' = \frac{-F_d\cos\varphi_{pg}}{m_p} - \frac{F_f\sin\varphi_{pg}}{m_p} - \frac{c_{px}\dot{x}_p}{m_p} - \frac{k_{px}x_p}{m_p} \\ x_7' = x_8 \\ x_8' = \frac{F_d\cos\varphi_{pg}}{m_g} + \frac{F_f\sin\varphi_{pg}}{m_g} - \frac{c_{gx}\dot{x}_g}{m_g} - \frac{k_{gx}x_g}{m_g} \\ x_9' = x_{10} \\ x_{10}' = \frac{-F_d\sin\varphi_{pg}}{m_p} + \frac{F_f\cos\varphi_{pg}}{m_p} - \frac{c_{py}\dot{y}_p}{m_p} - \frac{k_{py}y_p}{m_p} \\ x_{11}' = x_{12} \\ x_{12}' = \frac{F_d\sin\varphi_{pg}}{m_g} - \frac{F_f\cos\varphi_{pg}}{m_g} - \frac{c_{gy}\dot{y}_g}{m_g} - \frac{k_{gy}y_g}{m_g} \end{cases} \tag{34}$$

4. Numerical Simulation and Discussion

The high-contact-ratio gear system is non-linear. Table 2 lists its main parameters. The parameters in blue boxes are gear basic parameters. Gear basic parameters were designed via the KISSsoft software. Mesh damping ratio was obtained from the reference [32]. Additionally, the other parameters were calculated from the gear system via the Solidworks software. The fourth-order Runge–Kutta numerical integration method is used to solve the above differential equations of the system through the MATLAB software. The simulated data are processed for generating a bifurcation diagram, three-dimensional frequency spectrum, Poincaré map, phase map, etc.

Table 2. Main parameters of the high-contact-ratio gear.

Parameters	Pinion/Gear	Parameters	Pinion/Gear
Tooth Number z_p/z_g	27/31	Designed contact ratio	2.135
Transverse modulus (mm)	5	Tooth width (mm)	20
Pressure angle (°)	19	Moments of inertia (kg·m^2) I_p/I_g	0.0051/0.0089
Addendum coefficient	1.32	Mass (kg) m_p/m_g	2.13/2.84
Modification coefficient	0	Mesh damping ratio	0.06
Hub bore radius (mm)	14	Input power (kW)	20

4.1. Effect of Excitation Frequency

Gear excitation frequency often changes with working conditions. It is one of the key parameters affecting the dynamic characteristics of the gear system and is often used as a variable parameter to compare the systematic dynamical performance. Here, surface roughness R_a is taken equal to 0.8 μm, and fractal dimensions D_1 and D_2 are equivalent to 1.1. Figure 10 shows the bifurcation characteristics of the lateral displacement (x_p) of the pinion changing with the excitation frequency Ω. Its three-dimensional frequency map is illustrated in Figure 11. Both figures reach an agreement on systematic bifurcation behaviors.

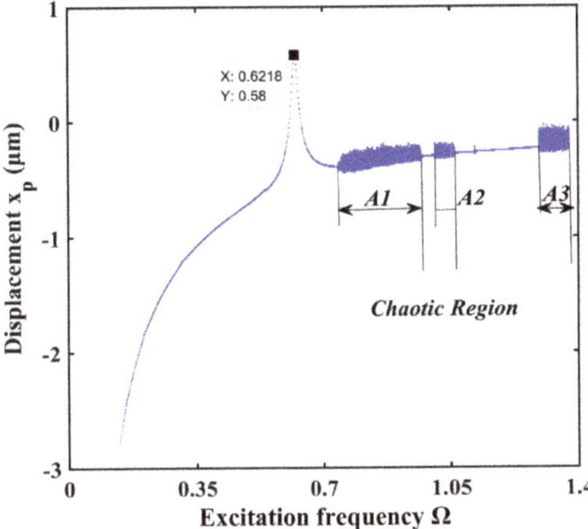

Figure 10. Bifurcation diagram under different excitation frequency.

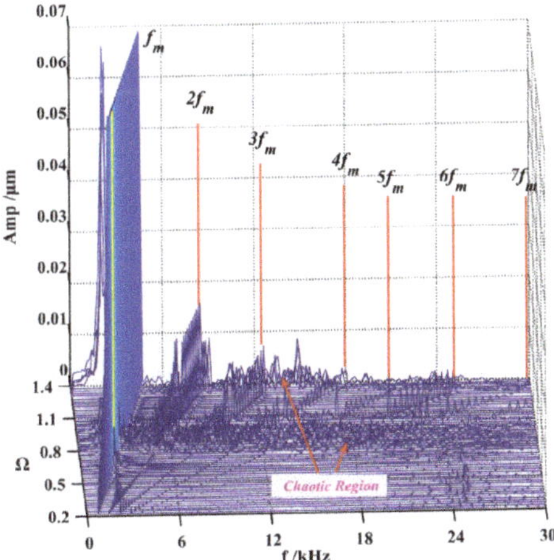

Figure 11. Three−dimensional frequency spectrum under different excitation.

The system is in periodic motion at low excitation frequency, as shown in Figure 12a,b. When Ω equals 0.6218, resonance occurs in the system. The system enters chaotic regions A1, A2, and A3 sequentially with the increasing excitation frequency. The system follows quasi-periodic motion between the chaotic regions, as shown in Figure 12e to Figure 12h. How the systems enter chaos is different. The system enters chaotic region A1 through quasi-periodic motion, as shown in Figure 12c,d. However, the system enters chaotic regions A2 and A3 by way of crisis, as shown in Figure 12i–l.

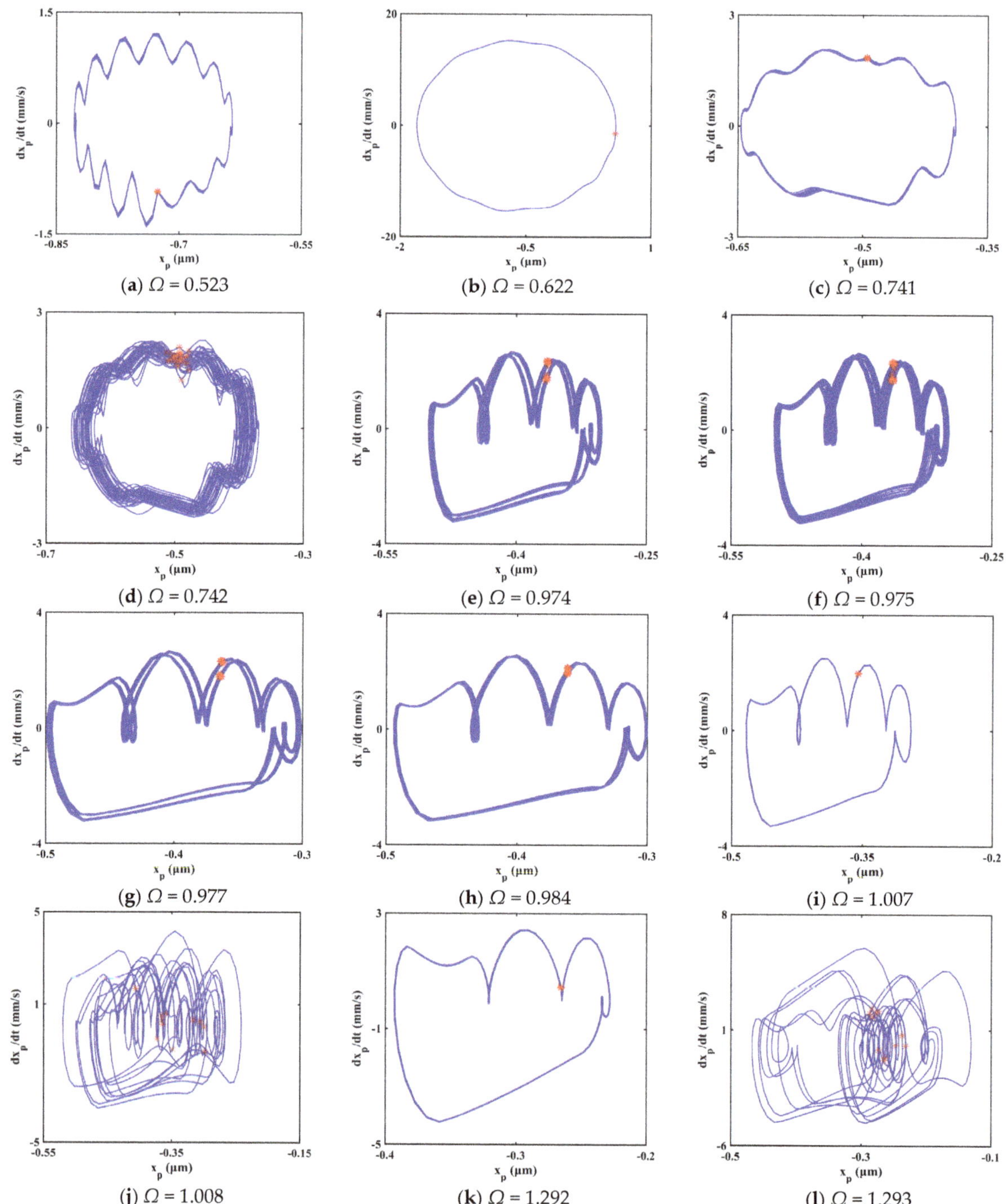

Figure 12. Evolution process of system response for $\Omega \in [0.5, 1.3]$.

As shown in Figure 11, the main components of the spectrum of gear system are composed of the meshing frequency and its harmonics. In the chaotic region, the frequency spectrum is a continuous line. The value range of the chaotic regions is also inconsistent with Figure 10.

Chaos means the uncontrollability and unpredictability of the system movement, which aggravates the vibration and noise of the system. In practice, such movement is to be avoided by appropriate measures. Through the above analysis of bifurcation characteristics and frequency analysis, the frequency regions and critical bifurcation values of chaotic motions and motions of different periods are obtained. Hence, the desired motion state can be obtained artificially.

4.2. Effect of Gear Backlash

Gear backlash is one of the main nonlinearities of the system. It generally changes the wear and the deformation of components of the system. The excitation frequency was set as a constant value ($\Omega = 1.2$), and the other parameters remained unchanged in the study of bifurcation characteristics of the system.

The bifurcation diagram and three-dimensional frequency spectrum are shown in Figures 13 and 14, respectively. Both cases have the same trend in motion state transitions. Several frequency jumps exist in the bifurcation diagram, for example, at b equals to 11.3. Various motion patterns such as single periodic motion, multi-periodic motion, quasi-periodic motion, and chaotic motions are seen within the rotational speed range. There are mainly four chaotic regions in the bifurcation diagram, namely B1, B2, B3, and B4. The way the system enters a chaotic region is different. The system enters B1, B2, and B4 through quasi-periodic motion, as shown in Figure 15b,c,e,f,j,k. However, the system enters B3 through the period-doubling route, as shown in Figure 15h,i.

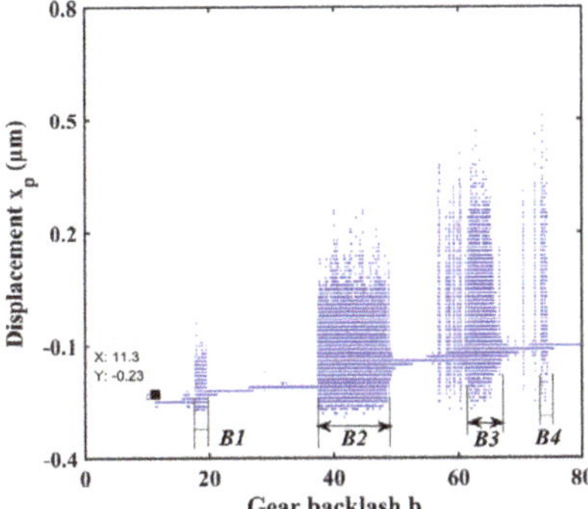

Figure 13. Bifurcation diagram under different gear backlash.

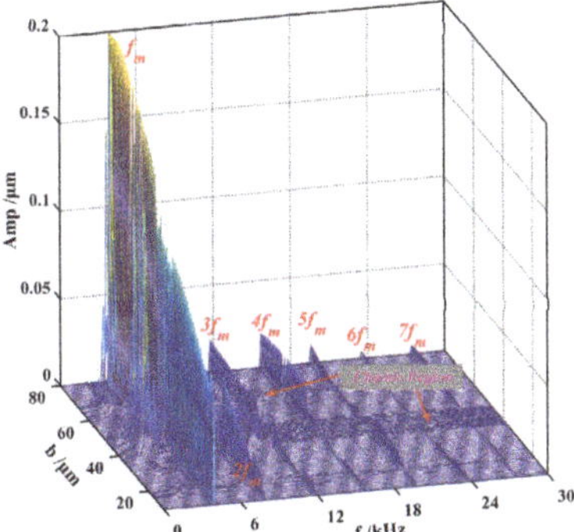

Figure 14. Three-dimensional frequency spectrum under different gear backlash.

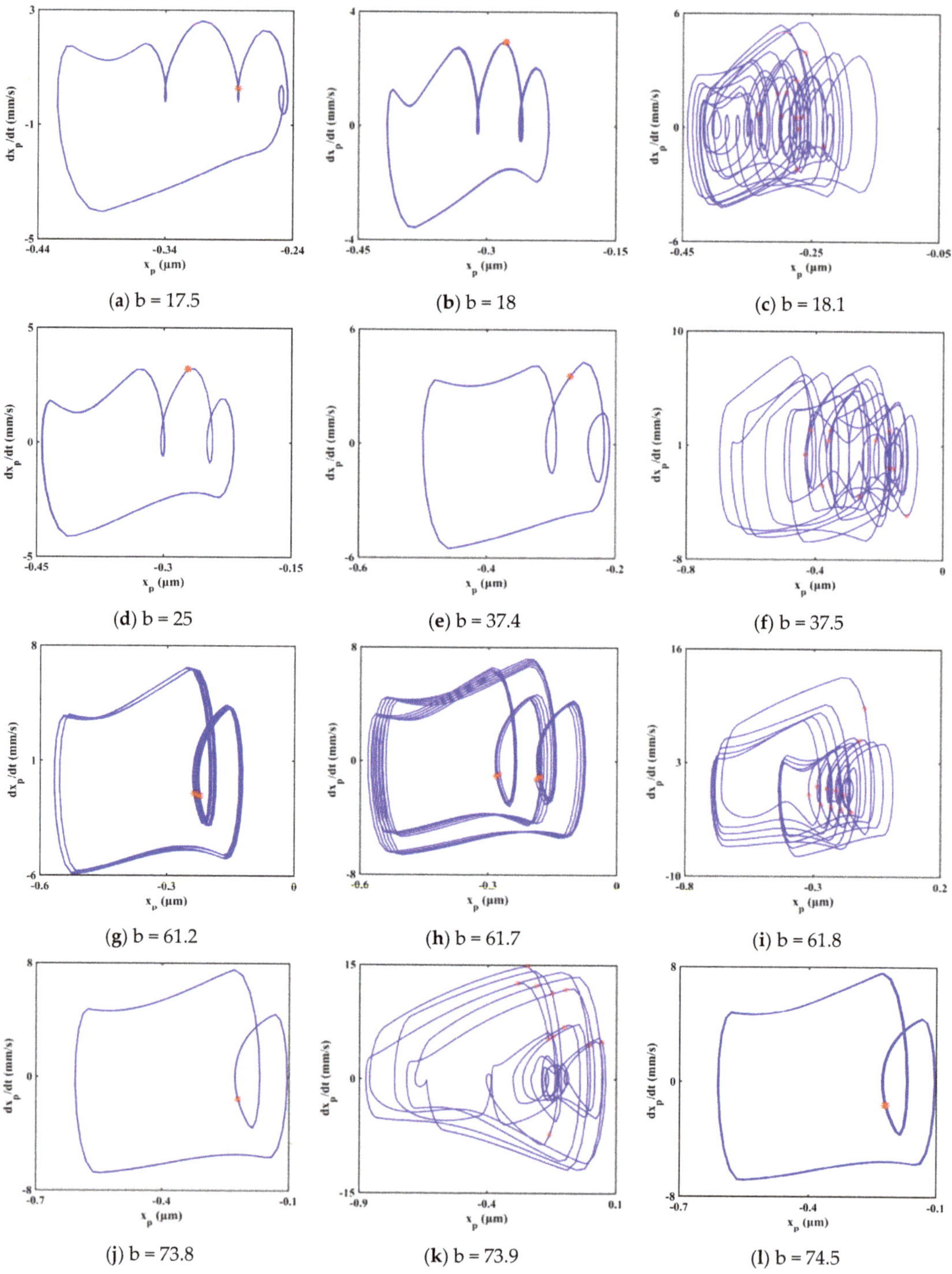

Figure 15. Evolution process of system response at the range of b∈ [17, 75].

5. Experimental Validation

A single-stage high-contact-ratio gear test rig is designed and developed to validate the dynamic model. Dynamic characteristics under different speed conditions to imitate real-life applications are studied. The test rig is set up at the Anhui Digital Design and manufacturing laboratory, Hefei University of Technology, China. It can measure the acceleration, acting load, and rotational speed of the system. The gearbox is coupled to the servo motor, which has a maximum speed of 3000 rpm. The prime mover is Delta make with a power capacity of 0.75 kW. ZHY-6001 piezoelectric accelerometers measure the vibration signals of the system. Two DYN-200 torque transducers track the rotational speed and torque of the gear system. A data acquisition system ZHKJ-1001 with six channels is used. The sampling rate for the experimental trials is set at 10 kHz. The experimental setup and test gears are shown in Figure 16. High-contact-ratio gear is a non-standard part. Thus, the test gears are machined by slow wire cutting. Tooth spalling defects in the gears are generated with the electric mill.

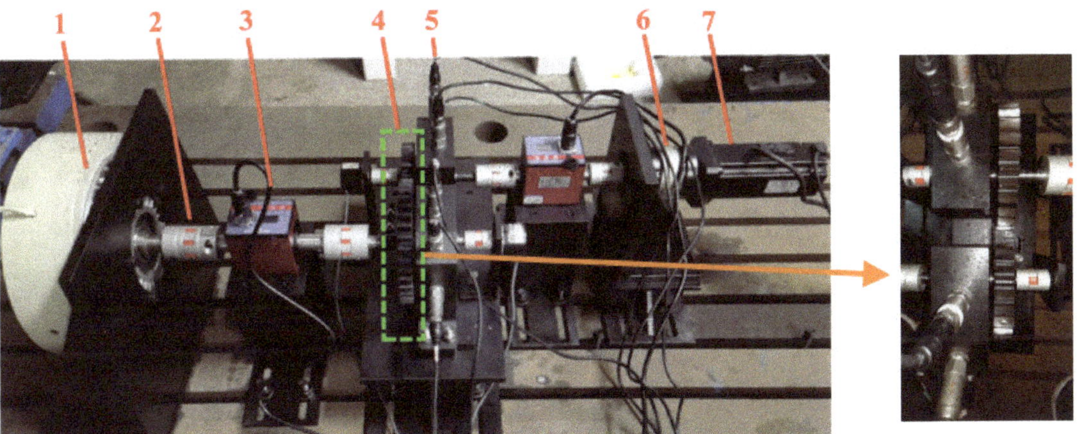

Figure 16. Experimental setup.

The comparison between the experimental and simulated signals in the time-domain and frequency-domain is presented in Figure 17. The gear vibration signal contains a lot of interference noise caused by the operation of mechanical equipment. The white noise of the ambient background appears in its full frequency band. In this paper, the moving average method is used to denoise the collected experimental signals. The rotational frequency of pinion and gear is f_p ($f_p = n_p/60 = 10$ Hz) and f_g ($f_g = 8.7$ Hz), respectively. The pinion's meshing frequency and meshing are f_m ($f_m = z_p * f_p = 270$ Hz) and T_p ($T_p = 1/f_m = 0.0037$ s). The characteristic fault frequency is f_s. It is the reciprocal of the failure period and can expressed as follows.

$$fs = \frac{1}{x_{p2} - x_{p1}} \; or \; fs = \frac{1}{x_{p3} - x_{p2}} \tag{35}$$

As shown in Table 3, there is little difference between the fault frequency calculated from the abscissa data of three points (P_1, P_2, P_3) in Figure 17a and the data of the three points (P_1, P_2, P_3) in Figure 17b.

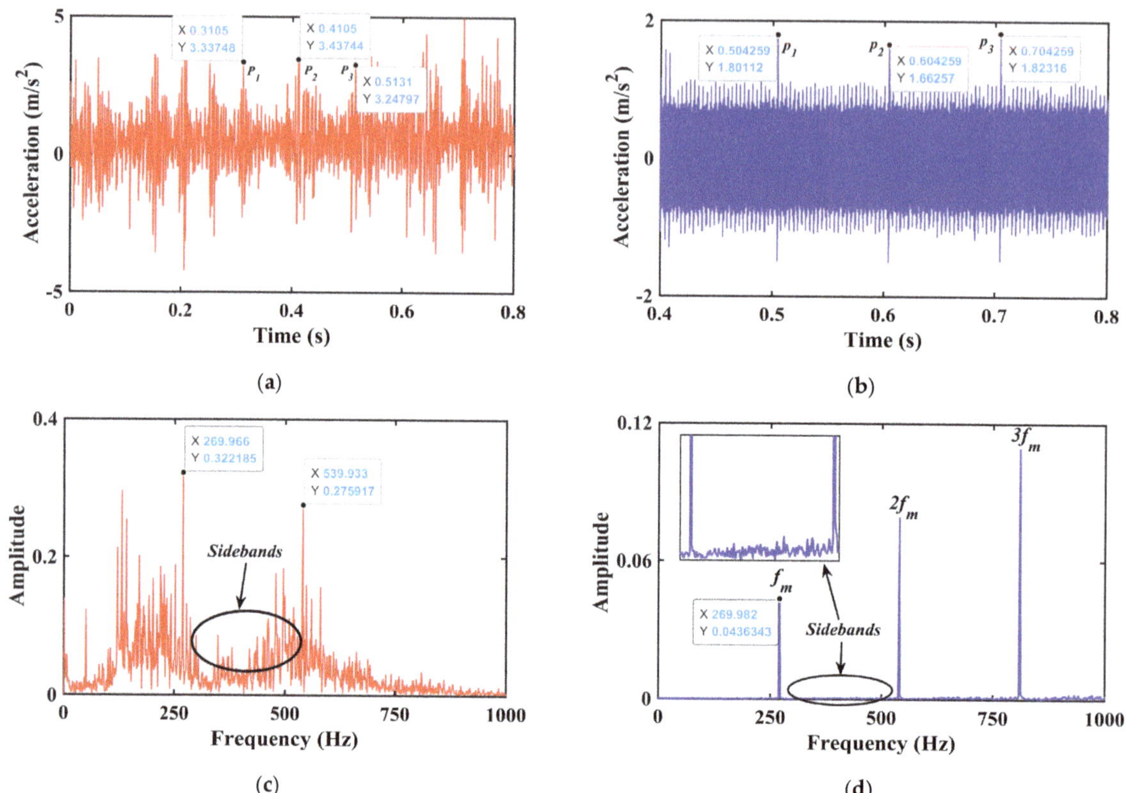

Figure 17. Comparison between the experimental and simulated signals in horizontal direction. (**a**) Time−domain of experimental signal, (**b**) Time−domain of simulated signal, (**c**) Frequency spectrum of the experimental signal, (**d**) Frequency spectrum of the simulated signal.

Table 3. Fault frequency calculation.

Fault Frequency (f_s)	Simulation Results	Experimental Results	Error
$\frac{1}{x_{p2}-x_{p1}}$	10 Hz	10 Hz	0%
$\frac{1}{x_{p3}-x_{p2}}$	10 Hz	9.75 Hz	2.5%

Figure 17a,b show the time-domain experimental and simulated horizontal displacement signal. It can be observed that there is a certain difference in amplitude between the two cases, which is caused by experimental errors. Compared to experimental results, the simulated time-domain signal has a clear periodic impact to $Z_p * T_p$. The frequency response characteristic of the system under the two conditions is presented in Figure 17c,d. The frequency spectrum is mainly composed of gear mesh frequency and its harmonics. Peaks appear in the frequency spectrum at $n * f_m$, where n is a positive integer. However, the sideband structures can be seen clearly in both cases, as indicated by the arrow. The vibration spectrum of the spalled high-contact-ratio gear system is primarily characterized by the gear mesh frequency and its harmonics, and the sidebands induced by the modulation phenomenon. The results of simulated signal agree with that of experimental signals, which shows the reliability of the dynamic model proposed in this paper.

6. Conclusions

A novel approach for calculating the meshing stiffness of gear with tooth spalling defect is proposed. Simultaneously, a gear dynamic model is developed incorporating the time-varying mesh stiffness, fractal backlash and time-varying friction. The bifurcation characteristic of the gear system is acquired through the gear dynamic model. The vibration signatures of the gear system obtained experimentally are used to validate the dynamic model. Significant contributions of the study are summarized as follows:

(1) The system's motion with ellipsoid tooth spalling fault exhibits rich bifurcation and chaotic characteristics under the influence of excitation frequency and gear backlash. The system presents diverse motion states, including single periodic motion, multi-periodic motion, quasi-periodic motion, and chaotic motion. There are three typical routes to chaos in the response, i.e., crisis to chaos, quasi-period to chaos, and period-doubling bifurcation to chaos.

(2) The frequency spectrum of the gear system with localized spalling fault is mainly composed of the meshing frequency and its harmonic components. The fault frequency appears in the form of sidebands in the spectrum at low speed. The tooth spalling fault could lead to the periodic impulses in the time-domain waveform.

Author Contributions: Z.C. (Writing—original draft: Lead); K.H. (Project administration: Lead); Y.X., (Data processing: Lead); M.S. (Software: Lead). All authors have read and agreed to the published version of the manuscript.

Funding: This study was supported by the National Natural Science Foundation of China (51775156), Natural Science Foundation of Anhui Province of China (1908085QE228), Key Research and Development Project of Anhui Province (202004h07020013), the Fundamental Research Funds for the Central Universities of China (PA2020GDSK0091), and the Horizontal Cooperation Project (0045022007).

Institutional Review Board Statement: Not applicable.

Informed Consent Statement: Not applicable.

Data Availability Statement: The data that support the findings of this study are available from the corresponding author upon reasonable request.

Conflicts of Interest: The authors state that they have no conflicting interest.

References

1. Wang, Y.; Dou, D.; Wang, J. Comparison on torsional mesh stiffness and contact ratio of involute internal gear and high contact ratio internal gear. *Proc. Inst. Mech. Eng. Part C J. Mech. Eng. Sci.* **2020**, *234*, 1423–1437. [CrossRef]
2. Huang, K.; Xiong, Y.; Wang, T.; Chen, Q. Research on the dynamic response of high-contact-ratio spur gears influenced by surface roughness under EHL condition. *Appl. Surf. Sci.* **2017**, *392*, 8–18. [CrossRef]
3. Huang, K.; Yi, Y.; Xiong, Y.; Cheng, Z.; Chen, H. Nonlinear dynamics analysis of high contact ratio gears system with multiple clearances. *J. Braz. Soc. Mech. Sci. Eng.* **2020**, *42*, 98. [CrossRef]
4. Wang, J.; Howard, I. Finite element analysis of high contact ratio spur gears in mesh. *J. Tribol.* **2005**, *127*, 469–483. [CrossRef]
5. Zhan, J.; Fard, M.; Jazar, R. A CAD-FEM-QSA integration technique for determining the time-varying meshing stiffness of gear pairs. *Measurement* **2017**, *100*, 139–149. [CrossRef]
6. Liang, X.; Zhang, H.; Zuo, M.J.; Qin, Y. Three new models for evaluation of standard involute spur gear mesh stiffness. *Mech. Syst. Signal Process.* **2018**, *101*, 424–434. [CrossRef]
7. Sun, Y.; Ma, H.; Huangfu, Y.; Chen, K.; Che, L.; Wen, B. A revised time-varying mesh stiffness model of spur gear pairs with tooth modifications. *Mech. Mach. Theory* **2018**, *129*, 261–278. [CrossRef]
8. Chaari, F.; Baccar, W.; Abbes, M.S.; Haddar, M. Effect of spalling or tooth breakage on gear mesh stiffness and dynamic response of a one-stage spur gear transmission. *Eur. J. Mech. A Solids* **2008**, *27*, 691–705. [CrossRef]
9. Jiang, H.; Shao, Y.; Mechefske, C.K. Dynamic characteristics of helical gears under sliding friction with spalling defect. *Eng. Fail. Anal.* **2014**, *39*, 92–107. [CrossRef]
10. Han, L.; Qi, H. Influences of tooth spalling or local breakage on time-varying mesh stiffness of helical gears. *Eng. Fail. Anal.* **2017**, *79*, 75–88. [CrossRef]
11. Luo, W.; Qiao, B.; Shen, Z.; Yang, Z.; Chen, X. Time-varying mesh stiffness calculation of a planetary gear set with the spalling defect under sliding friction. *Meccanica* **2020**, *55*, 245–260. [CrossRef]

12. Meng, Z.; Shi, G.; Wang, F. Vibration response and fault characteristics analysis of gear based on time-varying mesh stiffness. *Mech. Mach. Theory* **2020**, *148*, 103786. [CrossRef]
13. Ma, H.; Li, Z.; Feng, M.; Feng, R.; Wen, B. Time-varying mesh stiffness calculation of spur gears with spalling defect. *Eng. Fail. Anal.* **2016**, *66*, 166–176. [CrossRef]
14. Saxena, A.; Parey, A.; Chouksey, M. Time varying mesh stiffness calculation of spur gear pair considering sliding friction and spalling defects. *Eng. Fail. Anal.* **2016**, *70*, 200–211. [CrossRef]
15. Yoshida, A.; Ohue, Y.; Ishikawa, H. Diagnosis of tooth surface failure by wavelet transform of dynamic characteristics. *Tribol. Int.* **2000**, *33*, 273–279. [CrossRef]
16. Jia, S.; Howard, I. Comparison of localised spalling and crack damage from dynamic modelling of spur gear vibrations. *Mech. Syst. Signal Process.* **2006**, *20*, 332–349. [CrossRef]
17. Ma, R.; Chen, Y.; Cao, Q. Research on dynamics and fault mechanism of spur gear pair with spalling defect. *J. Sound Vib.* **2012**, *331*, 2097–2109. [CrossRef]
18. Dadon, I.; Koren, N.; Klein, R.; Lipsett, M.G.; Bortman, J. Impact of gear tooth surface quality on detection of local faults. *Eng. Fail. Anal.* **2020**, *108*, 104291. [CrossRef]
19. Rui, M.; Chen, Y. Research on the dynamic mechanism of the gear system with local crack and spalling failure. *Eng. Fail. Anal.* **2012**, *26*, 12–20.
20. Yu, W.; Mechefske, C.K.; Timusk, M. A new dynamic model of a cylindrical gear pair with localized spalling defects. *Nonlinear Dyn.* **2017**, *2018*, 2077–2095. [CrossRef]
21. Chen, K.; Ma, H.; Che, L.; Li, Z.; Wen, B. Comparison of meshing characteristics of helical gears with spalling fault using analytical and finite-element methods. *Mech. Syst. Signal Process.* **2019**, *121*, 279–298. [CrossRef]
22. Huangfu, Y.; Chen, K.; Ma, H.; Li, X.; Han, H.; Zhao, Z. Meshing and dynamic characteristics analysis of spalled gear systems: A theoretical and experimental study. *Mech. Syst. Signal Process.* **2020**, *139*, 106640. [CrossRef]
23. Luo, W.; Qiao, B.; Shen, Z.; Yang, Z.; Cao, H.; Chen, X. Investigation on the influence of spalling defects on the dynamic performance of planetary gear sets with sliding friction. *Tribol. Int.* **2021**, *154*, 106639. [CrossRef]
24. Shi, L.; Wen, J.; Pan, B.; Xiang, Y.; Zhang, Q.; Lin, C. Dynamic Characteristics of a Gear System with Double-Teeth Spalling Fault and Its Fault Feature Analysis. *Appl. Sci.* **2020**, *10*, 7058. [CrossRef]
25. Ma, R.; Chen, Y. Bifurcation of multi-freedom gear system with spalling defect. *Appl. Math. Mech.* **2013**, *34*, 475–488. [CrossRef]
26. Wu, J. *Root Transition Curve and Root Stress*; National Defense Industry Press: Arlington, VA, USA, 1989.
27. Yang, L.; Baddour, N.; Ming, L. Dynamical modeling and experimental validation for tooth pitting and spalling in spur gears. *Mech. Syst. Signal Process.* **2018**, *119*, 155–181.
28. Chen, Z.; Zhou, Z.; Zhai, W.; Wang, K. Improved analytical calculation model of spur gear mesh excitations with tooth profile deviations. *Mech. Mach. Theory* **2020**, *149*, 103838. [CrossRef]
29. Chen, Q.; Wang, Y.; Tian, W.; Wu, Y.; Chen, Y. An improved nonlinear dynamic model of gear pair with tooth surface microscopic features. *Nonlinear Dyn.* **2019**, *96*, 1615–1634. [CrossRef]
30. Chen, Q.; Zhou, J.; Khushnood, A.; Wu, Y.; Zhang, Y. Modelling and nonlinear dynamic behavior of a geared rotor-bearing system using tooth surface microscopic features based on fractal theory. *AIP Adv.* **2019**, *9*, 015201. [CrossRef]
31. Xu, H. *Development of a Generalized Mechanical Efficiency Prediction Methodology for Gear Pairs*; The Ohio State University: Columbus, OH, USA, 2005.
32. Yi, Y.; Huang, K.; Xiong, Y.; Sang, M. Nonlinear dynamic modelling and analysis for a spur gear system with time-varying pressure angle and gear backlash. *Mech. Syst. Signal Process.* **2019**, *132*, 18–34. [CrossRef]

Article

Functional Investigation of Geometrically Scaled Drive Components by X-in-the-Loop Testing with Scaled Prototypes

Michael Steck, Sven Matthiesen * and Thomas Gwosch

Karlsruhe Institute for Technology (KIT), 76131 Karlsruhe, Germany; michael.steck@kit.edu (M.S.); thomas.gwosch@kit.edu (T.G.)
* Correspondence: sven.matthiesen@kit.edu; +49-721-608-47156

Abstract: Validation is important for a high product quality of drive components. An X-in-the-Loop test bench enables the integration of scaled prototypes through coupling systems and scaling models even before serial parts are available. In the context of X-in-the-loop investigations, it is still unclear whether a scaling model enables the early investigation of geometry variants in powertrain subsystems. In this paper, scaled geometry experiments taking into account the interacting system are considered to evaluate the scaling model in terms of early investigation of geometry variants. The aim of this paper is the functional investigation of geometrically scaled drive components by integrating scaled prototypes in an X-in-the-Loop test bench. Using an overload clutch with detents, component variants of different size levels are investigated in scaled experiments with a scaling model. The results confirm possibilities of X-in-the-Loop integration of scaled prototypes and their investigation on geometrically scaled drive components. The investigations show, therefore, the opportunities of integrating scaled drive components through the scaling model to support the investigation of geometry variants before serial parts are available. Scaled geometry investigations considering the interacting system can, thus, support product development.

Keywords: test bench; hardware-in-the-loop; drive component; scaled prototypes; scaling model; experiment; product development; validation; hand-held power tool

Citation: Steck, M.; Matthiesen, S.; Gwosch, T. Functional Investigation of Geometrically Scaled Drive Components by X-in-the-Loop Testing with Scaled Prototypes. *Machines* **2022**, *10*, 165. https://doi.org/10.3390/machines10030165

Academic Editor: Antonio J. Marques Cardoso

Received: 4 February 2022
Accepted: 18 February 2022
Published: 22 February 2022

1. Introduction

To reduce uncertainties early in product development, it is necessary to validate the product, which requires knowledge about the behavior of the system [1–3]. The necessary system knowledge can be determined by experimental investigations with prototypes in powertrain test benches, whereby non-existent components must be simulated [2,4].

Frontloading can shorten development times in product development and avoid late design changes [5,6]. In the industry, validation is very important, as is evident from the wide range of testing and validation activities.

For the validation of product series, problems result because, often, not all subsystems are available for early experimental investigations. Validation can, therefore, only be postponed to later phases in the development process, or be carried out via simulation.

Especially in early development phases, the consideration of interactions is a challenge, especially if not all components and subsystems of the product are available for experimental investigations (see Figure 1). Investigations into the overall system with realistic conditions are, thus, difficult.

A challenge in the early validation of powertrain components is the suitable integration of powertrain components into the overall system to map the interactions.

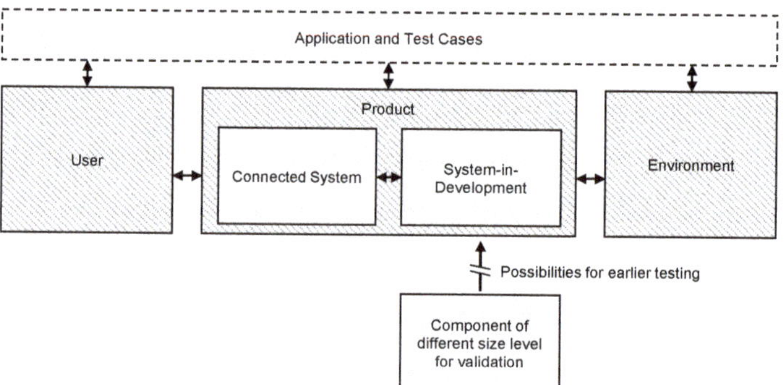

Figure 1. Early validation with XiL-approach by including available components of different size levels in a test bench.

For practical testing and validation activities, it is important to integrate the system under investigation into the overall system. The XiL approach can be used for this purpose [7–9]. The X-in-the-Loop (XiL) approach enables functional testing of a component or subsystem (System-in-Development, abbr. SiD) in test benches by integrating the remaining subsystems as physical or virtual models (as Connected Systems) [3,7,8,10]. Opportunities for early testing are provided by the use of early prototypes instead of using serial parts. Often, only subsystems of different size levels can be used for early validation as prototypes, for instance in wind tunnel experiments [11] (see Figure 1).

If only subsystems of a different size can be used as prototypes, an adaptation of the investigation test bench is necessary. For the integration of scaled prototypes in the context of X-in-the-Loop investigations, adapted coupling systems are required. The integration can be done by coupling systems (Figure 2). Coupling systems can allow the integration of geometrically scaled drive components in X-in-the-Loop investigations.

Coupling Systems and Scaling Models to Support Validation

Coupling systems are used for the integration of powertrain components. These coupling systems connect the component or subsystems under investigation in test benches with the interacting subsystems, which are either physical or virtual. The virtual coupling of subsystems on the test bench is used in distributed validation to couple several test benches across different locations.

Approaches also exist in the implementation of virtual shafts to connect testbeds. The theoretical background is well known in control engineering. A review of the last 20 years on the synchronization of multi-motor systems was presented by Perez–Pinal et al. [12]. It is important to consider interactions in the coupling of torque and speed between the interacting systems.

Andert et al. use an approach to connect different subsystems of drive trains by a virtual shaft [13]. The aim of this approach is to couple test facilities for components to a virtual powertrain test bench with a control logic that emulates a mechanical shaft connection. The superimposed controller synchronizes speed and torque like a rigid shaft connection with low inertia and high stiffness. This infrastructure allows a flexible configuration of hybrid and conventional powertrains with virtualized interaction [13].

In addition, there is an approach for the adaption of power quantities within the virtual coupling [14,15]. The integration of differently scaled subsystems, which are to be tested simultaneously on a test bench, is presented. Within the coupling systems, the power quantities are adapted to the scaled subsystem. Adaption can be made for the power quantities of a rotatory motion, for torque and angle of rotation [14].

The virtual coupling enables the modification of mechanical quantities in a rotational powertrain by scaling. The general structure of this virtual coupling is shown in Figure 2.

Studies with scaled subsystems also exist in the field of mechatronic systems and wind turbines [16–21].

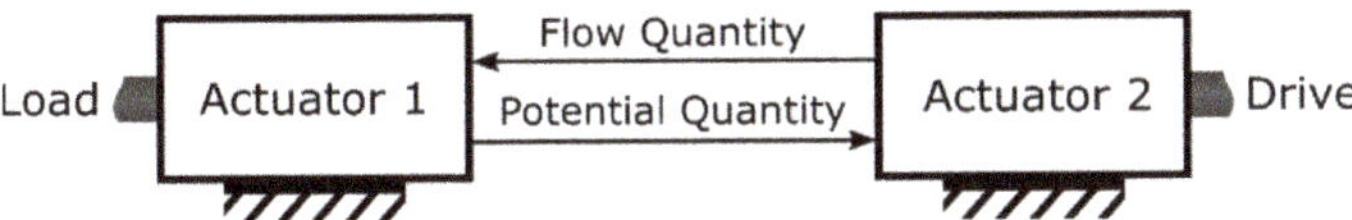

Figure 2. Virtual Coupling of mechanical subsystems [22].

Two examples of previous research [14,15] demonstrate virtual coupling and the integration of performance scaling within XiL test benches. In an initial investigation, scaling of rotational variables was used to enable simultaneous testing of powertrain subsystems with different scaling on a XiL test bench. This was demonstrated using the example of different performance scaling of the spring preload force of an overload clutch from a powertrain.

The results of the investigation [14] show that the rotational mechanical quantities can be adapted using scaling laws obtained from a similarity analysis. The scaling model makes it possible to adapt the system quantities between the subsystems and, thus, predict the behavior of the product, even if differently scaled subsystems were used on the XiL powertrain test bench.

In a further investigation [15], the scaling of the performance quantities was considered through a simulation using the example of an aerospace actuator. The aim was to take into account differences in size and, thus, different load capacities, and to compensate for them with a scaling model. Using the example of a torque limiting clutch from a geared rotary actuator out of an aircraft, the influences of the rotational inertia of the drive shaft on the scaling are analyzed in a simulation study. The determining influence is taken into account by the integration of virtual inertia within the scaling [15].

In previous XiL-investigations [14,15], the individual differing properties of the scaled powertrain components were already taken into account in the scaling model.

The problem is that an analysis of the scaling model for early investigation of geometry variations has not yet been performed in XiL-experiments. The following investigation addresses the question of whether the scaling enables a functional investigation of geometrically scaled drive components in X-in-the-Loop investigations.

The aim of the paper is the functional investigation of geometrically scaled drive components by integrating scaled prototypes in an X-in-the-Loop test bench. Using the example of an overload clutch with detents, component variants of different size levels are investigated in tests with performance scaling. For this purpose, an exemplary adaptation of the geometry of the overload clutch and its validation with the performance scaling is considered.

The following hypothesis is formulated: "The scaling enables the geometry testing of scaled components. Using a scaling model in X-in-the-Loop test bench, geometrically scaled drive components can be tested for functional properties, and geometry variants can be evaluated for product engineering."

For the investigation, a scaling model is considered concerning the early investigation of geometry variants in the XiL test stands. Using the example of an overload clutch, two different geometry variants are examined in scaled component tests, taking into account the interacting system.

2. Methods of Scaled Virtual Coupling for XiL-Test Bench

This chapter describes the approaches used to perform scaled component tests considering the interacting system. First, the used approach is shown in general, and implementation in a XiL test bench is presented. Based on this, the derivation of the scaling model to adapt the rotational quantities between size variants of a cordless screwdriver is presented.

2.1. Approach and Theory of Virtual Coupling

To integrate the scaled drive train components, an adaptation of the transmitted power within the coupling systems is necessary. The following approach (see Figure 3) can be used as a basis for integrating the scaled prototypes into the overall system and an investigation in XiL test benches.

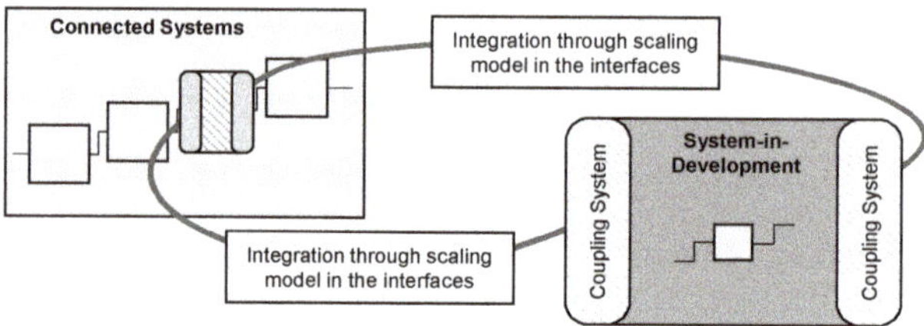

Figure 3. Approach to integrating scaled prototypes into the overall system.

The approach for adapting power between interconnected subsystems (Figure 3) is based on the virtual coupling [23] and virtual shaft concepts [13].

Virtual coupling refers to the decoupling of mechanically connected subsystems in a powertrain test bench and their coupling via virtual connections. The approach can be used to set up XiL test benches for early validation using reduced-size prototypes. The following subsystems shown in Figure 4 are necessary for the implementation:

- System-in-Development (SiD): Is the subsystem to be developed whose functionality is to be verified.
- Connected system represents the remaining subsystems of the system under development (SiD). It is physically present and forms the remaining system model. This is needed to represent the interactions with the SiD and to integrate it into the overall system.
- Coupling systems connect the Connected System and the System-in-Development. With the help of scaling models, they translate physical performance quantities into virtual system quantities and vice versa.

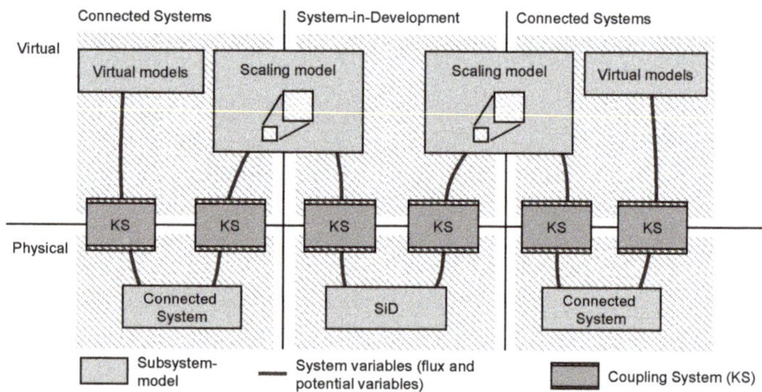

Figure 4. Example XiL architecture for scaled validation.

The structure (Figure 4) can be used for the implementation of system-specific test benches. From this, an example of XiL architecture for scaled tests (Figure 4 can be derived. The implementation for scaling the rotational mechanical quantities is shown below.

In general, the system-in-development (SiD) is connected to the remaining system only via coupling systems (KS) and the virtual domain. The coupling systems consist of sensor-actuator systems that adapt the system quantities from virtual to physical quantities and in reverse. Servo motors are used as actuators, which are supplemented by additional sensors to detect torque and speed. The flow and potential quantities are continuously exchanged between the coupling systems in real-time and can be adapted by the scaling models. These coupling systems allow the adaptation of the rotational power quantities through a scaling model. The integration of scaling via virtual coupling is shown in Figure 5.

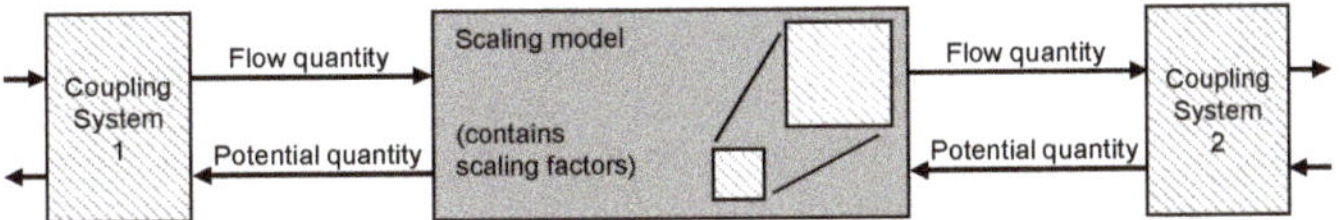

Figure 5. Integration of the scaling model via virtual coupling to adapt the mechanical quantities. [14].

The selection of actuators and control requirements depends on the dynamic requirements of the investigation. For dynamic systems with interactions between the interacting systems, the requirement of the real-time capability of the virtual coupling has to be considered when selecting the sensor-actuator systems. Virtual coupling can be evaluated using the criteria shown in Gwosch et al. [22]. A scaling model can then be integrated via the virtual coupling to adapt the mechanical performance variables.

2.2. Example System and Component Variants of the Overload Clutch

The scaling model is investigated in component tests to evaluate geometry variants of geometrically scaled drive components. The influence of the different geometry variants is investigated in both scaled and unscaled component tests. For this purpose, two prototypically implemented component variants of the clutch ring are examined concerning the release characteristics of the overload clutch on a XiL test bench.

To investigate the release characteristics, key values in the torque curve of the output shaft are considered. The scaled geometry variants are integrated via virtual coupling systems and scaling to enable scaled component tests, taking the rest of the system into account. Via the scaling, an adaptation of the torques and speeds between the considered subsystems is possible.

Two clutch systems from cordless screwdrivers are used as an example system. A cordless screwdriver in the 10.8-Volt version (Cordless screwdriver GSR 10 8 Li of the power tool manufacturer Robert Bosch GmbH [24]) and a cordless screwdriver in the 18-Volt version (Cordless screwdriver GSR 18-2 Li of the power tool manufacturer Robert Bosch GmbH [25]) is used here. Both clutch systems use the same operating principle but differ in terms of the size level, which is related to the battery voltage (size level 10 V/size level 18 V).

To test the formulated hypothesis, two variants of the clutch ring of a cordless screwdriver are investigated on a XiL test bench. The variants differ in the raceway geometry of the clutch ring (See Figure 6). Variant 1 corresponds approximately to the original geometry of the series component, and is used as a reference. Variant 2 was adapted in geometry and contains a pocket geometry. Comparable geometries of the component variants have already been used by Gwosch [26].

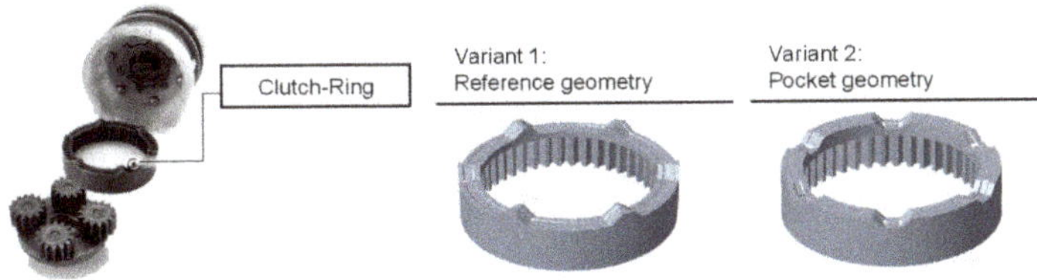

Figure 6. Differences in geometry for two different component variants of the clutch ring of a cordless screwdriver.

A theoretical consideration of the torque curve for active overload clutch is shown as a qualitative curve in Figure 7 based on investigations of Gwosch [26] and simulation results from Steck et al. [27].

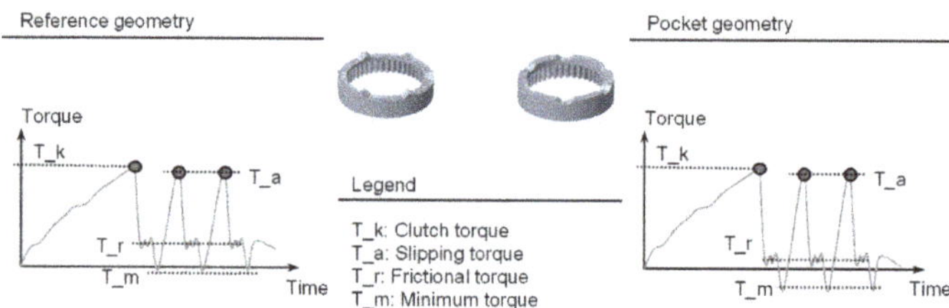

Figure 7. Qualitative torque curve of the overload clutch for two geometry variants. The torque characteristics are plotted to characterize the geometry variants. Adapted representation from Gwosch [26] (p. 148).

Description (based on research in [26]) of qualitative progression: If the applied torque exceeds the clutch torque, the balls move over the detents of the clutch ring and the clutch disengages. The torque then drops. Repeating contact between the detent balls and the driver of the clutch ring causes further torque increases. The functional behavior depends on the geometric properties of the clutch and the dynamic behavior of the driveline. Gwosch showed in investigations an influence of the behavior depending on the raceway geometry [26]. A deviating frictional torque is expected for the pocket geometry compared to the reference geometry.

2.3. Derivation of Scaling by Similarity Ratios

Necessary for the derivation of the scaling (based on research in [14,15]) is a sophisticated model understanding, which takes into account all relevant parameters of the investigation target. Relevant parameters are based on the system understanding of the components under consideration and experimental investigations in the reference system, as well as the requirements for the component currently in focus and under development.

Based on these model differences, similarity mechanics and dimensional analysis can be used to derive the scaling factors. The general procedure is shown in Figure 8.

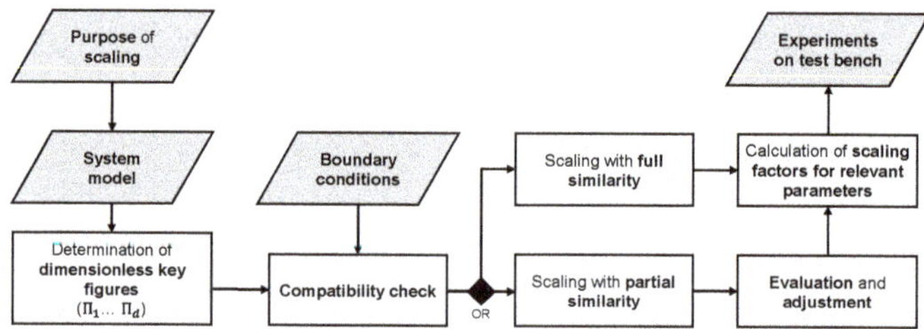

Figure 8. Procedure for deriving the scaling starting from a system model and considering the boundary conditions [14].

For rotational mechanical systems, the mechanical parameters M, $\dot{\varphi}$, and l can be used as the main variables of the system since they represent the model differences of the scaled and unscaled systems and the physical quantities in the virtual clutch. The rotational power variables are adapted depending on the scaling factors. Based on the scaling of the size step, an adaptation of the torques is performed, and additionally, the speeds between the subsystems are scaled.

Based on a modeling of the clutch, the relevant parameters, which are called the relevance list, were analyzed by dimensional analysis. The dimensions of the relevant parameters are classified according to the {MLT} system, the dimensions are mass (M), length (L), time (T) [28–30]. The parameters T, $\dot{\varphi}$ and l are used as main quantities of the system, since they represent the differences of the scaled and unscaled system.

The dimensionless ratios are calculated based on the dimensionless quantities and their relationships. The calculated ratios are then checked for consistency. In case of inconsistencies, not all dimensionless ratios can be considered, resulting in partial similarity.

For the coupling system study example, the scaling factors are derived below. The following Table 1 shows the result of the dimensional analysis for the investigation of both size variants of the cordless screwdrivers. In this study, friction effects were not considered in the scaling models.

Table 1. Result of the dimensional analysis. The parameters T, $\dot{\varphi}$, and d are used as the main quantities of the system, since they represent the difference of the scaled and unscaled system and the physical quantities in the virtual coupling.

Dimension \ Parameter	T	$\dot{\varphi}$	d	F	t	h	φ	$\ddot{\varphi}$	J	m
mass M [kg]	1	0	0	1	0	0	0	0	1	1
length L [m]	2	0	1	1	0	1	0	0	2	1
time T [s]	−2	−1	0	−2	1	0	0	−2	0	1

Result of the dimensional analysis. The parameters T, $\dot{\varphi}$, and d (torque, angular velocity, and diameter) are used as the main quantities of the system since they represent the difference of the scaled and unscaled system and the physical quantities in the virtual coupling.

The two size levels of the clutches do not differ in terms of the used materials and the manufacturing processes. Due to the characteristics of the XiL test bench, the properties of the prototype, and the purpose of the scaling, the following boundary conditions are relevant:

1. The diameter of the clutch (d) is scaled by the factor a. Thus, the ratio between the diameter (δd) of the scaled and unscaled system is described by $\delta d = \frac{d_S}{d_U} = a$.
2. The height of the clutch ring (h) is scaled with b. There is no complete geometric scaling of the size level. The height of the clutch ring is responsible for the preload of the spring and, thus, determines the clutch torque. This length is scaled with $\delta h = b$.
3. The torque is given by the spring force multiplied by the diameter of the clutch ring. As a result, the torque is scaled with $\delta T = \delta d \cdot \delta h$
4. The transmission ratio is already a dimensionless key figure that describes the ratio of the number of teeth of the gears.
5. The speed is scaled by the factor $\delta n = n$. The differences in the transmission ratio determine the scaling of the speed.
6. Time scaling is not permitted. This is necessary to be able to represent the interactions between the subsystems. Therefore, a ratio between the time of the scaled and unscaled system is defined such that there is no scaling ($\delta t = 1$).
7. The same material should be used, i.e., there is no scaling for the material: $\delta\rho = 1$. Since the same material is used for all tests, any differences that may occur in the friction condition are not taken into account.

The boundary conditions do not allow a complete geometric scaling, because the diameter and the height of the clutch cannot be scaled with the same factor. The verification of the ratios shows a discrepancy for the moments of inertia. Due to the partial similarity of the geometric dimension, the inertia of the clutch ring is not scaled to the same degree. Additionally, there is a discrepancy between the scaling of the inertia and the scaling of the moments of inertia. The external torque is scaled by the factor ($\delta M = a{\cdot}b$) while the moment of inertia is scaled by a different factor due to the different geometry. To account for the two states of the planetary gear in interaction with the clutch, the scaling of the speed is split. There are adaptations of the speed before and after the clutch. The check with the boundary conditions results in a partial similarity for these component tests in this publication. Finally, the following scaling factors result (see Table 2).

Table 2. Summary of the scaling factors.

Dimension	Mass*M* [kg]	Length *L* [m]	Time *T* [s]	Scaling Factor
time (δt)	0	0	1	1
diameter (δd)	0	1	0	a
height (δh)	0	1	0	b
angular velocity ($\delta\dot{\varphi}$)	0	0	−1	n_1/n_2
mass (δm)	1	0	0	1
external force (δF_e)	1	1	−2	1
internal force (δF_i)	1	1	−2	1
applied torque (δT)	1	2	−2	$a{\cdot}b$
moment of inertia (δT_T)	1	2	−2	1
stiffness (δc)	1	0	−2	1
mass inertia (δJ)	1	2	0	1

The scaling factors δT and $\delta\dot{\varphi}$ are used to calculate the power quantities (flow and potential quantities) on the test bench:

- The scaling of the applied torque depends on the ratio of the coupling dimensions. From this follows the scaling of the torque with $(\delta T) = a{\cdot}b$
- Scaling of the angular velocity is according to: $(\delta\dot{\varphi}) = n_1/n_2$

The calculation of the performance variables (see Figure 9) is performed within the control system in real-time during the operation of the test bench. The adapted mechanical quantities are then applied to the interacting subsystems in the coupling systems.

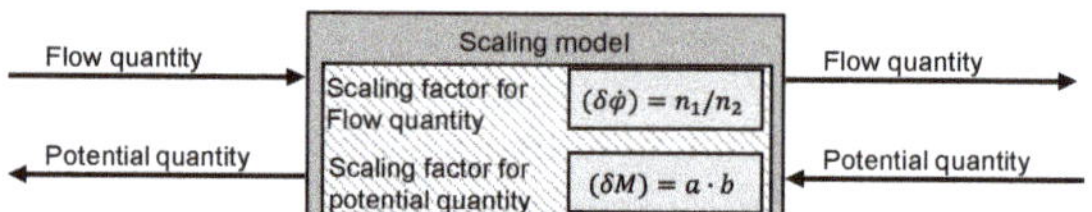

Figure 9. Scaling factors to compensate for size level.

The scaling factors (given in Table 2) are used to calculate the power quantities (flow and potential quantities) on the test bench. For implementation, the scaling factors are integrated into the virtual coupling at the test bench.

3. Experimental Setup

In the following, the test bench setup (experimental setup) for evaluating the scaling for geometry variation is shown. The implementation of the test bench is based on the approach presented in Chapter 2.1 Using clutches from two cordless screwdrivers as an example, the scaling model for investigating geometry variation is considered. In individual investigations, the clutch from one cordless screwdriver was used to evaluate the influence of geometry on the release behavior in scaled investigations.

This chapter presents the experimental setup of the scaled component tests to evaluate scaling. In addition, the component variations of the overload clutch are described. The series of tests for performing the scaled component tests on a XiL test bench is then shown. Based on the scaled component tests with different geometries, the analysis of the approach for the investigation of scaled geometries is carried out.

3.1. Test Setup

The test setup is based on the approach shown in Chapter 2.1. The component variants of the clutch rings can be inserted into the overload clutch on the test bench

(Figure 10 The investigation is exemplified on the scaled-components-in-the-loop (sCiL) test bench [14,15,22] with the investigation setup shown in Figure 11. The test bench is operated in closed-loop mode. The individual drive train systems can be positioned and aligned via standardized base plates. The accuracy and the real-time capability of the signal transmission must be taken into account. It is important that the influence of the coupling systems on the signal transmission is as low as possible, and that no negative deviations occur.

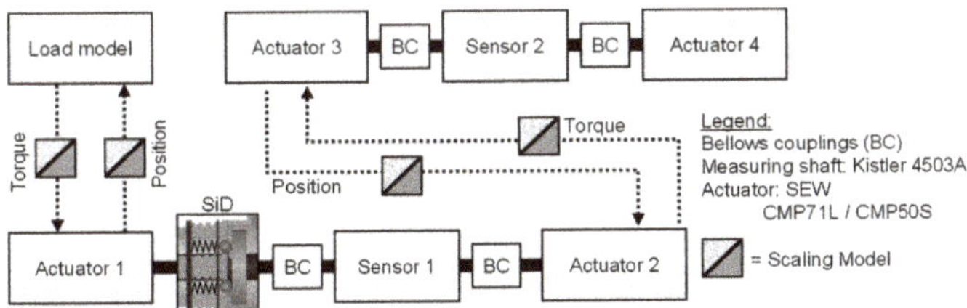

Figure 10. Investigation setup for the variant investigation of the overload clutch in scaled test bench investigations (schematic representation). Actuator 1 is torque controlled.

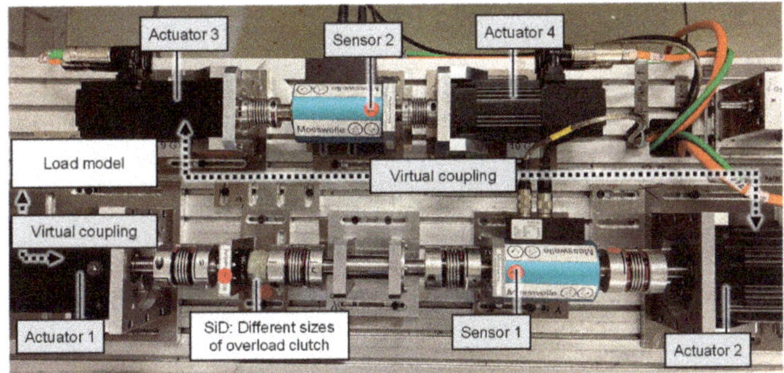

Figure 11. sCiL test bench for performing the verification of the scaled examination. The sCiL test bench contains the mechanical components of the drive train of a cordless screwdriver. The virtual coupling is integrated between actuator 2 and actuator 3 as well as between actuator 1 and the virtual load model in the drive train.

The investigations are carried out with a generic load model. The load torque is obtained from the angle of rotation via a constant load factor. The load model and the virtual models of the XiL-architecture can be considered as a digital twin. The speed of the drive shaft is used as the default value, which is kept at a constant value of 100 rpm after a ramp-up phase.

The scaling models are integrated within the virtual coupling between the Systems under Development. The coupling systems are realized with sensor-actuator systems and a centralized or decentralized control system. The implemented test setup is shown in Figure 11.

3.2. Execution of the Component Tests

In Gwosch [26], the component variants were manufactured from the material AlSi10Mg using an additive manufacturing process. In this publication, additively manufactured components made of polymer were used for geometry variants due to manufacturing restrictions. To be able to investigate the influence of the geometry variation, both geometry variants are produced using the same manufacturing process and the same material.

The clutch rings used in this investigation are shown in Figure 12. Both component variants of the clutch ring are made of photopolymer (Photopolymer VeroWhitePlus from manufacturer Stratasys. Material properties are given in [31].) by an additive manufacturing process. The raceway of the clutch ring was oriented upward during additive printing to realize a smooth surface. The geometry of the clutch rings was measured in terms of overall height before testing. Consideration of different manufacturing processes and material properties in the scaling is not part of this investigation.

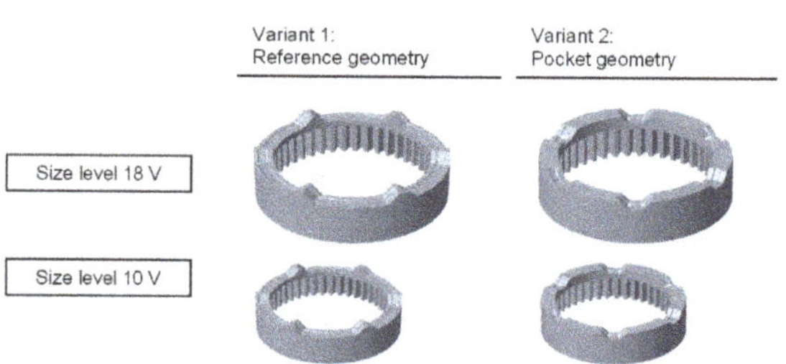

Figure 12. Geometries and size level of the used components of the clutch.

The test series differ in terms of the used geometry variant and the used scaling model to adapt the performance quantities between the different scaled clutch to the environment. For the scaled tests (10 V Scaled), the small clutch (size level 10 V) is included in the scaled environment. The differences in power quantities are compensated for by the scaling model. This is compared to two experiments without a scaling model at two different size levels (size level 10 V and 18 V). The release characteristics of the two geometry variants are compared both in the scaled (size level 10 V Scaled) and unscaled experiments (size level 10 V and 18 V).

To evaluate the scaling model for the investigation of geometry variants, the following test series are carried out and compared concerning the release characteristics:

- Unscaled component in unscaled environment (Standard case in size level 10 V)
- Unscaled component in performance scaled environment (relevant case in scaled size level 10 V)
- Size scaled component in scaled environment (Case for evaluation in size level 18 V)

These experiments are performed for two different geometry variants, so in total 6 different experiments. The coupling torque and the frictional torque are considered as evaluation variables of the release characteristics.

The characteristic values are plotted in Figure 7. The selected evaluation variables are relevant for the derivation of target variables in the development since the safety function of the clutch depends on them. The filtered torque at measuring shaft 2 in Figure 10 is selected as the measured variable. The test is repeated 21 times for each variant. A total of 126 tests were performed.

4. Results of the X-in-the-Loop Integration of Scaled Prototypes

In this chapter, scaled component testing is presented considering the rest of the system. This is then followed by the evaluation of the scaling for the early evaluation of design decisions.

The results for the evaluation of the geometry variants are summarized in Table 3. It can be seen that the scaling model within the coupling systems allows an adaptation of the clutch torques of the different variants. For the frictional torque, the results show a difference between reference geometry and the pocket geometry. Table 3 shows the mean values of the clutch torque and the frictional torque of the two geometry variants for measuring shaft 2.

Table 3. Parameters for the evaluation of geometry variants. Mean values of the clutch torque and frictional torque were determined from 21 test repetitions in each case.

Characteristic Value (Mean Values)		10 V	10 V Scaled	18 V
Clutch torque	Reference geometry	0.554 Nm (σ = 0.029 Nm)	0.838 Nm (σ = 0.028 Nm)	0.865 Nm (σ = 0.029 Nm)
	Pocket geometry	0.518 Nm (σ = 0.021 Nm)	0.787 Nm (σ = 0.035 Nm)	0.834 Nm (σ = 0.030 Nm)
Characteristic value (mean values)		**10 V**	**10 V Scaled**	**18 V**
frictional torque	Reference geometry	0.457 Nm (σ = 0.025 Nm)	0.697 Nm (σ = 0.036 Nm)	0.729 Nm (σ = 0.25 Nm)
	Pocket geometry	0.194 Nm (σ = 0.034 Nm)	0.329 Nm (σ = 0.086 Nm)	0.110 Nm (σ = 0.063 Nm)

To check the functionality of the scaling, the coupling torques were evaluated. The clutch torques of the reference geometry and pocket geometry are shown in Figure 13. The mean value of the clutch torques is at a comparable level for the two variants. Here, there is a good agreement between the clutch torques of the scaled tests (10 V Scaled) and the tests of the size level 18 V. It is shown that it is indeed possible to adapt the coupling torques by scaling.

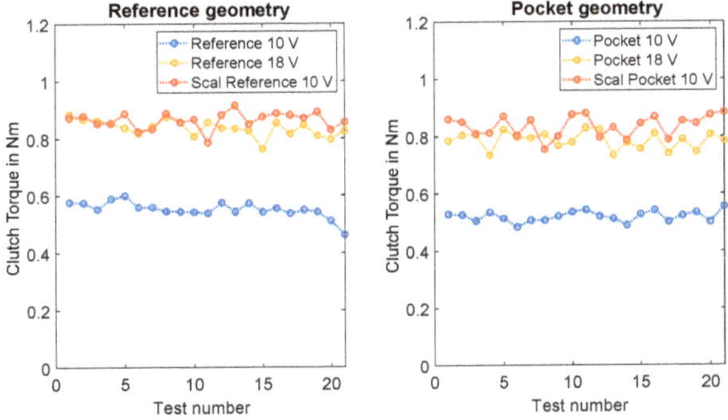

Figure 13. Comparison of clutch torques between reference and pocket geometry at measuring shaft 2.

In addition, the characteristic value of the frictional torque was considered in the investigations. For the reference geometry, there is good agreement between the frictional torques of the scaled tests (10 V Scaled) and the tests of the size level 18 V. For the pocket geometry, the frictional torque is at a lower level for all size levels (see Figure 14). For both geometry variants, some outliers can be identified (see Figure 14).

In summary, there is a difference between the mean values of the frictional torque parameter for the reference geometry and the pocket geometry. At a comparable clutch torque level, the frictional torque of the pocket geometry is lower than that of the reference geometry. In the scaled tests, the frictional torque of the pocket geometry is also lower compared to the reference geometry. For the mean values of the frictional torque, there is a deviation in the comparison of the scaled investigation (10 V scaled) with the unscaled variant (18 V).

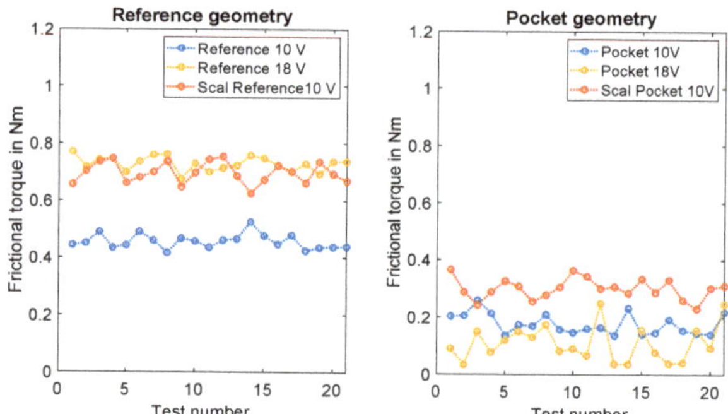

Figure 14. Frictional torque of the two geometry variants. The frictional torques at measuring shaft 2 are plotted above the test number.

5. Discussion

In the following, the scaled component tests and the approach with a scaling model within the coupling system are discussed.

The hypothesis formulated at the beginning of this paper: "Scaling enables geometry investigations of scaled components. By using a scaling model in an X-in-the-Loop test bench, geometrically scaled drive components can be tested for functional properties and geometry variants can be evaluated for product development." This is reviewed below.

The results in this paper show how scaling models support the XiL-investigation of geometry variants for scaled components.

For the example system of a clutch out of an electric screwdriver, the match between scaled investigations and non-scaled investigations is good for the evaluation variable of the clutch torque. By changing the raceway geometry of the clutch ring, the release characteristic of the overload clutch can be adapted and investigated in scaled experiments. These results confirm a dependence of the behavior on the raceway geometry similar to investigation results presented by Gwosch [26].

It is not necessary to adapt the scaling to the geometry variants, provided that no additional physical effects (e.g., friction) become relevant as a result of the geometric changes, which are not currently taken into account. In scaled investigations, subsystems of different size levels can be used to test variants with different geometries for their functional performance.

The results of the scaled component tests confirmed an influence of the geometry on the frictional torque at the output shaft of the cordless screwdriver drive train. The scaled component tests show a deviation in the frictional torque of the pocket geometry compared with the reference geometry. The frictional torque here is well below the clutch torque. These differences show that not all effects responsible for this characteristic point in the torque behavior are taken into account in the scaling model. Due to the relatively low frictional torque, effects may be present here that are less important to the clutch torque. In the case of changing friction conditions during the rotational motion, it is therefore difficult to map the friction conditions in the scaling model.

To represent the friction behavior for the present clutch in the scaling model, further investigations are necessary to analyze the friction conditions as a function of the material pairings and the position-dependent effective surface pairs. The friction ratios determined can then be taken into account in the scaling models using known friction models. To integrate the friction models into the scaling, the friction models must take into account the scaling parameters and map effects on the performance variables. For an evaluation of the friction-dependent function, adaptation to the scaling model is necessary.

With the use of the sCiL test bench and the developed scaling, the influence of a modified raceway geometry on the torque curve in the drive train of the cordless screwdriver can be investigated under defined boundary conditions even before prototypes of the follow-on product are available.

The coupling systems of the sCiL environment allow the power quantities of the rotational motion to be adapted according to the scaling models. The coupling systems consist of servo motors, rotational position sensors, and torque sensors (sensor-actuator systems). The challenge regarding the sensor-actuator systems is the accuracy with which the power quantities are determined and then reapplied. For this purpose, the accuracy and the real-time capability of the signal transmission must be taken into account. It is necessary to check whether the signal transmission is fast enough or not.

Further developments of the control system at the sCiL test rig offer the potential to further improve the quality of the mapping of the interactions between the subsystems under consideration and to reduce the influence of the coupling systems. In particular, the mechanical influences of the actuators are to be reduced so those investigations can also be carried out in more demanding dynamic ranges.

The presented component tests are suitable for estimating the influence of geometry adaption on the function with scaled investigations. The scaling model allows the influence of geometry on the frictional torque and the clutch torque to be predicted. The ratios of the parameters are comparable for geometry adaptation. This means that the scaling model and the approach of the test bench can be used for the early validation of development variants regarding geometry adaptation.

The scaled investigations enable the influence of the geometry change to be evaluated on the powertrain test bench in the early phases of product development. The scaled experiments can also be used to build a model and derive a digital twin. The digital twin can then support during the product life cycle. The investigations and the models derived from them can also be used to validate errors and investigate possible improvements.

When deriving the scaling, it is important to consider the relevant parameters and properties. If relevant effects are neglected, deviations in the evaluation of the scaled systems and the transfer to systems of other sizes level may result. Existing system knowledge, experiments, and research results can be used to identify the relevant parameters.

The internal system conditions and loads on the components in the drive train of the cordless screwdriver can differ despite the same clutch torque. An evaluation of the applied loads is already possible to some extent through the virtual coupling on the sCiL test bench. The system behavior at the selected interfaces can, thus, be recorded in detail.

The investigated component variants differ from the series components in terms of material and manufacturing process. This can lead to deviations if the findings are transferred directly to the series component. The influence of tolerances and the material must be verified in further investigations concerning the scaling models. The evaluation of the torque at the drive shaft enables the safety function of the clutch to be assessed.

If the relevant effects are taken into account in the scaling model, an evaluation is possible in the early phases of product development. In scaled investigations, the influence of geometry adaptation can already be estimated. If additional effects occur that are not taken into account in the scaling model, then only limited statements can be made about the behavior that depends on them. If the derivation of quantitative target values is necessary for validation, then the relevant effects must be taken into account in the scaling.

The scaling model makes it possible to investigate the influence of geometry variants on the functionality already with scaled components.

Quantitative evaluation of geometry parameters in early development phases can ensure the functionality of the subsequent product. The further development and design of the clutch ring are, thus, supported at an early stage and validated by objective measurements.

With the help of scaled variant testing, target values can already be derived from subsystems of different scales before serial parts can be used in investigations on test benches. The results of the scaled component tests confirmed that the derived scaling model enables the evaluation of geometry variants in scaled investigations for the considered effects.

6. Conclusions

In the context of scaled component tests, the possible applications of the scaled components-in-the-loop test bench were demonstrated as an example for the early validation of two geometry variants. As a result of the investigation, system knowledge is available regarding geometry changes for the components of the overload clutch. The results of the experiment confirmed the scaling model and also the entire approach. The approach shown in chapter 2 can serve as a basis for further scaled investigations also with other systems.

The paper shows that a scaling model in an X-in-the-Loop test bench can be used to test geometrically scaled drive components concerning functional properties and to evaluate geometry variants for product engineering. Scaling models allow for the influence of geometry variants to be evaluated in scaled tests at an early stage.

This study shows the possibilities of scaling for investigating geometry adaption for scaled components on XiL test stands.

Differences concerning friction ratios are not currently taken into account in the scaling model. In further studies, the influence of other effects such as friction could be investigated and taken into account in the scaling. Therefore, it is necessary to analyze further effects such as different stiffness, friction, and other differences of the components concerning design and function. The differences between scaled and unscaled XiL-based experiments could be addressed so that an even more accurate prediction of real behavior is possible.

Extending the scaling model to include friction models to adapt the performance variables could lead to improved validity in the future. The proposed approach offers potential for powertrain re-design. By adjusting the performance variables in the interfaces, different subsystems can also be studied together; for example, novel powertrain systems. Further studies on other subsystems are necessary to consolidate the approach and support frontloading in more applications with scaled experiments.

Author Contributions: Conceptualization, M.S. and T.G.; Data curation, M.S.; Formal analysis, M.S.; Funding acquisition, S.M.; Investigation, M.S.; Project administration, S.M.; Supervision, S.M. and T.G.; Validation, M.S.; Visualization, M.S.; Writing—original draft, M.S.; Writing—review & editing, M.S., S.M. and T.G. All authors have read and agreed to the published version of the manuscript.

Funding: The project upon which this publication is based was funded by Federal Ministry for Economic Affairs and Energy under the funding number 20y1509b. The authors of this publication are responsible for its contents. We acknowledge support by the KIT-Publication Fund of the Karlsruhe Institute of Technology.

Institutional Review Board Statement: Ethical review and approval were waived for this study due to the fact that it was not a clinical study and did not bear the characteristics of a medical experiment. Only measurement data was recorded non-invasively during an ordinary routine work procedure of one subject.

Informed Consent Statement: Informed consent was obtained from the subject involved in the study.

Data Availability Statement: The data that support the findings of this study are available from the corresponding author upon reasonable request.

Conflicts of Interest: The authors declare no conflict of interest.

References

1. Božek, P.; Turygin, Y. Measurement of the Operating Parameters and Numerical Analysis of the Mechanical Subsystem. *Meas. Sci. Rev.* **2014**, *14*, 198–203. [CrossRef]
2. Albers, A.; Behrendt, M.; Ott, S. Validation-Central Activity to Ensure Individual Mobility. In Proceedings of the FISITA 2010 World Automotive Congress, Budapest, Hungary, 30 May–4 June 2010.
3. Albers, A.; Pinner, T.; Yan, S.; Hettel, R.; Behrendt, M. Koppelsystems: Obligatory Elements within Validation Setups. In *Proceedings of the DESIGN 2016, the 14th International Design Conference*; Marjanović, D., Štorga, M., Pavković;, N., Bojčetić, N., Škec, S., Eds.; The Design Society: Dubrovnik, Kroatien, 2016; pp. 109–118. ISBN 2220-4334.
4. Albers, A.; Düser, T. Implementation of a Vehicle-in-the-Loop Development and Validation Platform. In Proceedings of the FISITA 2010 World Automotive Congress, Budapest, Hungary, 30 May–4 June 2010.

5. Thomke, S.; Fujimoto, T. The effect of "front-loading" problem-solving on product development performance. *J. Prod. Innov. Manag.* **2000**, *17*, 128–142. [CrossRef]
6. Thomke, S.; Bell, D.E. Sequential Testing in Product Development. *Manag. Sci.* **2001**, *47*, 308–323. [CrossRef]
7. Albers, A.; Düser, T. Implementation of a Vehicle-in-the-Loop Development and Validation Platform. In *Automobiles and Sustainable Mobility, Proceedings of the FISITA 2010 World Automotive Congress, Budapest, Hungary, 30 May–4 June 2010*; Scientific Association for Mechanical Engineering (GTE): Budapest, Hungary, 2010; ISBN 978-963-905829-3.
8. Albers, A.; Behrendt, M.; Klingler, S.; Matros, K. Verifikation und validierung im produktentstehungsprozess. In *Handbuch Produktentwicklung*; Lindemann, U., Ed.; Hanser-Verlag: München, Germany, 2016; ISBN 978-3-446-44518-5.
9. Fathy, H.K.; Filipi, Z.S.; Hagena, J.; Stein, J.L. Review of hardware-in-the-loop simulation and its prospects in the automotive area. In *Modeling and Simulation for Military Applications*; Schum, K., Sisti, A.F., Eds.; Defense and Security Symposium; SPIE: Orlando, FL, USA, 2006; p. 62280E.
10. Krammer, M.; Benedikt, M.; Blochwitz, T.; Alekeish, K.; Amringer, N.; Kater, C.; Materne, S.; Ruvalcaba, R.; Schuch, K.; Zehetner, J.; et al. *The Distributed Co-Simulation Protocol for the Integration of Real-Time Systems and Simulation Environments*; Society for Computer Simulation International: San Diego, CA, USA, 2018.
11. Szwedziak, K.; Łusiak, T.; Bąbel, R.; Winiarski, P.; Podsędek, S.; Doležal, P.; Niedbała, G. Wind Tunnel Experiments on an Aircraft Model Fabricated Using a 3D Printing Technique. *JMMP* **2022**, *6*, 12. [CrossRef]
12. Perez-Pinal, F.J.; Nunez, C.; Alvarez, R.; Cervantes, I. Comparison of multi-motor synchronization techniques. In Proceedings of the 30th Annual Conference of IEEE Industrial Electronics Society, 2004. IECON 2004, Busan, Korea, 2–6 November 2004; IEEE: Manhattan, NY, USA, 2004; pp. 1670–1675, ISBN 0-7803-8730-9.
13. Andert, J.; Klein, S.; Savelsberg, R.; Pischinger, S.; Hameyer, K. Virtual shaft: Synchronized motion control for real time testing of automotive powertrains. *Control. Eng. Pract.* **2016**, *56*, 101–110. [CrossRef]
14. Steck, M.; Gwosch, T.; Matthiesen, S. Scaling of Rotational Quantities for Simultaneous Testing of Powertrain Subsystems with Different Scaling on a X-in-the-Loop Test Bench. *Mechatronics* **2020**, *71*, 102425. [CrossRef]
15. Steck, M.; Gwosch, T.; Matthiesen, S. Compensation of mass-based effects in component scaling on a hardware-in-the-loop test bench by virtual inertia. *Mechatronics* **2021**, *78*, 102622. [CrossRef]
16. Budinger, M.; Passieux, J.-C.; Gogu, C.; Fraj, A. Scaling-law-based metamodels for the sizing of mechatronic systems. *Mechatronics* **2014**, *24*, 775–787. [CrossRef]
17. Cho, U.; Wood, K.L.; Crawford, R.H. Novel Empirical Similarity Method for the Reliable Product Test with Rapid Prototypes. Proceedings of DETC'98: ASME 1998 Design Engineering Technical Conferences, Atlanta, GA, USA, 13–16 September 1998. [CrossRef]
18. Castellani, F.; Astolfi, D.; Peppoloni, M.; Natili, F.; Buttà, D.; Hirschl, A. Experimental Vibration Analysis of a Small Scale Vertical Wind Energy System for Residential Use. *Machines* **2019**, *7*, 35. [CrossRef]
19. Tan, Q.-M. *Dimensional Analysis: With Case Studies in Mechanics*; Springer: Berlin/Heidelberg, Germany, 2011; ISBN 978-3-642-19234-0.
20. Petersheim, M.D.; Brennan, S.N. Scaling of hybrid-electric vehicle powertrain components for Hardware-in-the-loop simulation. *Mechatronics* **2009**, *19*, 1078–1090. [CrossRef]
21. Castellani, F.; Astolfi, D.; Becchetti, M.; Berno, F. Experimental and Numerical Analysis of the Dynamical Behavior of a Small Horizontal-Axis Wind Turbine under Unsteady Conditions: Part I. *Machines* **2018**, *6*, 52. [CrossRef]
22. Gwosch, T.; Steck, M.; Matthiesen, S. Virtual coupling of powertrain components: New applications in testing. In *ASIM-Workshop Simulation Technischer Systeme: Grundlagen und Methoden in Modellbildung und Simulation, Braunschweig*; Durak, U., Deatcu, C., Hettwer, J., Eds.; ARGESIM Verlag: Wien, Austria, 2019; pp. 163–168. ISBN 978-3-901608-06-3.
23. Nickel, D.; Behrendt, M.; Bause, K.; Albers, A. Connected testbeds—Early validation in a distributed development environment. In *18. Internationales Stuttgarter Symposium*; Bargende, M., Reuss, H.-C., Wiedemann, J., Eds.; Springer Fachmedien Wiesbaden: Wiesbaden, Germany, 2018; pp. 1173–1185. ISBN 978-3-658-21193-6.
24. Robert Bosch Power Tools GmbH. Product Datasheet GSR 10 8-2-LI. Available online: https://www.bosch-professional.com/de/de/archive/gsr-10-8-v-li-2-16834-p/ (accessed on 7 July 2021).
25. Robert Bosch Power Tools GmbH. Product Datasheet GSR 18-2-LI. Available online: https://www.bosch-professional.com/ch/de/pdf/productdata/gsr-18-2-li-sheet.pdf (accessed on 7 July 2021).
26. Gwosch, T. Antriebsstrangprüfstände zur ableitung von konstruktionszielgrößen in der produktentwicklung handgehaltener power-tools. In *Forschungsberichte*; Albers, A., Matthiesen, S., Eds.; Karlsruher Institut für Technologie (KIT): Karlsruhe, Germany, 2019.
27. Steck, M.; Paland, D.; Gwosch, T.; Matthiesen, S. Simulation model of a torque-limiting clutch with adjustable design parameters to investigate the release behavior. In *Tagungsband ASIM Workshop STS/GMMS/EDU 2021*; Workshop ASIM Fachgruppen STS, GMMS und EDU 2021; Liu-Henke, X., Durak, U., Eds.; ARGESIM Verlag: Wien, Austria, 2021; ISBN 9783960018698.
28. Gibbings, J.C. *Dimensional Analysis*; Springer: London, UK; New York, NY, USA, 2011; ISBN 978-1-84996-316-9.
29. Buckingham, E. On Physically Similar Systems; Illustrations of the Use of Dimensional Equations. *Phys. Rev.* **1914**, *4*, 345–376. [CrossRef]

30. Christophe, F.; Sell, R.; Coatanéa, E.; Tamre, M. System modeling combined with dimensional analysis for conceptual design. In Proceedings of the 8th International Workshop on Research and Education in Mechatronics 2007, Tallinn, Estonia, 14–15 June 2007; pp. 242–247.
31. Stratasys. Vero Material Data Sheet. Available online: https://www.stratasys.com/de/materials/search/vero (accessed on 6 September 2021).

MDPI

Article

Analysis of Load Inhomogeneity of Two-Tooth Difference Swing-Rod Movable Teeth Transmission System under External Excitation

Rui Wei [1], Yali Yi [1], Menglei Wu [1], Meiyu Chen [1] and Herong Jin [2,*]

[1] School of Mechanical Engineering, Yanshan University, Qinhuangdao 066004, China; weirui@stumail.ysu.edu.cn (R.W.); yiyali@ysu.edu.cn (Y.Y.); 496550306@stumail.ysu.edu.cn (M.W.); mychen@stumail.ysu.edu.cn (M.C.)

[2] Parallel Robot and Mechatronic System Laboratory of Hebei Province, Yanshan University, Qinhuangdao 066004, China

* Correspondence: ysujhr@ysu.edu.cn; Tel.: +86-139-3359-9493

Abstract: In order to improve the load state of the two-tooth difference swing-rod movable teeth transmission system, in this paper, a dynamic equivalent calculation model of the transmission system is established based on lumped parameter theory, and then a calculation method of system dynamic load is derived. The influence of external excitation on load inhomogeneity of the transmission system is analyzed from a dynamic point of view. The theoretical results are verified by Adams dynamic load simulation analysis and strain test based on a test bench. The results show that when errors of the transmission system are fixed, the system load inhomogeneity is improved effectively with the increase of load torque, while the system load inhomogeneity becomes worse as input speed increases. This study provides a theoretical reference for improving the load inhomogeneity of the two-tooth difference swing-rod movable teeth transmission system.

Keywords: swing-rod movable teeth transmission system; external excitation; load inhomogeneity; simulation analysis; strain test

Citation: Wei, R.; Yi, Y.; Wu, M.; Chen, M.; Jin, H. Analysis of Load Inhomogeneity of Two-Tooth Difference Swing-Rod Movable Teeth Transmission System under External Excitation. *Machines* **2022**, *10*, 502. https://doi.org/10.3390/machines10070502

Academic Editors: Sven Matthiesen and Thomas Gwosch

Received: 6 May 2022
Accepted: 20 June 2022
Published: 22 June 2022

1. Introduction

A two-tooth difference swing-rod movable teeth transmission system has strong bearing capacity because of the multi-tooth meshing characteristics [1]. The structure of wave generator is symmetrical so that the symmetrical meshing pairs have strong self-balance and stable dynamic performance. In addition, the symmetrical meshing pair forces are balanced in the multi-tooth meshing process [2]. Under real conditions, the load inhomogeneity of symmetrical meshing pairs is obvious due to various excitation factors. Furthermore, the excitation factors can cause system vibration, fatigue failure of tooth surface, and damage to movable teeth. The consequences seriously affect the performance of the transmission system [3]. Under the condition of certain system errors, the influence of external excitation on the load inhomogeneity of the two-tooth difference swing-rod movable teeth transmission system cannot be ignored.

At present, research papers on the swing-rod movable teeth transmission mainly focus on tooth shape analysis [4], tolerance design [5], dynamic characteristics [6–8], strain analysis [9], and so on. Most of the studies on load inhomogeneity of transmission system are about planetary gear transmission. Taking the concentric face gear split-torque transmission system (CFGSTTS) as the study object, Dong et al. [10] analyzed the characteristics of dynamic load sharing by establishing a lumped parameter model according to the Newton theorem. Dong et al. [11] studied the load-sharing characteristics of face-gear four-branching split-torque transmission system by establishing a static load-sharing mechanical analysis model. Using the surface gear flow system as the research object, Mo et al. [12]

studied the change curves of meshing force and load-sharing coefficient, and analyzed the effects of input power and input speed on the load-sharing coefficient of the system. Hu et al. [13] analyzed the influence of meshing impact on the load-sharing coefficients and dynamic load factors of the planetary transmission system. Taking the large-scale wind power planetary gear system as the research object, Xu et al. [14] studied the effect of external load changes on the system load sharing through finite element simulation analysis and experimental tests. Zhang et al. [15] introduced single factor analysis to investigate the influence of each nonlinear internal excitation on the load-sharing coefficient (LSC) and determined the most significant control factors affecting LSC. On this basis, Zhang et al. [16] proposed a dynamic tooth wear prediction model and studied the influence of tooth wear on load sharing. Bodas et al. [17] simulated the effects of carrier and gear errors associated with manufacturing and assembly on load sharing between planetary gears by using the finite element method. Sanchez-Espiga et al. [18] proposed a numerical method to calculate the load-sharing problem of the planetary transmission by measuring the strains of the root of the sun gear teeth. Taking two-stage planetary transmission as the study object, Sun et al. [19] proposed a numerical method to calculate dynamic sensitivity of the load-sharing coefficient to errors. The above studies systematically studied the influence of external excitation on the load inhomogeneity of the planetary transmission system, wherein the important influencing factors of external excitation were clarified and the simulation and experimental techniques were analyzed. The two-tooth difference swing-rod movable teeth transmission belongs to the planetary transmission with small tooth difference [20], but its internal structure is more complex. Therefore, it is more difficult to establish a dynamic model, conduct simulation modeling, and carry out test measurements.

In this paper, the two-tooth difference swing-rod movable teeth transmission system is taken as the research object. A dynamic equivalent calculation model is established. Combined with the relative displacement relationship of each component, dynamic differential equations of the system are derived to obtain the system dynamic load. Load proportional coefficients are used to evaluate the degree of the system load inhomogeneity. The influence of external excitation on system load inhomogeneity is investigated by theoretical calculation. Subsequently, the results are verified by simulation analysis and experimental tests, which provides a theoretical reference for improving the system load inhomogeneity.

2. Equivalent Calculation Model

The two-tooth difference swing-rod movable teeth transmission system is shown in Figure 1. As a typical planetary transmission system with small tooth difference, the two-tooth difference swing-rod movable teeth transmission system consists of a wave generator *H*, a ring gear *K*, a separator *G*, swing rods, and movable teeth. In this case, the ring gear *K* is fixed. The wave generator *H* is fixed to the input shaft. The separator *G* is fixed to the output shaft. The input power of the wave generator *H* is transmitted to the separator *G* through multiple movable teeth which mesh with the wave generator *H* and the ring gear *K* at the same time.

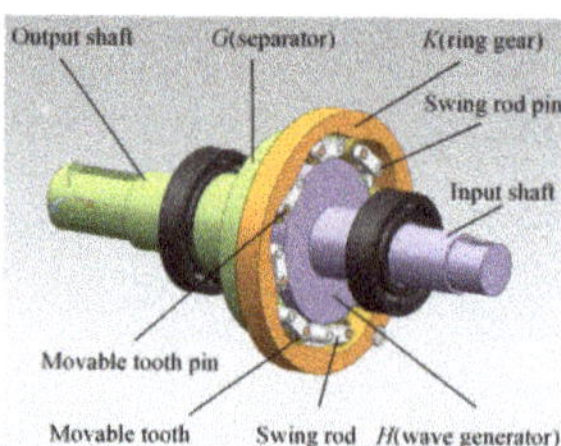

Figure 1. The two-tooth difference swing-rod movable teeth transmission system.

In order to obtain the micro-displacement of each component in the transmission process, a dynamic equivalent calculation model (shown in Figure 2) is established based on the lumped parameter theory. The following assumptions are made. Firstly, only small elastic deformation occurs in the meshing pairs and bearings in the system. Secondly, all the components are regarded as units composed of dampers and springs; the flexible deformation caused by components can be ignored. Thirdly, each swing rod, movable tooth, swing rod pin, and movable tooth pin group is regarded as a whole, and the deformation and errors between them are ignored. In addition, the mass body is treated as rigid, and deformation occurs only in the spring elements and dampers.

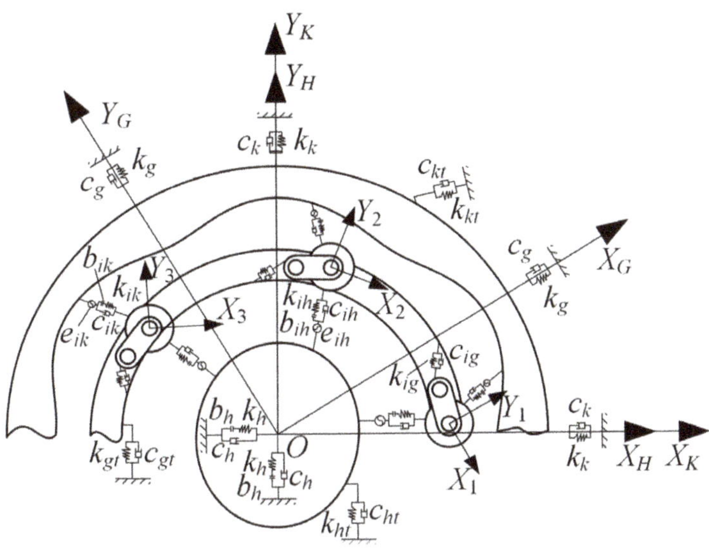

Figure 2. Dynamic equivalent calculation model of the two-tooth difference swing-rod movable teeth transmission system.

The coordinate systems X_HOY_H, X_KOY_K, and X_GOY_G are fixedly connected with the wave generator, the ring gear, and the separator, respectively. X_KOY_K is the fixed absolute coordinate system; X_HOY_H and X_GOY_G are the relative coordinate systems rotating with the wave generator and separator, respectively. X_i and Y_i are fixedly attached to the *i*th movable tooth ($i = 1, 2, 3, \ldots, n$, where n represents the number of the movable teeth).

In the dynamic equivalent calculation model of the two-tooth difference swing-rod movable teeth transmission system, there are $(9 + 3n)$ degrees of freedom.

3. Dynamic Load Calculation of the Transmission System

3.1. Relative Displacement and Differential Equation of Motion of the Transmission System

The dynamic load of the movable teeth transmission system is caused by the relative elastic displacements of the meshing points between the movable teeth and the wave generator and the meshing points between the movable teeth and the ring gear in their normal direction. They can be calculated according to the relationship between force and deformation.

To obtain the relative displacement relationship between the components in Figure 2, a relative displacement analysis model of the transmission system is established, as shown in Figure 3.

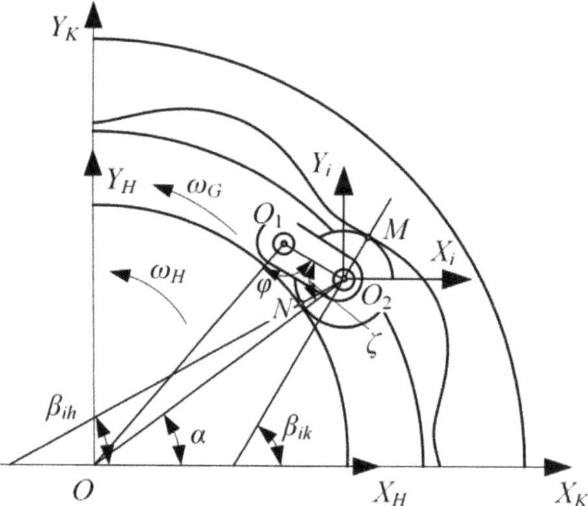

Figure 3. Relative displacement relation of components.

The direction of the meshing point pointing to the movable tooth center O_2 is defined as the positive direction of relative displacement, and the displacements of the wave generator and the ring gear relative to the movable tooth are projected on the positive direction. Combined with Figures 2 and 3, the relative displacements on each meshing line can be obtained as follows:

$$\Delta_{ih} = (x_h - x_i)\cos\beta_{ih} + (y_h - y_i)\sin\beta_{ih} - u_h\sin(\alpha - \beta_{ih}) - u_i\cos\left(\frac{\pi}{2} - \zeta + \alpha - \beta_{ih}\right) - e_{ih}, \quad (1)$$

$$\Delta_{ik} = (x_i - x_k)\cos\beta_{ik} + (y_i - y_k)\sin\beta_{ik} + u_k\sin(\alpha - \beta_{ik}) + u_i\cos\left(\frac{\pi}{2} - \zeta + \alpha - \beta_{ik}\right) - e_{ik}, \quad (2)$$

$$\Delta_{ig} = -(x_i - x_g)\cos(\zeta - \alpha) + (y_i - y_g)\sin(\zeta - \alpha) + u_g\cos\left(\frac{\pi}{2} - \varphi\right), \quad (3)$$

The subscripts h, g, and k represent the wave generator, the separator, and the ring gear, respectively. i is the serial number of the movable tooth. Based on a transmission error model of the two-tooth difference swing-rod movable teeth transmission system, the equivalent meshing errors can be obtained by using the action line increment method [21].

In the working process of the two-tooth difference swing-rod movable teeth transmission system, the whole system is in a state of force balance. According to Newton's second law, combined with Figures 2 and 3, the wave generator, the ring gear, the separator, and the ith movable tooth subsystem are analyzed.

The dynamic differential equation of the wave generator, the separator, and the ring gear can be uniformly expressed as follows:

$$\begin{cases} m_p\ddot{x}_p + c_p\dot{x}_p + k_p f(x_p, b_p) + \sum\limits_{i=1}^{n}\left(c_{ip}\dot{\Delta}_{ip} + k_{ip} f(\Delta_{ip}, b_{ip})\right)a_{px} = 0 \\ m_p\ddot{y}_p + c_p\dot{y}_p + k_p f(y_p, b_p) + \sum\limits_{i=1}^{n}\left(c_{ip}\dot{\Delta}_{ip} + k_{ip} f(\Delta_{ip}, b_{ip})\right)a_{py} = 0 \\ \left(I_p/r_p^2\right)\ddot{u}_p + c_{pt}\dot{u}_p + k_{pt}u_p + \sum\limits_{i=1}^{n}\left(c_{ip}\dot{\Delta}_{ip} + k_{ip} f(\Delta_{ip}, b_{ip})\right)a_{pu} = F_p \end{cases}, \quad (4)$$

In Equation (4), $f(\Delta,b)$ is a symbolic function, which represents the relationship between the tooth clearance and the meshing displacement of the two-tooth difference swing-rod movable teeth transmission system. The equation is as follows:

$$f(\Delta,b) = \begin{cases} \Delta - b & \Delta > b \\ 0 & |\Delta| \le b \\ \Delta + b & \Delta < -b' \\ \Delta & b = 0 \end{cases} \tag{5}$$

The external force acting on the part p is

$$F_p = \begin{cases} T_{in}/r_h & p = h \\ -T_{out}/r_g & p = g, \\ 0 & p = k \end{cases} \tag{6}$$

The projection vector of the error excitation on the ring gear coordinate system is

$$(a_{px}, a_{py}, a_{pu})^T = \begin{cases} (\cos\beta_{ih}, \sin\beta_{ih}, -\sin(\alpha-\beta_{ih}))^T & p = h \\ (\cos(\zeta-\alpha), -\sin(\zeta-\alpha), \cos(\frac{\pi}{2}-\varphi))^T & p = g, \\ (-\cos\beta_{ik}, -\sin\beta_{ik}, \sin(\alpha-\beta_{ik}))^T & p = k \end{cases} \tag{7}$$

The dynamic differential equation of the ith movable tooth is

$$\begin{cases} m_i\ddot{x}_i - \left(c_{ih}\dot{\Delta}_{ih} + k_{ih}f(\Delta_{ih}, b_{ih})\right)\cos\beta_{ih} + \left(c_{ik}\dot{\Delta}_{ik} + k_{ik}f(\Delta_{ik}, b_{ik})\right)\cos\beta_{ik} - \left(c_{ig}\dot{\Delta}_{ig} + k_{ig}\Delta_{ig}\right)\cos(\zeta-\alpha) = 0 \\ m_i\ddot{y}_i - \left(c_{ih}\dot{\Delta}_{ih} + k_{ih}f(\Delta_{ih}, b_{ih})\right)\sin\beta_{ih} + \left(c_{ik}\dot{\Delta}_{ik} + k_{ig}f(\Delta_{ik}, b_{ik})\right)\sin\beta_{ik} + \left(c_{ig}\dot{\Delta}_{ig} + k_{ig}\Delta_{ig}\right)\sin(\zeta-\alpha) = 0 \\ \left(I_i/r_i^2\right)\ddot{u}_i - \left(c_{ih}\dot{\Delta}_{ih} + k_{ih}f(\Delta_{ih}, b_{ih})\right)\cos\left(\frac{\pi}{2} - \xi + \alpha - \beta_{ih}\right) + \left(c_{ik}\dot{\Delta}_{ik} + k_{ik}f(\Delta_{ik}, b_{ik})\right)\sin\left(\frac{\pi}{2} - \zeta + \alpha - \beta_{ik}\right) = 0 \end{cases}, \tag{8}$$

According to the dynamic differential equation and the relative displacement of each component, considering the stiffness and damping of the system, the transverse and longitudinal vibration displacements of each component can be obtained via the Newmark method, which lays a foundation for the calculation of dynamic loads and load proportional coefficients between the movable teeth and the ring gear.

3.2. Calculation of Dynamic Load and Load Proportional Coefficient

In the two-tooth difference swing-rod movable teeth transmission system, it is assumed that the dynamic load between the ith movable tooth and the ring gear is F_{ki}. According to Equations (2) and (5), and combined with the meshing stiffness between the ith movable tooth and the ring gear, the dynamic load F_{ki} can be derived as

$$F_{ki} = k_{ik}f(\Delta_{ik}, b_{ik}), \tag{9}$$

where k_{ik} can be calculated from the system deformation coordination conditions and the Palmgren deformation equation [22].

In the two-tooth difference swing-rod movable teeth transmission system, the load proportional coefficient is defined to measure the load inhomogeneity of the system, which means the ratio of the actual dynamic load borne by the two center-symmetric movable teeth in the load distribution. Under ideal conditions, the load proportional coefficient of two symmetrical movable teeth is 1. The larger the load proportional coefficient, the more serious the system load inhomogeneity. The instantaneous load proportional coefficient when the ith and jth movable teeth mesh with the ring gear can be expressed as

$$\Omega_{kij} = \frac{2F_{ki}}{F_{ki} + F_{kj}}(i = 1, 2, 3, \ldots, n/2), \tag{10}$$

where $j = i + n/2$.

In the transmission system, a movable tooth undergoes a complete lift and return motion in one meshing cycle. In order to characterize the extreme value of the load proportional coefficient, the maximum instantaneous load proportional coefficient in each meshing cycle is taken to represent the load proportional coefficient in this cycle. The maximum load proportional coefficient when the ith and jth movable teeth mesh with the ring gear in a meshing period can be expressed as

$$B_{ki} = \left|\Omega_{kij} - 1\right|_{\max} + 1, (i = 1, 2, 3, \ldots, n) \tag{11}$$

4. Analysis of the Influence of External Excitation on System Load Inhomogeneity

In order to study the influence of external excitation on the system load inhomogeneity, a prototype of the two-tooth difference swing-rod movable teeth transmission system was designed. Here, 45 steel was selected as the material of the wave generator, the separator, the ring gear, and the movable teeth. According to the design requirements, the machining accuracy of each component was selected as grade 7, and the clearances of the meshing pairs are 5 μm. The basic parameters of the transmission system are shown in Table 1.

Table 1. Basic parameters of the transmission system.

Name of Component	Number of Teeth	Quality (kg)	Moment of Inertia (kg·m^2)
H (wave generator)	2	0.4258	8.525×10^{-5}
K (ring gear)	6	0.1832	3.868×10^{-4}
G (separator)		0.9871	4.420×10^{-4}
movable tooth	8	0.0103	1.193×10^{-5}

In reality, the influence of external excitation on the system load inhomogeneity cannot be ignored. Therefore, it is necessary to explore the influence of load torque and input speed on the load inhomogeneity of the system, so as to guide the adjustment of the working state of the transmission system and make the load distribution of the system more uniform.

4.1. Influence of Load Torque on System Load Inhomogeneity

In order to study the influence of system load torque on the system load inhomogeneity, the input speed of the wave generator was set as 800 r/min, whilst other parameters remained unchanged. Using the first movable tooth and the symmetrical fifth movable tooth in the system as an example, the load proportional coefficients of the movable tooth with the ring gear under different load torques can be calculated by Equation (11). Making the separator rotate a circle, the load proportional coefficient curves of the movable tooth in six meshing cycles under the load torques of 2 N·m, 4 N·m, 6 N·m, 8 N·m, and 10 N·m can be obtained, as shown in Figure 4.

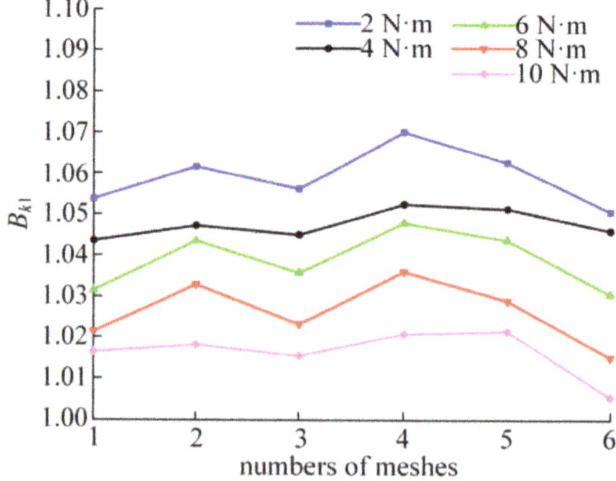

Figure 4. The variation curves of the load proportional coefficient under different load torques.

In Figure 4, with the increase of the load, the load proportional coefficient presents a downward trend. Under the same load, the load proportional coefficient fluctuates in a rotation cycle of the separator, which is caused by errors. However, due to the coupling effect of various errors and deformation, the variation of the load proportional coefficient does not show obvious periodicity. The average values of the load proportional coefficients B_{k1} under different load torques are 1.0592, 1.0477, 1.039, 1.0265, and 1.0166, respectively. With the increase of system load torque, the average value of the load proportional coefficient B_{k1} decreases. The average value can reflect the central tendency of the load proportional coefficient at a certain level. The larger the average value, the more serious the system load inhomogeneity. Therefore, it can be seen from Figure 4 that increasing the system load torque can make the system load distribution more uniform and improve the stability of the system.

4.2. Influence of Input Speed on System Load Inhomogeneity

On the premise of keeping other basic parameters unchanged, the load torque of the separator was set as 10 N·m, and the influence of speed on system load inhomogeneity was studied. Using the first movable tooth and the symmetrical fifth movable tooth in the system as an example, the load proportional coefficient curves under the input speeds of 400 r/min, 600 r/min, 800 r/min, 1000 r/min, and 1200 r/min can be obtained, as shown in Figure 5.

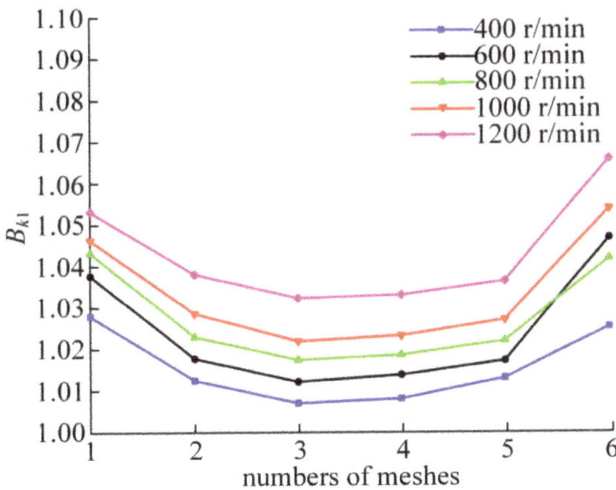

Figure 5. The variation curves of the load proportional coefficient under different input speeds.

In Figure 5, with the increase of the input speed, the load proportional coefficient presents an upward trend. In addition, it can be seen from the figure that the load proportional coefficient curve at 600 r/min intersects with that at 800 r/min. This may be due to the existence of error factors in the system, so that the system is not completely centrosymmetric. When the rotation speed reaches 600 r/min, the vibration displacement reaches the maximum in the sixth engagement, resulting in a large load proportional coefficient. The average values of the load proportional coefficients B_{k1} are 1.0159, 1.0244, 1.0279, 1.0337, and 1.0433, respectively. With the increase in input speed, the average value of B_{k1} also increases, and the system load inhomogeneity becomes more serious. This shows that the load uniformity of the two-tooth difference swing-rod movable teeth transmission system is better when working under low speed conditions.

5. Dynamic Load Simulation Analysis

In order to verify the influence of the load torque and input speed on system load inhomogeneity, a virtual prototype model of the transmission system was established in ADAMS to analyze the load proportional coefficient. Considering the actual working state of the prototype, the error accuracy level of the main components of the prototype model was selected as grade 7, and the error accuracy level of the other components was selected as grade 9.

The dynamic simulation model of the two-tooth difference swing-rod movable teeth transmission system is presented in Figure 6. The constraint conditions of the major components are shown in Table 2. The impact function is used in the contact parameters setting of the movable teeth with the wave generator and the ring gear. According to the Hertz collision theory, the contact stiffness was set as 5×10^7 N/m, the damping coefficient was set as 0.1% of the contact stiffness, the force exponents are 1.5, and the penetration depths of the meshing pairs are 0.1 mm.

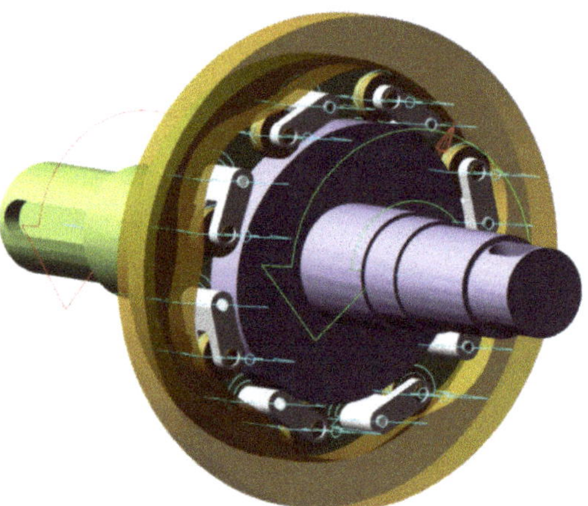

Figure 6. Virtual prototype simulation model.

Table 2. Constraints of each component.

Parts	Reference	Constraint Type
ring gear	the earth	fixed pair
wave generator	the earth	revolute pair
separator	the earth	revolute pair
swing rod	separator	revolute pair
movable tooth	swing rod	fixed pair
movable tooth	wave generator	contact pair
movable tooth	ring gear	contact pair

On this basis, by referring to the working conditions in Section 4, control groups were established respectively to compare and analyze the effect of load and rotational speed on system load inhomogeneity.

5.1. Simulation Analysis of Load Torque

The input speed of the wave generator was set as 800 r/min. Using the first movable tooth and the symmetrical fifth movable tooth as an example, the change curves of the load proportional coefficient B_{k1} under different load torques have been analyzed, as shown in Figure 7.

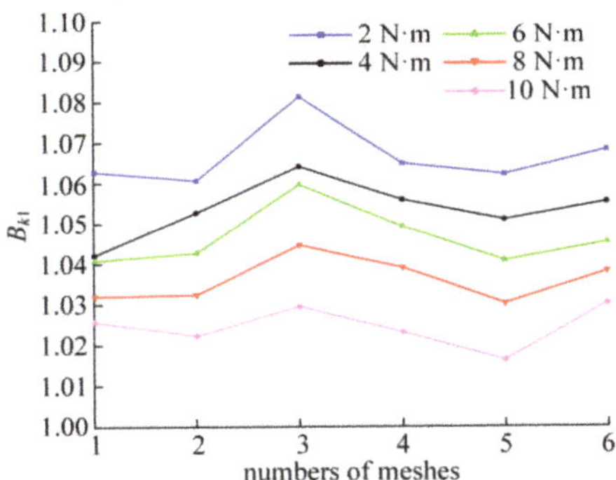

Figure 7. Load proportional coefficient curves under different load torques.

In Figure 7, the average values of the load proportional coefficients under each load condition are 1.0669, 1.0538, 1.0468, 1.0364, and 1.0249, respectively. It can be seen that as the load torque increases, the average value of the system load proportional coefficient B_{k1} decreases.

Figure 8 shows a comparison between the simulation results and the theoretical analysis results under different loads. Although there are differences in the variation rules between the theoretical and simulation load proportional coefficient curves, the difference is within 2%. The main reason is that the interaction between components and the impact load of meshing pairs cannot be analyzed through the theoretical model.

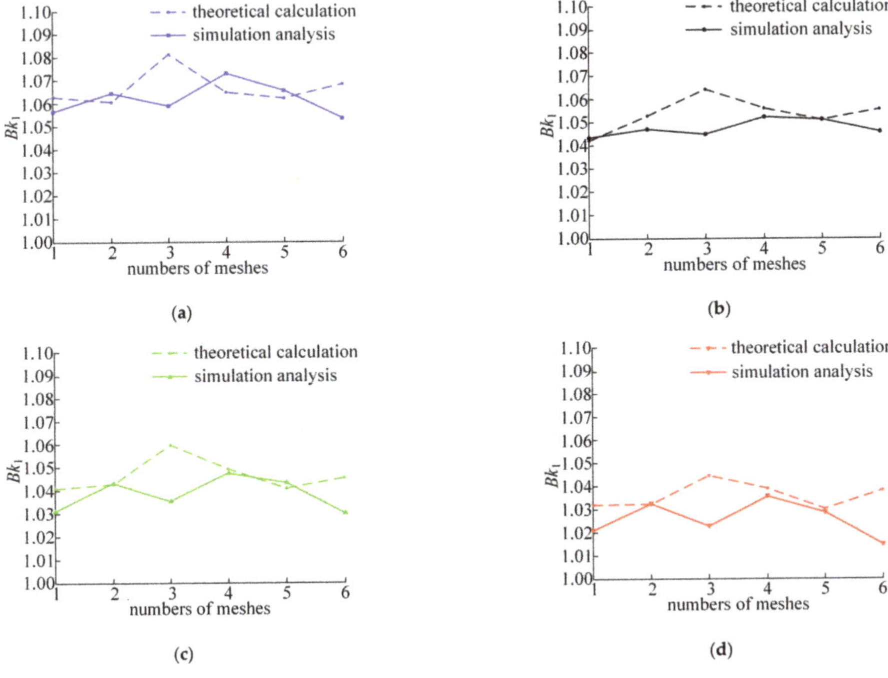

Figure 8. *Cont.*

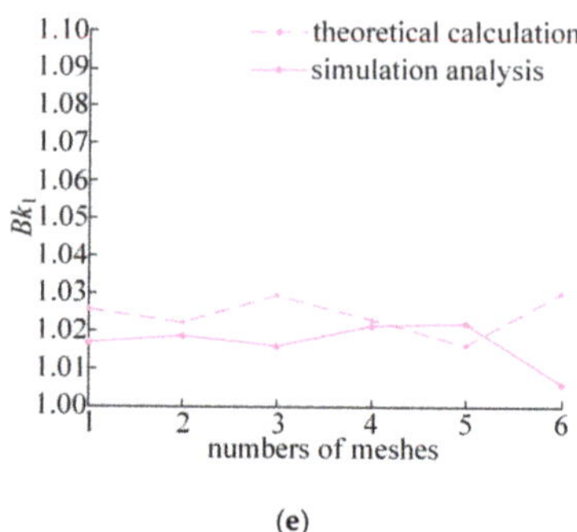

Figure 8. Comparison of load proportional coefficient curves under different loads: (**a**) 2 N·m; (**b**) 4 N·m; (**c**) 6 N·m; (**d**) 8 N·m; (**e**) 10 N·m.

Table 3 lists the average value of B_{k1} by theoretical calculation and simulation method under different load, respectively. The average values of the load proportional coefficient obtained by simulation are only 0.6–1.0% larger than those obtained by theoretical calculation. By comparing the changes of the average value of load proportional coefficients under different load conditions, it is obvious that increasing load torque can improve system load inhomogeneity.

Table 3. Average value comparison of load proportional coefficient under different loads.

Load	Theoretical Calculation Value	Simulation Value	Error (%)
2 N·m	1.0592	1.0669	0.7
4 N·m	1.0477	1.0538	0.6
6 N·m	1.039	1.0468	0.8
8 N·m	1.0265	1.0364	1.0
10 N·m	1.0166	1.0249	0.8

5.2. Simulation Analysis of Input Speed

The load torque of the separator was set at 10Nm. Under the condition of keeping other basic parameters unchanged, the load proportional coefficients under each input speed condition have been analyzed. Using the first movable tooth and the symmetrical fifth movable tooth as an example, the load proportional coefficient curves under different speed conditions can be obtained, as shown in Figure 9.

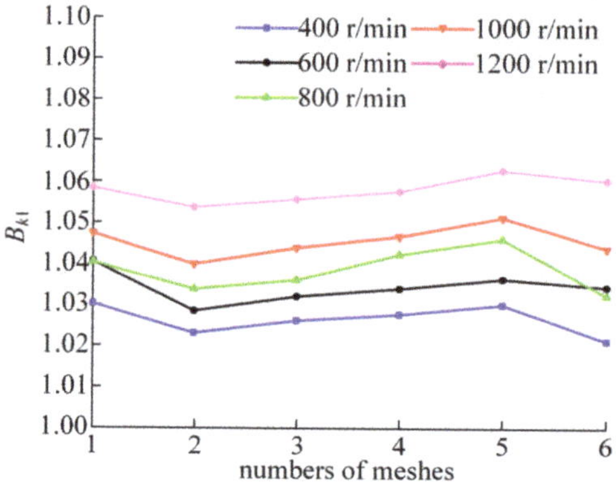

Figure 9. Load proportional coefficient curves under different input speeds.

In Figure 9, the average values of the load proportional coefficients under each speed condition are 1.027, 1.0348, 1.0389, 1.046, and 1.0585, respectively. It can be seen that the average value of the system load proportional coefficient B_{k1} increases with the increase in input speed. As shown in Figure 10, the theoretical calculation results of load proportional coefficients corresponding to different input speeds are smaller than the simulation results, but the difference is within 3%.

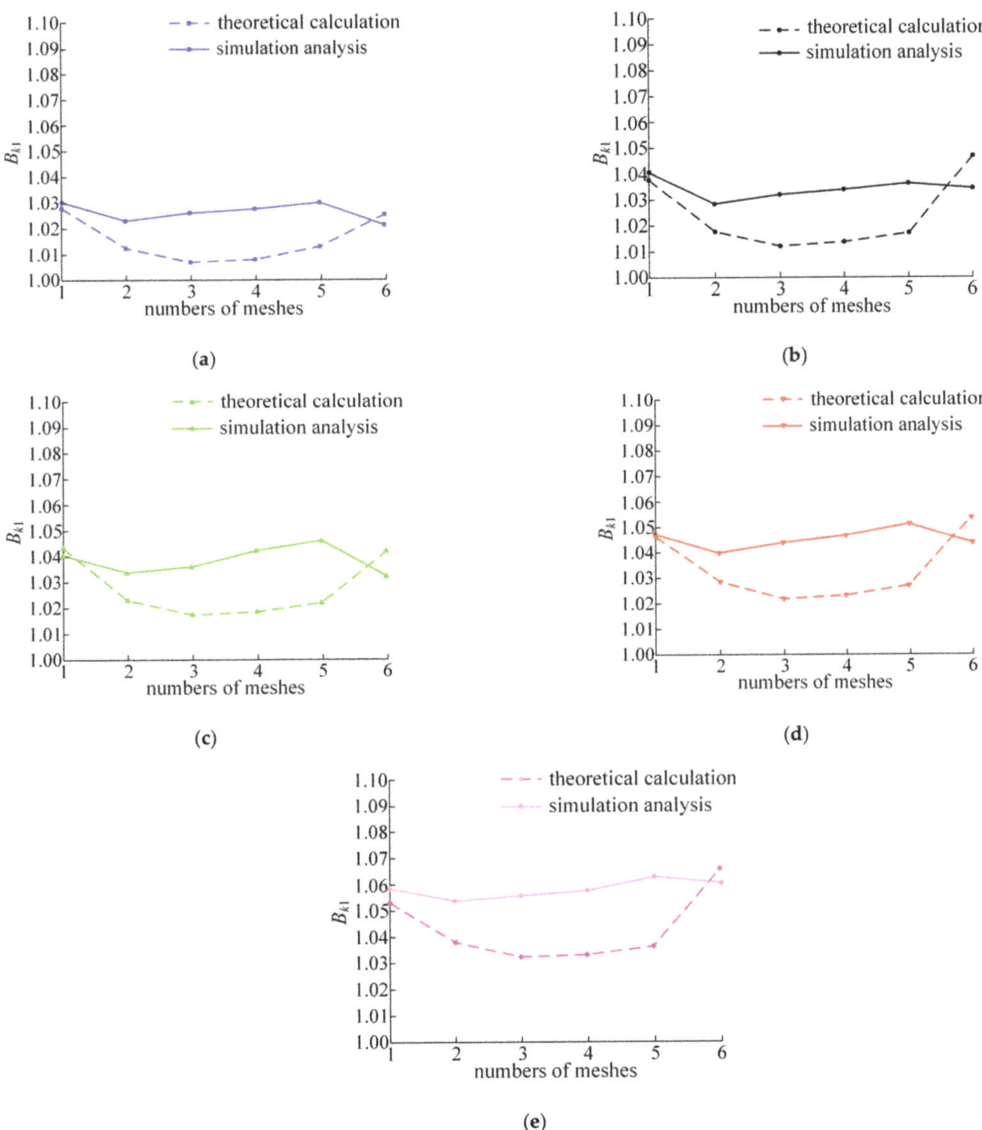

Figure 10. Comparison of load proportional coefficient curves at different input speeds: (**a**) 400 r/min; (**b**) 600 r/min; (**c**) 800 r/min; (**d**) 1000 r/min; (**e**) 1200 r/min.

Table 4 lists the average value of B_{k1} by theoretical calculation and simulation method under different input speeds, respectively. Compared with the average value of the load proportional coefficient obtained from the theoretical calculation, the simulation results increase by 1.0–1.5%. It can be seen that there is little difference between theoretical calculation and simulation results. Both theoretical calculation and simulation analysis prove that the system load distribution is more uniform under low speed conditions.

Table 4. Average value comparison of load proportional coefficient under different input speeds.

Input Speed	Theoretical Calculation Value	Simulation Value	Error (%)
400 r/min	1.0159	1.027	1.1
600 r/min	1.0244	1.0348	1.0
800 r/min	1.0279	1.0389	1.1
1000 r/min	1.0337	1.046	1.2
1200 r/min	1.0433	1.0585	1.5

6. Experimental Test and Analysis

In order to verify the influence of the load torque and the input speed on the load inhomogeneity of the two-tooth difference swing-rod movable teeth transmission system, a dynamic load test bench was designed and built to study the distribution uniformity of the system dynamic load under different working conditions. Because the meshing position between the movable teeth and the ring gear changes constantly, it is impossible to measure the dynamic load of the meshing pairs directly. Therefore, in this experiment, the strain generated by dynamic load on the ring gear was measured to complete the study. Then, according to the elastic modulus of the material, the measurement results were transformed into dynamic load between the meshing pairs to study the system load inhomogeneity.

Due to the low power of the prototype to be tested and the need for load conversion during operation, an open power flow mode was used for the test bench. The power required by the system was provided by the servo motor and then transferred to the loading device by the transmission parts, and the remaining energy was consumed by the loading device. The whole power transfer direction forms a closed circuit. Its basic schematic diagram is shown in Figure 11.

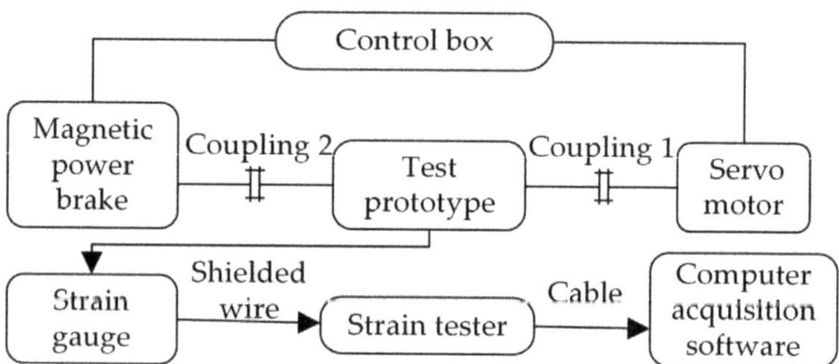

Figure 11. Basic schematic diagram of test bench.

The test bench mainly consists of two parts. One is the measurement system, the other one is the dynamic strain testing and analysis system. The measurement system is mainly composed of a drive motor, a control box, a test prototype, a magnetic powder brake, a DHDAS dynamic strain tester, and strain sensors. There are two symmetrical measuring points on the prototype; the corresponding pasted positions of strain gauges are shown in Figure 12a. The strain gauges required for the test were 120 Ω resistance unidirectional strain gauges. We used a half-bridge circuit composed of test strain gauges, temperature compensation strain gauges, and two 120 Ω resistors built in the dynamic strain tester, which can improve the accuracy of the test system. The temperature compensation strain gauges are pasted as shown in Figure 12b.

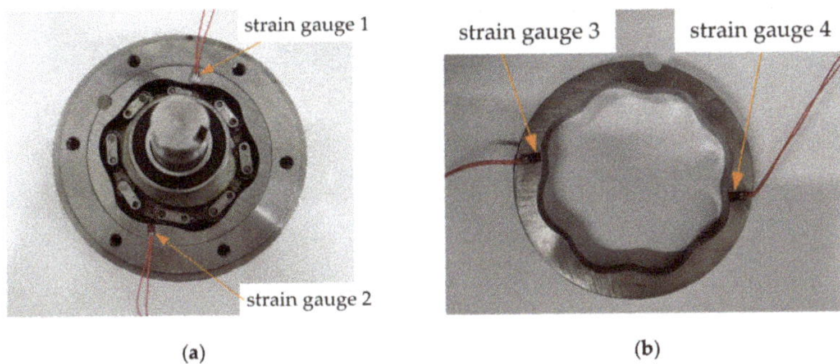

Figure 12. Strain gauges for test: (**a**) Ring gear pasted strain gauges; (**b**) Temperature compensated strain gauges.

The dynamic strain testing and analysis system was mainly accomplished via the test software, whose function is to receive, store, and analyze the data signals transmitted by the dynamic strain tester. The overall installation of the strain test system is shown in Figure 13.

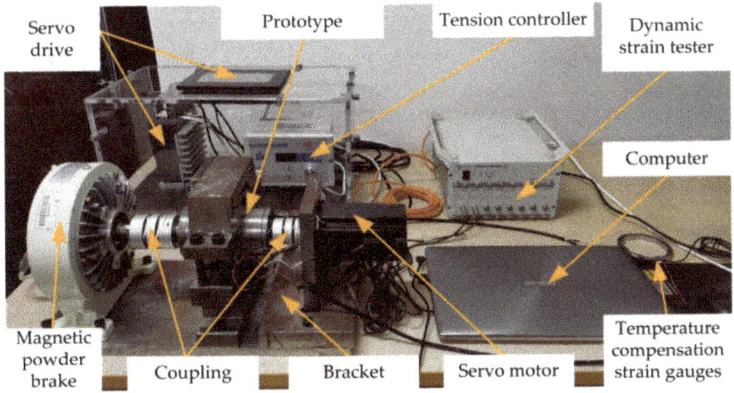

Figure 13. Figure of the overall installation of strain test system.

6.1. *Influence of Load on Dynamic Load of Meshing Pairs*

By controlling the tension controller to change the load torque in the test system, the variation curves of strain values ε are shown in Figure 14. The values ε were measured at two strain measuring points in the direction of center symmetry under different load torques.

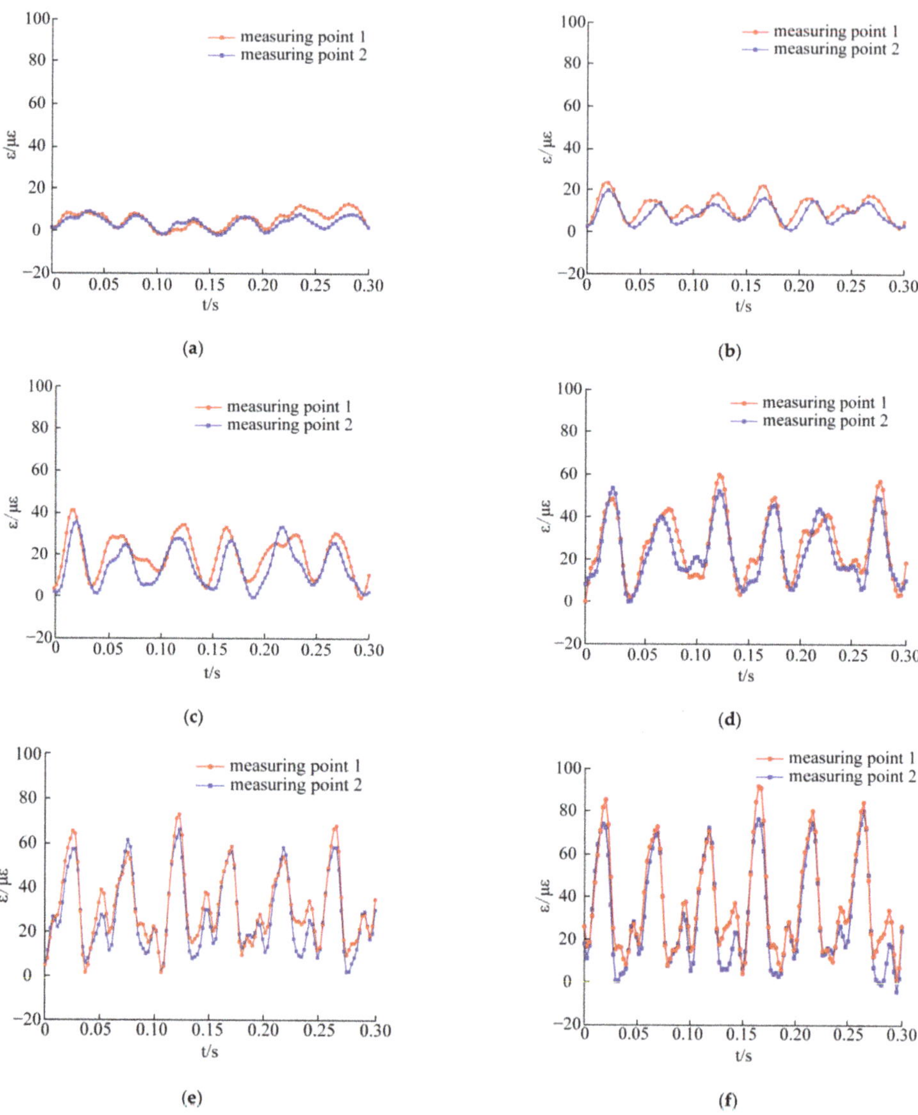

Figure 14. Strain curves under different loads: (**a**) Strain curves without load; (**b**) 2 N·m; (**c**) 4 N·m; (**d**) 6 N·m; (**e**) 8 N·m; (**f**) 10 N·m.

There are six peaks and troughs at the measuring points when the separator rotates for one cycle. As the load torque increases, the periodicity becomes more obvious. In Figure 14a, the strain signal detected by the dynamic strain test system under no-load conditions is mainly due to the vibration of the system, the sinusoidal AC signal, and the electromagnetic interference of the servo motor, which causes a zero shift of the strain values ε in the test. Therefore, alternating current should be avoided as much as possible in the strain detection test, and electromagnetic interference to test results should be shielded.

In Figure 14b, when a 2 N·m load is applied to the magnetic powder brake, the amplitude fluctuations of the strain curves are larger than that without load. The maximum strain values detected by the two measuring points are 23.24 με and 19.86 με, respectively.

Since the strain value generated after loading is smaller than the strain value of the basic error, the strain value is relatively sensitive to the change of external factors under the load of 2 N·m and is easily influenced by the change of external factors. In order to ensure the accuracy of the test data, the load conditions above 4 N·m were selected for comparative analysis.

As shown in Figure 14c, the maximum strain values of the two measuring points are 40.74 με and 35.34 με. The maximum strain values of the two measuring points in Figure 14d are 60.15 με and 53.82 με. The maximum strain values of the two measuring points in Figure 14e are 74.40 με and 66.70 με. In Figure 14f, the maximum strain values of the two measuring points are 91.08 με and 84.34 με. It can be seen from the comparison that with the increase in load torque, the strain amplitude of the meshing pairs also increases, the interference of external errors on the detection results decreases, and the periodicity of the strain curves becomes more obvious. The strain curves of the two measuring points in the above figures are obviously different, which indicates that the load is not uniform in the system.

As shown in Figure 15, test data with loads above 4 N·m were selected and transformed into dynamic load between the meshing pairs for comparative analysis. Using Equations (10) and (11), the load proportional coefficient curves under different load conditions at the same speed can be obtained.

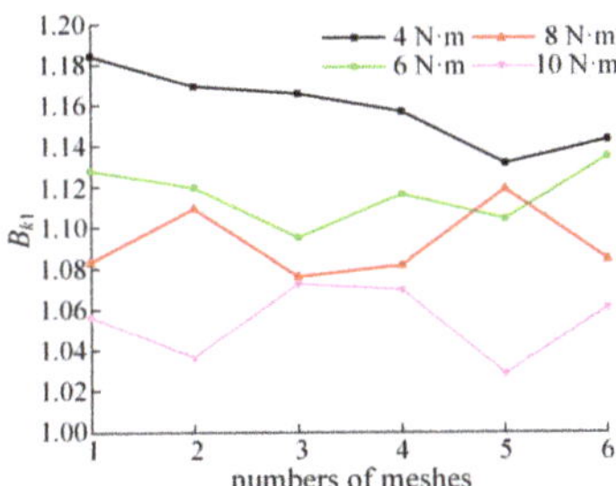

Figure 15. Load proportional coefficient curves under different loads.

Figure 15 shows the change curves of the load proportional coefficient under different load conditions. The average values corresponding to each curve are 1.1582, 1.1167, 1.0928, and 1.0550, respectively. Compared with the average values of the load proportional coefficient analyzed in Figure 8, the result is shown in Figure 16.

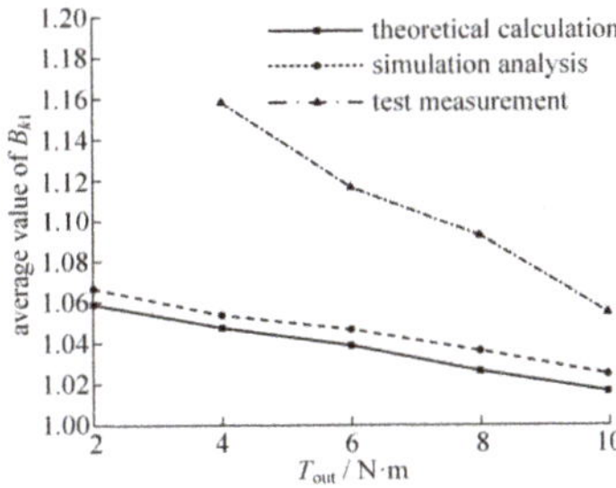

Figure 16. Variation curves of the average value of B_{k1} under different loads.

Figure 16 shows the variation curves of the average value of the load proportional coefficient under different loads obtained from theoretical calculation, simulation analysis, and test measurement. The test measurement results are 3.78–10.5% larger than the theoretical analysis results and 2.93–9.91% larger than the simulation analysis results. The main reason is that only the fixed errors of some components are considered in the theoretical and simulation analysis, while all the errors of the prototype are involved in the experiment. Moreover, the vibration caused by impact load, the fluctuation of input speed, and load torque have not been considered in the theoretical calculation or the simulation analysis. In addition, the effect of external magnetic field and current on the strain system during the test leads to a zero shift of the strain value, resulting in larger overall measurement results. Therefore, there are some numerical differences between the test results and those of the theoretical calculation and the simulation analysis.

In addition, it can be seen from Figure 16 that with the increase in load, the decrease rate of the load proportional coefficient of theoretical calculation and simulation analysis is basically the same and relatively gentle. However, the rate of decrease rate of the load proportional coefficient obtained from the test is obviously greater. This is because in the actual prototype, the transmission components will undergo flexible deformation under the action of load, and the load will cause the redistribution of assembly errors, so as to reduce the impact of initial errors on the load distribution. Therefore, as the load increases, the load proportional coefficient decreases faster in the test than in theoretical calculation or simulation analysis.

6.2. Influence of Rotational Speed on Dynamic Load of Meshing Pairs

The load torque of the magnetic powder brake was controlled to be 10 N·m, and the motor speed was set to 400 r/min, 600 r/min, 800 r/min, 1000 r/min, and 1200 r/min. The rotation periods of the separator corresponding to these rotation speeds are 0.6 s, 0.4 s, 0.3 s, 0.24 s, and 0.2 s, respectively. As shown in Figure 17, under the load of 10 N·m, the strain curves of the two measuring points in one rotation period of the separator at different speeds were recorded.

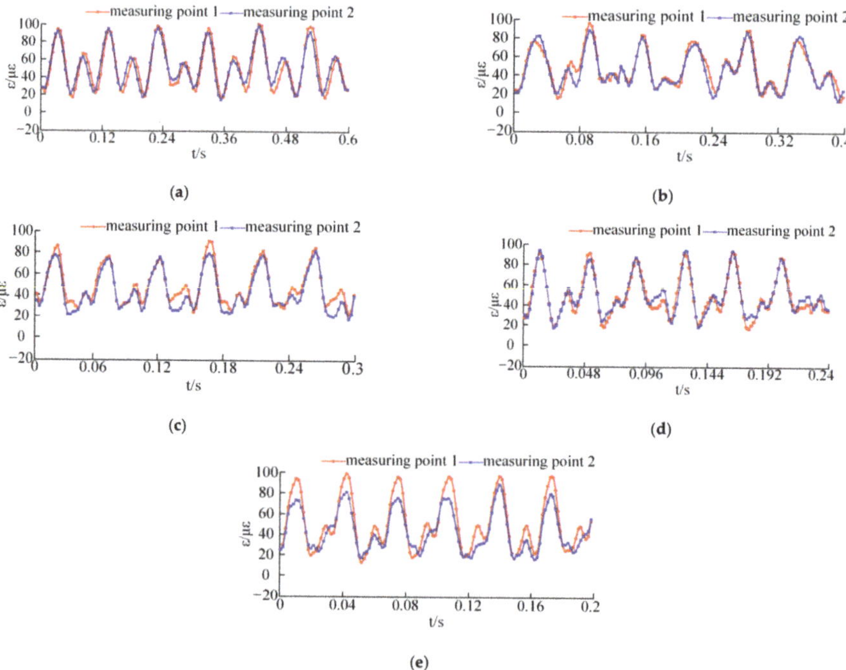

Figure 17. Strain curves under different input speeds: (**a**) 400 r/min; (**b**) 600 r/min; (**c**) 800 r/min; (**d**) 1000 r/min; (**e**) 1200 r/min.

It can be seen from the amplitudes that the change of the input speed has little effect on the maximum value of the strain amplitude. However, with the increase in input speed, the difference of strain values measured at the two measuring points becomes obvious.

According to the strain data at different input speeds, the dynamic loads between the corresponding meshing pairs are calculated, and then the load proportion coefficient curves can be obtained, as shown in Figure 18.

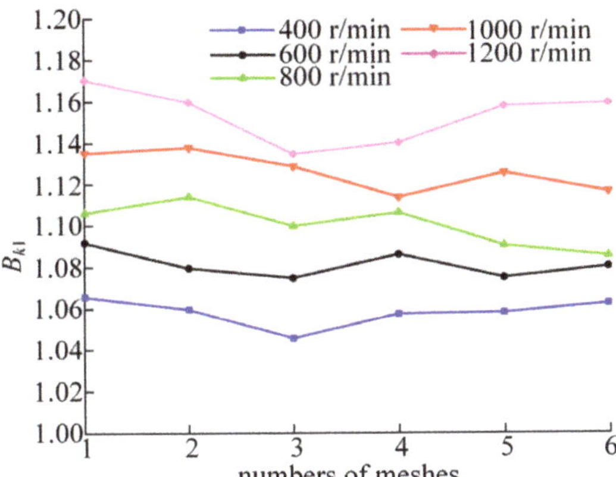

Figure 18. Load proportion coefficient curves under different speeds.

According to the load proportion coefficient curves in Figure 18, the average values of B_{k1} under different speeds are 1.0602, 1.0831, 1.1019, 1.1276, and 1.1549, respectively. Compared with the average values of the load proportional coefficient analyzed in Figure 10, the variation curves of the load proportional coefficient obtained through theoretical calculation, simulation analysis, and test measurement are shown in Figure 19.

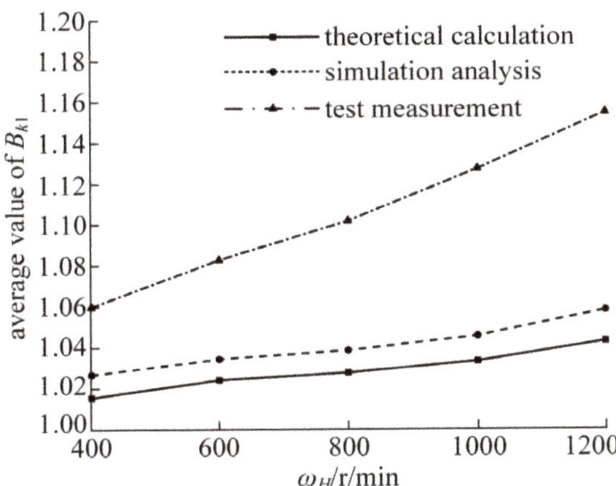

Figure 19. Variation curves of the average value of B_{K1} under different speeds.

By comparing the values, it can be found that the experimental test results are 4.4–10.7% larger than that of the theoretical analysis and 3.2–9.1% larger than that of the simulation analysis. In addition, it can be seen from Figure 19 that as the input speed increases, the difference between the results obtained by theoretical calculation and those obtained by simulation analysis does not change much, whilst the gap with the experimental results gradually widens. The main reason is that due to the existence of clearance, the system vibration caused by impact load will become more obvious with the increase in input speed. Therefore, although the average value of the load proportional coefficient in theoretical calculation, simulation analysis, and test increases with the increase in input speed, the increase rate of test results is obviously faster.

7. Conclusions

In this study, a lumped parameter dynamic equivalent model of the two-tooth difference swing-rod movable teeth transmission system is established to study the influence of external excitation on load inhomogeneity, and the theoretical analysis method is verified by simulation and experiments. The main conclusions are as follows:

(1) Under different working conditions, the difference rate between the average value of the load proportional coefficients measured by experiment and that obtained by theoretical calculation and simulation analysis is within 10.7%. In addition, the general variation trend of the average value of the load proportional coefficients obtained from theoretical calculation, simulation analysis, and test measurement are consistent, which verifies the correctness of the theoretical model, the virtual prototype model, and the feasibility of the test method.

(2) The deformation of components can ameliorate the uneven load distribution caused by errors to a certain extent. The increase of the load torque increases the strain amplitude of the meshing pair, thereby reducing the interference of the errors on the uniform distribution of the load, which makes the load distribution more uniform with the increase in load.

(3) The change in input speed has little effect on the maximum value of the strain amplitude. However, the increase in input speed leads to more obvious load difference between the two movable teeth with center symmetry, which makes the system load non-uniformity more serious. This shows that the load uniformity of the two-tooth difference swing-rod movable teeth transmission system is better when working under low speed conditions.

Author Contributions: Conceptualization, R.W.; data curation, M.W. and M.C.; formal analysis, R.W.; investigation, R.W.; methodology, R.W. and M.W.; resources, H.J.; supervision, Y.Y.; validation, R.W., M.W. and M.C.; writing—original draft, R.W.; writing—review and editing, Y.Y. and H.J. All authors have read and agreed to the published version of the manuscript.

Funding: This work was supported by the National Natural Science Foundation of China (Grant No. 51605416).

Institutional Review Board Statement: Not applicable.

Informed Consent Statement: Not applicable.

Data Availability Statement: Not applicable.

Conflicts of Interest: The authors declare no conflict of interest.

Nomenclature

Notation	Unit	Description
k_p	N/m	the support stiffness of the part p; $p = h, g, k$
c_p	N·s/m	the support damping of the part p; $p = h, g, k$
k_{pt}	N/m	the torsional stiffness of the part p; $p = h, g, k$
c_{pt}	N·s/m	the torsional damping of the part p; $p = h, g, k$
k_{ip}	N/m	the meshing stiffness between the ith movable tooth and the part p; $p = h, g, k$
c_{ip}	N·s/m	the meshing damping between the ith movable tooth and the part p; $p = h, g, k$
x_p	m	the transverse displacement of the part p; $p = h, g, k$
y_p	m	the longitudinal displacement of the part p; $p = h, g, k$
u_p	m	The torsional displacement of the part p; $p = h, g, k$
x_i	m	the transverse displacement of the ith movable tooth

y_i	m	the longitudinal displacement of the *i*th movable tooth
u_i	m	the torsional displacement of the *i*th movable tooth
b_h	m	The support clearance of the wave generator
b_{ih}	m	half of the meshing gap between the *i*th movable tooth and the wave generator
b_{ik}	m	half of the meshing gap between the *i*th movable tooth and the ring gear
M		the meshing point of the *i*th movable tooth and the ring gear
N		the meshing point of the *i*th movable tooth and the wave generator
β_{ih}	rad	the angle between the normal line at the meshing point N relative to the fixed horizontal axis X_K
β_{ik}	rad	the angle between the normal line at the meshing point M relative to the fixed horizontal axis X_K
α	rad	the angle between the horizontal axis X_K and the line connecting the coordinate origin O and the movable tooth center O_2
φ	rad	the angle between the length direction of the swing rod and the line connecting the coordinate origin O and the swing rod rotation center O_1
ξ	rad	the angle between the length direction of the swing rod and the line connecting the movable tooth center O_2 and the coordinate origin O
Δ_{ih}	m	the relative displacement of the wave generator to the *i*th movable tooth in the normal direction of their meshing point N
Δ_{ik}	m	the relative displacement of the ring gear to the *i*th movable tooth in the normal direction of their meshing point M
Δ_{ig}	m	the relative displacement of the separator to the *i*th movable tooth in the length direction of the swing rod
e_{ih}	m	the equivalent error of the *i*th movable tooth meshing with the wave generator
e_{ik}	m	the equivalent error of the *i*th movable tooth meshing with the ring gear
m_p	kg	the mass of the part p; $p = h, g, k$
I_p	kg·m^2	the moment of inertia of the part p; $p = h, g, k$
r_p	m	the equivalent radius of the part p; $p = h, g, k$
T_{in}	N·m	the input torque of the wave generator
T_{out}	N·m	the output torque of the separator
m_i	kg	the mass of the *i*th movable tooth
I_i	kg·m^2	the moment of inertia of the *i*th movable tooth
r_i	m	the equivalent radius of the *i*th movable tooth
F_{ki}	N	the dynamic load between the *i*th movable tooth and the ring gear
Ω_{kij}		the load proportional coefficient when the *i*th movable tooth meshes with the ring gear
B_{ki}		the load proportional coefficient of the *i*th movable tooth with the ring gear in the meshing period
ω_H	r/min	the input speed of the wave generator
ω_G	r/min	the output speed of the separator

References

1. Wang, Y.F.; Zhang, Q.P. Study on Virtual Prototype Modeling of Swing Movable Teeth Transmission. *Appl. Mech. Mater.* **2014**, *607*, 325–328. [CrossRef]
2. Qu, J.F. *Mechanism Innovation Principle*; Science Press: Beijing, China, 2001; pp. 1–20.
3. Qin, B.; Yin, H.; Wang, Z.; Zhang, J.Q.; Li, Z.J.; Wang, J.G. Application of EMPE and KP-KELM in Fault Diagnosis of Planetary Gearbox. *J. Mech. Transm.* **2019**, *43*, 146–151.
4. Dong, X.R.; Li, J.F.; Wang, X.H.; Liu, Z.F. Structural and Tooth Profile Analysis on Cam Profile Compound Teeth Transmission. *China Mech. Eng.* **2006**, *16*, 1661–1665.
5. Wei, R.; Jin, H.R.; Yi, Y.L. Research on the transmission error of swing-rod movable teeth transmission system. *Mech. Ind.* **2020**, *21*, 409. [CrossRef]
6. Liang, S.M.; Zhang, J.F.; Xu, L.J. Study on elasto-dynamic model of swing movable teeth transmission system. *J. Mech. Eng.* **2002**, *38*, 142–146. [CrossRef]
7. Yang, R.G.; An, Z.J.; Duan, L.Y. Analysis of Free Vibration of Cycloid Ball Planetary Transmission. *China Mech. Eng.* **2016**, *27*, 1883–1891.

8. Yi, Y.L.; Gao, Y.F.; He, L.; Jin, H.R. Thermoelastic stress-strain analysis of the meshing pairs of a two-tooth difference swing-rod movable toothed drive. *Strength Mater.* **2019**, *51*, 633–645. [CrossRef]
9. Xu, L.Z.; Wang, W.P. Quasi-static analysis of forces and stress for a novel two-step movable tooth drive. *Mech. Des. Struct. Mach.* **2018**, *46*, 285–295. [CrossRef]
10. Dong, J.X.; Wang, Q.B.; Tang, J.Y.; Hu, Z.H.; Li, X.Q. Dynamic characteristics and load-sharing performance of concentric face gear split-torque transmission systems with time-varying mesh stiffness, flexible supports and deformable shafts. *Meccanica* **2021**, *56*, 2893–2918. [CrossRef]
11. Dong, H.; Zhang, H.Q.; Zhao, X.L.; Duan, L.L. Study on the load-sharing characteristics of face-gear four-branching split-torque transmission system. *Adv. Mech. Eng.* **2021**, *13*, 1–15.
12. Mo, S.; Yue, Z.X.; Feng, Z.Y.; Gao, H.J. Analytical investigation on load sharing characteristics for face gear split flow system. *J. Huazhong Univ. Sci. Technol.* **2020**, *48*, 23–28. [CrossRef]
13. Hu, S.Y.; Fang, Z.D.; Xu, Y.Q.; Guan, Y.B.; Shen, R. Meshing impact analysis of planetary transmission system considering the influence of multiple errors and its effect on the load sharing and dynamic load factor characteristics of the system. *J. Multi-Body Dyn.* **2021**, *235*, 57–74. [CrossRef]
14. Xu, X.Y.; Yang, W.; Diao, P.; Tao, Y.C. Research of the Dynamic Load Sharing of Heavy Load Planetary Gear System with Multi-floating Component. *J. Mech. Transm.* **2016**, *40*, 6–11.
15. Zhang, H.B.; Wu, S.J.; Peng, Z.M. A nonlinear dynamic model for analysis of the combined influences of nonlinear internal excitations on the load sharing behavior of a compound planetary gear set. *J. Mech. Eng. Sci.* **2016**, *230*, 1048–1068. [CrossRef]
16. Zhang, H.B.; Shen, X.F. A dynamic tooth wear prediction model for reflecting "two-sides" coupling relation between tooth wear accumulation and load sharing behavior in compound planetary gear set. *J. Mech. Eng. Sci.* **2020**, *234*, 1746–1763. [CrossRef]
17. Bodas, A.; Kahraman, A. Influence of Carrier and Gear Manufacturing Errors on the Static Load Sharing Behavior of Planetary Gear Sets. *JSME Int. J. Ser. C* **2004**, *47*, 908–915. [CrossRef]
18. Sanchez-Espiga, J.; Fernandez-del-Rincon, A.; Iglesias, M.; Viadero, F. Planetary gear transmissions load sharing measurement from tooth root strains: Numerical evaluation of mesh phasing influence. *Mech. Mach. Theory* **2021**, *163*, 104370. [CrossRef]
19. Sun, W.; Li, X.; Wei, J.; Zhang, A.Q.; Ding, X.; Hu, X.L. A study on load-sharing structure of multi-stage planetary transmission system. *J. Mech. Sci. Technol.* **2015**, *29*, 1501–1511. [CrossRef]
20. Wang, Y.F.; Yang, B. Study on Dynamic Simulation of Swing Movable Teeth Reducer. *Appl. Mech. Mater.* **2014**, *496–500*, 749–753. [CrossRef]
21. Yi, Y.L.; Guo, H.; Wei, R.; Jin, H.R. Transmission error analysis of swing-rod movable teeth drive based on functionary line increment means. *Manuf. Technol. Mach. Tool* **2018**, *8*, 64–70.
22. Yi, Y.L.; Dou, L.R.; Guo, H.; Jin, H.R. Coupling Stiffness of Double Outer Generator Swing Rod Movable Teeth Transmission. *China Mech. Eng.* **2018**, *29*, 644–649.

Article

Investigation of the Voltage-Induced Damage Progression on the Raceway Surfaces of Thrust Ball Bearings

André Harder, Anatoly Zaiat *, Florian Michael Becker-Dombrowsky, Steffen Puchtler and Eckhard Kirchner

Department of Mechanical Engineering, Institute for Product Development and Machine Elements, Technical University of Darmstadt, Otto-Berndt-Straße 2, 64287 Darmstadt, Germany
* Correspondence: zaiat@pmd.tu-darmstadt.de

Abstract: In the course of the electrification of powertrains, rolling element bearings are increasingly subject to electrical damage. In contrast to mechanically generated pittings, voltage-induced surface damage is a continuous process. Though several approaches for the description of the damage state of a bearing are known, a generally accepted quantification for the bearing damage has not been established yet. This paper investigates surface properties, which can be used as a metric damage scale for the quantification of the electric bearing damage progression. For this purpose, the requirements for suitable surface properties are defined. Afterwards, thrust ball bearings are installed on a test rig, with constantly loaded mechanically and periodically damaged electrically in multiple phases. After each phase, the bearings are disassembled, the bearing surfaces are graded and measured for 45 different standardized surface properties. These properties are evaluated with the defined requirements. For the ones meeting the requirements, critical levels are presented, which allow for a quantified distinction between grey frosting and corrugation surfaces. These values are compared with measurements presented in the literature showing that the identified surface properties are suitable for the quantification of electrical bearing damages.

Keywords: electric bearing damages; thrust ball bearings; electric powertrains; corrugation pattern; electric damage progression

Citation: Harder, A.; Zaiat, A.; Becker-Dombrowsky, F.M.; Puchtler, S.; Kirchner, E. Investigation of the Voltage-Induced Damage Progression on the Raceway Surfaces of Thrust Ball Bearings. *Machines* **2022**, *10*, 832. https://doi.org/10.3390/machines10100832

Academic Editor: Vincenzo Niola

Received: 31 August 2022
Accepted: 16 September 2022
Published: 21 September 2022

1. Introduction

Harmful electric bearing currents like electric discharge machining currents (EDM-currents) or rotor ground currents have been known for several decades [1–4]. Modern e-drive systems integrate electric motors and transmissions in one housing [5,6]. This increases the likeliness for currents appearing not only in motor bearings, but also in transmission bearings, especially in the case of insufficient isolation between motor and transmission [7,8]. Electric rolling bearing damages are responsible for a large amount of failures in e-drive systems [9,10]. Due to the new mobility concepts focusing on electric vehicles bearing faults caused by electric damages, these become more important within the scope of research [11,12]. The focus of research is the mitigation of the damage occurrence, the modelling of the occurring voltages and the monitoring of the damages [1]. Although the voltage induced bearing damages are known for some time, there is still no model established for the calculation of the lifetime of bearings under electric load.

Typical scales to describe the harmfulness of bearing currents are the apparent bearing current density [10] and the virtual electric power [13]. The bearing current density is the electric current passing through the bearing divided by the heartzian contact area. Muetze defines threshold values below which no harmful electric surface damages occur ($J < 0.1\ \mathrm{A/mm^2}$) [10]. Although White Etching Cracks (WEC) can occur at current densities of $J \geq 10^{-6}\ \mathrm{A/mm^2}$ [14], typical electric bearing damages like grey frosting and corrugation patterns occur at apparent current density levels above $J > 0.1\ \mathrm{A/mm^2}$. Therefore, it is possible to apply unharmful electric signals on bearings for a sensory utilization [15,16]. The virtual electric power is the product of the peak value of the voltage and the current applied on a bearing. It indicates the electric power, which is induced in the bearing and can also be used to quantify the likeliness of electric bearing damages [17]. Both values have the same disadvantages:

- They describe if a corrugation progression is likely, but they do not provide information about the point in time when a damage occurs or the rapidity of its progression.
- They do not quantify the bearing damage.
- The bearing current is a quantity difficult to measure in real e-motor applications. Therefore, these two values of damage progression and damage value respective to the bearing current are difficult to obtain.

An approach to overcome these disadvantages is to monitor the bearing surface, which in turn lead to increasing bearing vibrations in case of surface damages. The state of the surface can be related to vibration measurement data allowing for a condition monitoring of the bearing [17]. Tischmacher [17] presented a scale to quantify bearing surface damages (cf. Figure 1). It differentiates between six different surface states, starting with zero as first grade for grey frosting damaged surfaces. Severe surface damages, including fatigue damages, are assigned to grade six. The advantage of this bearing surface evaluation is the availability of a scale which allows for a comparison of different bearing damages based on the degree of damage. Thus, the bearing surface is considered suitable to describe the electric bearing damage and works as a starting point for the development of a bearing lifetime calculation. Unfortunately, the scale presented by Tischmacher has several disadvantages:

- The scale is ordinal, meaning it cannot be assumed that the differences between two different grades are equal. This leads to several obstacles like the inability to apply basic calculus operators like plus or minus, which makes calculations based on this scale impossible.
- There is no objective quantification of the different scale grades. The grade of a specific bearing surface depends on the person evaluating it.

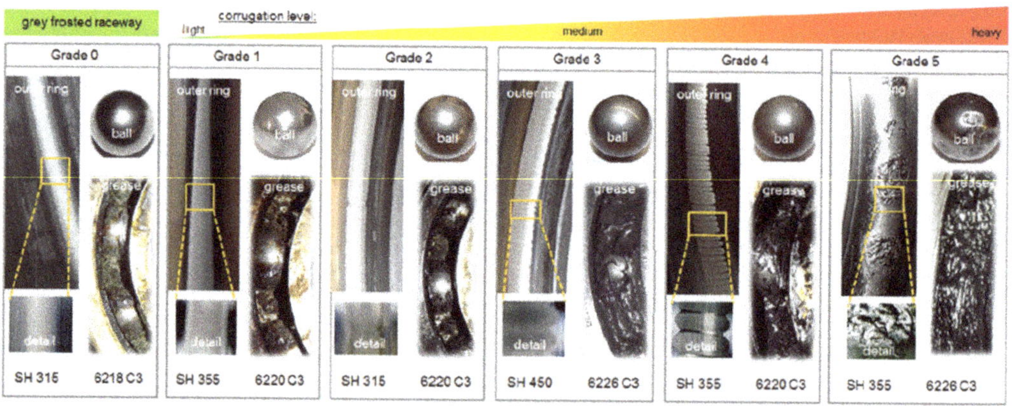

Figure 1. Different grades of bearing damages. ©2018 IEEE. Reprinted, with permission, from [17].

While the scale seems to be suitable for a qualitative and comparative description of the electric bearing damages, these disadvantages make this scale unsuitable for a quantitative description of the bearing damage. Several authors use different approaches to quantify the surface properties of electrically damaged bearing surfaces [18–23]. The shown results are measurements of standardized one- or two-dimensional surface properties, described by DIN ISO 4287 or ISO 25178-2 [24,25]. This surface properties are by design metrical scales suitable for quantitative descriptions of surfaces and independent of the evaluating person. Since each of the aforementioned publication uses different surface properties, the question is which surface property describes the electrical bearing damages sufficiently and is therefore suitable for the description of the electrical bearing damage.

To investigate this question, this paper uses the following approach: A test series on electrically damaged bearings under test rig condition is performed. The experiments are carried out on thrust ball bearings at different operation conditions (c.f. Section 2). After specific operation times, the subjective condition of the bearings is evaluated in a study and the surface properties are measured. Based on these data, the following requirements are defined, and a suitable surface property has to comply with the following:

- Requirement 1: At least a moderate correlation of $|R| \geq 0.3$ of the Spearman Rank correlation coefficient has to exist between a suitable surface property and the subjective evaluation.
- Requirement 2: A suitable surface property should allow for the distinction between corrugations and crater surfaces. Hence, the probability that the corrugation and grey frosting surfaces are equally distributed should be $p < 0.05$ in a Wilcoxon rank-sum test [26].
- Requirement 3: The variation of the suitable surface property should be lower than the variation of the subjective surface evaluation.
- Requirement 4: The variation of the suitable surface property should be independent from the value of the surface property. The surface property and its variation should have a Spearman rank correlation coefficient of $|R| < 0.8$.

The first two requirements thereby evaluate the quality a surface property describing the investigated effect. The last two requirements define the necessary reliability of the measured data.

In the following, the experimental setup is described first in Section 2. Then, the study to visually evaluate the degree of damage on the basis of Tischmachers grading [17], which is introduced in Section 3. Afterwards, the deduced surface properties that meet the defined requirements are presented in the same chapter. For these suitable surface properties, the threshold values are defined at the end of Section 3, allowing for a distinction between grey frosting and corrugation patterns. These values are discussed in Section 4 based on the literature data. The most relevant findings are summed up, and a conclusion and an outlook on future work are given in Section 5.

2. Materials and Methods

The experiments are conducted on the bearing test rig of the Institute for Product Development and Machine Elements (pmd) of the Technical University of Darmstadt [27]. The rest rig provides four testing cells, which can be operated separately (cf. Figure 2, left). In prior research, it was used for the investigation of the electric impedance of radial rolling element bearings [27]. One of the testing cells is modified for the experimental investigations of thrust ball bearings, like shown in Figure 2 on the right. Though thrust ball bearings are rarely used in electric drives, for investigations of the electric damage progression, this bearing type is used because of the possibility to investigate the bearing surface without destroying the bearing [19,20].

The setup consists of two thrust ball bearings supported by two roller bearings. The right thrust ball bearing (highlighted by the yellow box) is the investigated bearing, and the other ones are supporting. These are electrically isolated (red line) to ensure a definite current path. The shaft is connected to ground using a slip ring. A signal generator and an amplifier provide the voltage between housing and collector ring. A horizontal cylinder applies an axial force F_{Ax} onto the bearing. The shaft in turn is connected to an external electric motor with a claw coupling. Internal force, sensors, acceleration and temperature sensors monitor the test cell. The test cell allows for a disassembly of the bearings without harming the contact surfaces of the bearings. Prior to the here-presented experiments, a test run was carried out, showing no significant changes on the bearing surfaces due to the assembly and disassembly of the test cell.

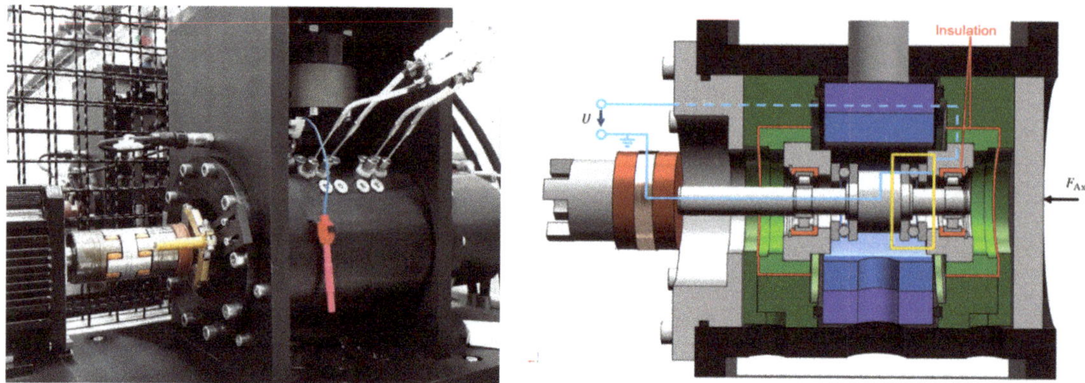

Figure 2. (**left**): Testing cell of the bearing test rig at the Technical University of Darmstadt; (**right**): sectioning and current path of the testing cell.

The experiment is designed as fractional factorial with five factors at two levels, resulting in a total of 16 tests. It includes five factors with two patterns. The factor combinations of axial load F_{Ax}, revolution speed n, voltage amplitude $\hat{U}$, frequency f and the signal form are investigated, as seen in Table 1. The lubricant, oil temperature, bearing type and bearing size are constant during the investigation and listed in Table 2. In contrast to the grease-lubricated thrust ball bearings used in the industry, the tests in this paper are carried out with oil. The oil circuit is operated with a volume of 5 L per minute and 20 L in total. The oil is filtered before entering the test cell. The large amount of lubricant in comparison to grease-lubricated bearings as well as its constant circulation and filtration reduce the effect of lubricant deterioration on the bearing surfaces.

Table 1. Experimental factors.

Factor	Unit	-	+
Axial Load F_{Ax}	kN	1	3.5
Rotation Speed n	rpm	500	2500
Voltage $\hat{U}$	V	2.5	5
Frequency f	Hz	5000	20,000
Signal Form	-	Square Wave	Sine Wave

Table 2. Constant parameters.

Parameter	Value
Lubricant	FVA3 reference oil
Oil Temperature T	40 °C
Bearing Type	Thrust ball bearing
Bearing Size	51,305

The experiment investigates the connection between the surface properties, vibration data of the test rig and the electric impedance of the damaged bearing, although this work focuses primarily on the bearing surfaces. The evaluation of the vibration and impedance data will be presented in later work. Figure 3 summarizes the experimental design.

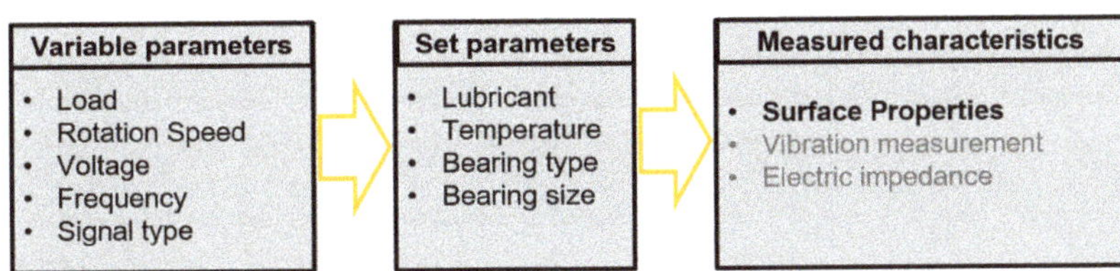

Figure 3. Design of the experiment.

Each test sequence is grouped in five different phases, as shown in Figure 4. The first phase is a run-in of the bearings for six hours under an axial load $F_{Ax} = 3500$ N and a rotation speed $n = 2500$ rpm. The aim is to reduce surface peaks and obtain comparable starting conditions for the damaging phase. It is followed by three electric damaging phases of three hours each with disassembly and a surface evaluation at the end of each phase. The fifth and last phase is an electric damaging period of 12 hours. The operating conditions at each of the damaging phases are constant and given for all tests in Table A1.

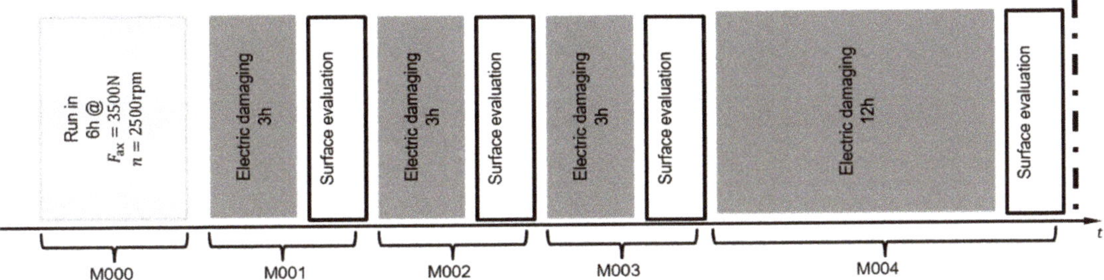

Figure 4. Test setup.

For measurements of the runway surfaces, the white light interferometer Smart WLI from GBS is used. It enables a three-dimensional gauging of the surfaces to detect micro geometries and surface roughness. Using the Smart WLI, each bearing ring is measured at four marked positions on each bearing ring to ensure the comparability of the measurements after each damaging period. These areas are designated as "N", "E", S" and "W". The area "N" is marked by two engraved dots, the other areas by one dot (c.f. Figure 5). Each scanned area has the size of about 2 mm × 0.7 mm.

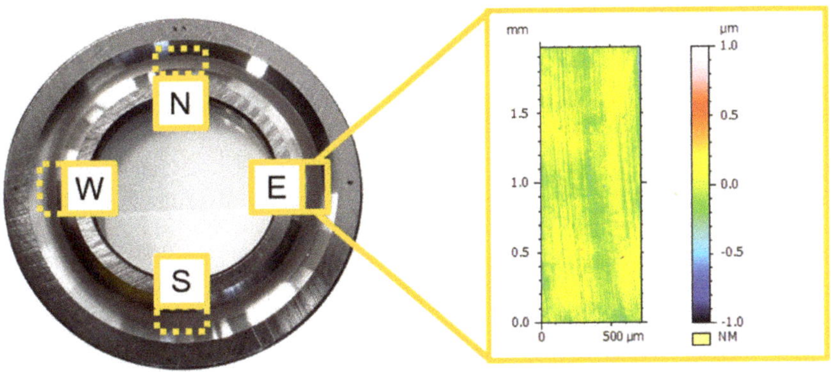

Figure 5. Damaged bearing surface with the scanned areas.

The measurement principle is based on the objectives, which cause interference patterns during scanning of the surfaces. The patterns are detected by optical sensors and transformed into electrical signals. From these signals, topographical data are derived, and the raceway geometry and possible outliers are removed, leaving the raceway surface. An exemplary scan of the raceway surface in the course of voltage-induced damage can be taken from Figure 6. After the first damage period, it is observable that there are some remains of the honing process visible on the bearing surface. After the damage periods M002 and M003, these patterns reduce and a crater pattern is observable. After M004, there is a clear corrugation pattern observable. The black line at M003 resembles a contamination during the scanning process. Such contaminations were excluded from evaluation.

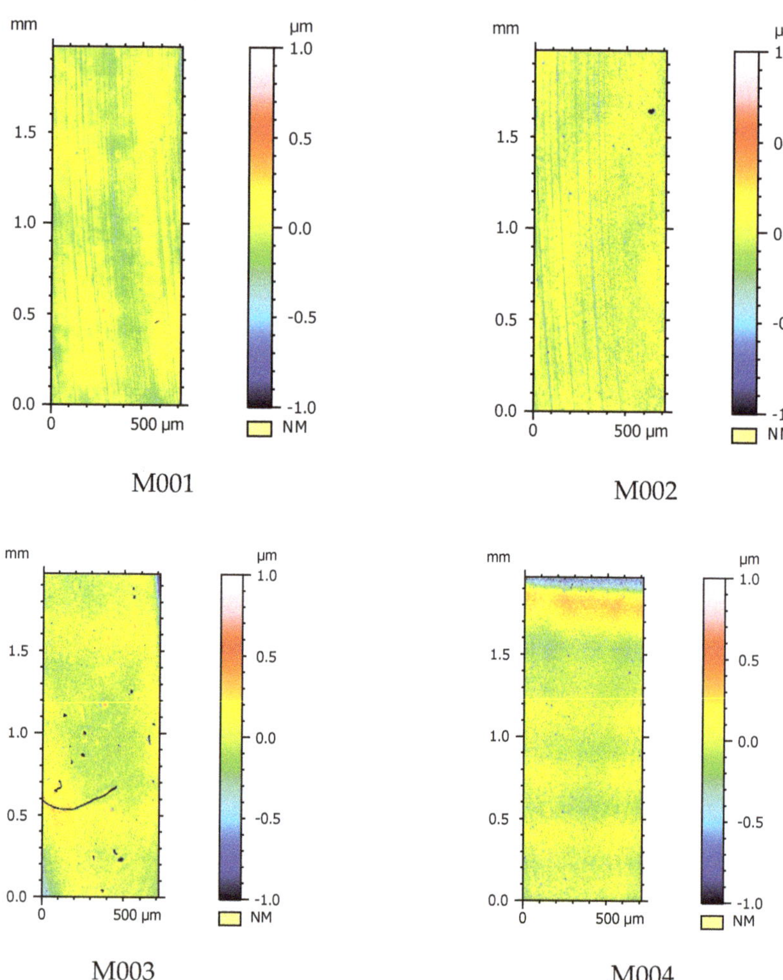

Figure 6. Exemplary scan of the V09 raceway surface after 3 h, 6 h, 9 h and 21 h of voltage-induced surface damaging.

The surface properties extracted from the scans are listed in Table 3. The Variable X is used for properties, which are evaluated in different dimensions and filtered at different wavelengths, leaving different surfaces, such as the roughness or the waviness. For example, the maximum profile height is evaluated for the 2D surface at different extracted wavelengths. Therefore, the maximum profile height is calculated for the primary profile as well as for the roughness and waviness profile. The same procedure is applied at 1D surface properties. In total this leads to 45 different surface properties.

Table 3. Characteristic surface parameters.

Variable	Parameter	Variable	Parameter
X_z	Maximum Profile Height	V_m	Material Volume
X_t	Total Profile Height	V_v	Void Volume
X_a	Arithmetic Mean of Profile Ordinate	V_{mp}	Material Volume of peaks
X_q	Quadratic Mean of Profile Ordinate	V_{mc}	Material Volume at the core
X_{sk}	Skewness of Profile	V_{vc}	Void Volume at the core
X_{ku}	Steepness of Profile	V_{vv}	Void Volume at the valleys
X_{Sm}	Median Groove Width of Profile Elements	S_k	Core height
X_{dq}	Quadratic Mean of Profile Pitch	S_{pk}	Average peak height above the core
X_{mr}	Material Fraction of Profile	S_{vk}	Average valley depth below the core
X_{dc}	Height Difference between two Intersection Lines		

In addition to the surface property measurement the bearing damage scale presented by Tischmacher is used for the evaluation of the bearing damage in a study. For this purpose, digital photographs of the individual raceways are created in the course of increasing surface damage, as already shown in Figure 4. These are shown to eight persons, all research assistants at pmd who have a Master's degree in mechatronic or mechanical engineering. Prior to conducting the assessment, all participating individuals were introduced to the damage scale. For the purpose of the study, the scale was extended to take unharmed bearing surfaces into account. Unharmed surfaces are therefore the new damage grade 0, and all other grades are increased by +1. Additionally, the participants were allowed to evaluate the surfaces with intermediate steps (e.g., 2.5) if a surface seems to be between to damage grades. To ensure an unbiased surface evaluation, the raceway images were evaluated by the participants in a random order.

The damage scale used in this study, like the evaluation scale according to Tischmacher, is an ordinal scale. Therefore, a mean value or standard deviation may not be calculated for the surveyed study results. Thus, the median of the degree of damage DG and the first quartile $DG_{0.25}$, the third quartile $DG_{0.75}$ and the interquartile range Q_{DG} were evaluated. In addition, the number N_{eval} is indicated, describing the order of evaluation. An adjustment of the results, e.g., by removing outliers, was omitted.

3. Results

In the following section, the results of the damage assessment study are presented. Based on these results, the surface properties are evaluated and compared to the defined requirements for a suitable surface property. Due to the large amount of measured data obtained during the study, this paper focuses on the evaluated data leading to suitable surface properties. The complete set of data, including all study results of the surface evaluation, the surface scans and the evaluated surface properties and the monitoring data of the test rig are published in [28].

3.1. Degree of Damage Assessment Study

An investigation on the influence of the rating order shows that there is only a small rank correlation between the order and the damage grade ($R_{DG-N} = 0.2$), and there is no rank correlation between the rating order and the interquartile range ($R_{Q-N} = -0.016$). Furthermore, the rank correlation between the damage grade and interquartile range is also neglectable ($R_{DG-Q} = 0.008$). This means that neither the rating order nor the interquartile range influence the results of the study—and therefore, the variation of the degree of damage depends on the degree of damage itself.

Figure 7 shows the course of the assessed degree of damage after each damage period of test V09 as a boxplot. The medians DG are encircled and connected by a line to describe the time course of the assessed damage. The interquartile range Q_{DG} is shown as a rectangle, and readings outside the interquartile range are shown as a thin vertical line. Outliers with a deviation of more than 1.5 Q_{DG} below the first quartile or above the third quartile are marked with a cross. Test V09 shows a clear tendency of an increasing damage grade DG with an increasing time of electrical damaging. In the other tests, which are shown in Figure A1, it is though observable that the increase of the damage grade over different tests is mostly insignificant, and in some cases, there is even a decrease observable. Furthermore, the median of the interquartile range is at $\widetilde{Q}_{DG} = 1$, meaning that evaluated damage grade varies around one grade of the damage scale. Due to this variation, a precise description of the damage state is unlikely. It is possible though to use this scale to distinguish between corrugation damages and crater surfaces. All tests with observable corrugation patterns were evaluated with a damage grade of $DG = 3$ or higher.

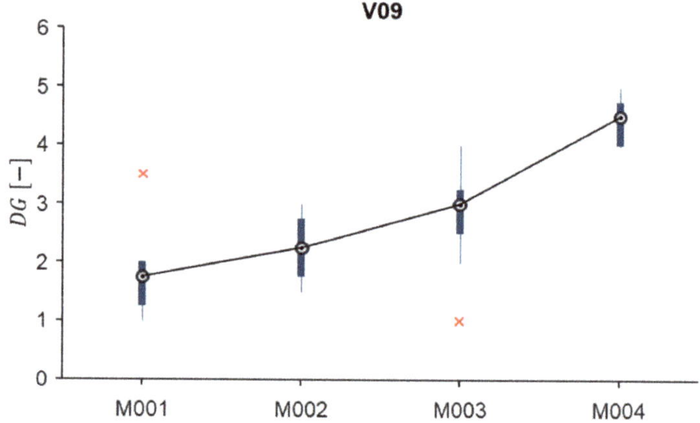

Figure 7. Assessed damage grade (DG) of test V09 of a 51305 axial ball bearing with axial load $F_A = 1$ kN and rotational speed $n = 2500$ rpm with an applied square wave voltage $\hat{V} = 2.5$ V at $f = 5$ kHz after 3 h (M001), 6 h (M002), 9 h (M003) and 21 h (M004) of electric damaging. Circles mark the median, boxes mark the interquartile range, vertical lines mark the overall range of values and crosses mark outliers.

As the results show, the damage grade assessment study is able to identify corrugation damages and to differentiate them from grey frosting. Thus, it is suitable for a qualitative evaluation of rolling bearing damages. Because the high median of the interquartile range is at $\widetilde{Q}_{DG} = 1$, the use of this scale for a comparative damage description is limited. Reasons for the variability can be the study design and the design of the scale itself. Another aspect is that single deep craters in the surface are not considered in the scale, which lead to a high variability in the assessment.

3.2. *Testing Surface Property Requirements*

Based on the results of the surface assessment study the measured surface properties are compared with the requirements for properties suitable for the quantification of electric bearing damages.

3.2.1. Correlation with the Damage Grade

First, the correlation between each surface property and the assessed damage grade is examined. The condition of each raceway after each damage period is comparatively described by the DG. If a surface property is related to the electrical bearing damage, it should correlate with the damage grade evaluated. The rank correlation between all damage grade evaluations after each damage period with the corresponding surface properties is calculated. Figure 8 shows the rank correlation coefficients between the damage grade of the surface and all investigated surface properties as a bar chart. In addition, the bounds for the significant correlation $|R| \geq 0.3$ are plotted. A higher requirement for the correlation is not defined, because it would overestimate the results of the damage grade evaluation. It is observable that 19 of 45 surface properties investigated correlate with the damage-assessed damage grade.

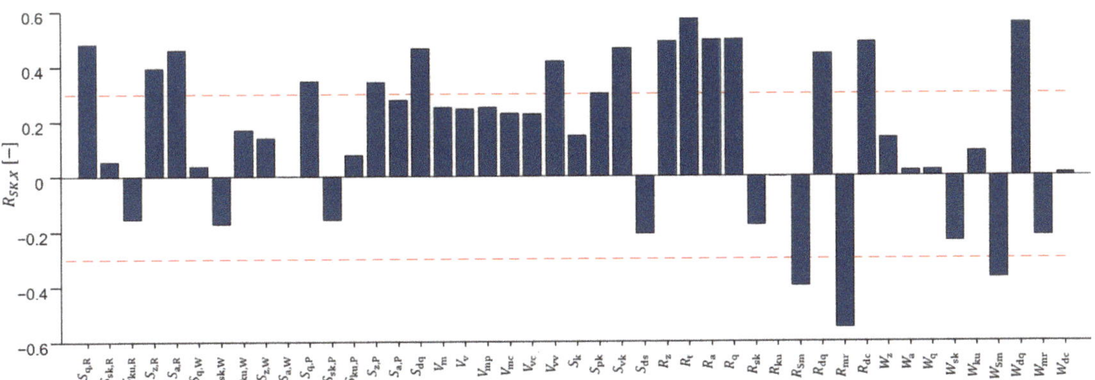

Figure 8. Rank correlation coefficients of the investigated surface properties with the assessed damage grade *DG*. All properties outside the marked boundary satisfy the requirement of significant correlation.

3.2.2. Identification of Corrugation

In this section, it is examined whether bearings with pronounced corrugation can be distinguished with the aid of the measured surface properties. For this purpose, the data of the raceways examined are divided into two groups. Bearings that received a median rating of $DG < 3$ belong to set $A = \{DG < 3\}$ of bearings without corrugations, and the complement set $B = \{DG \geq 3\}$ describes bearings with corrugation. Thus, there are 45 bearing surfaces in set A and 15 bearing surfaces in set B. Using the Wilcoxon rank-sum test, it can be quantified whether a certain threshold of a measured surface property is able to classify both sets. This procedure tests the hypothesis that two sets of data have the same median and therefore the same distribution. If the probability calculated in this process is below a critical value, this hypothesis can be rejected, and the two groups are different. $P(A = B) < 0.05$ is defined as the critical probability.

The result of the Wilcoxon rank-sum test for all investigated surface properties is shown in Figure 9. The dashed line indicates the threshold value above which the hypothesis of equal quantities can be rejected. It can be seen that 26 of the 45 properties studied meet the requirement.

3.2.3. Investigation of the Variation of the Surface Properties

Subsequently, the standard deviation of all investigated surface properties is compared to the interquartile range of the assessed damage grade. Therefore, the damage grade scale is assumed as metric, i.e., the grades are equidistant, and both the assessment and measured surface properties are normally distributed. To convert all quantities in relative values, a reference value for the quantity X is calculated from the 75% quartile $X_{0.75}$ and the corresponding interquartile range Q_X [29],

$$z_{u,X} = X_{0.75} + 1.5 \cdot Q_X. \quad (1)$$

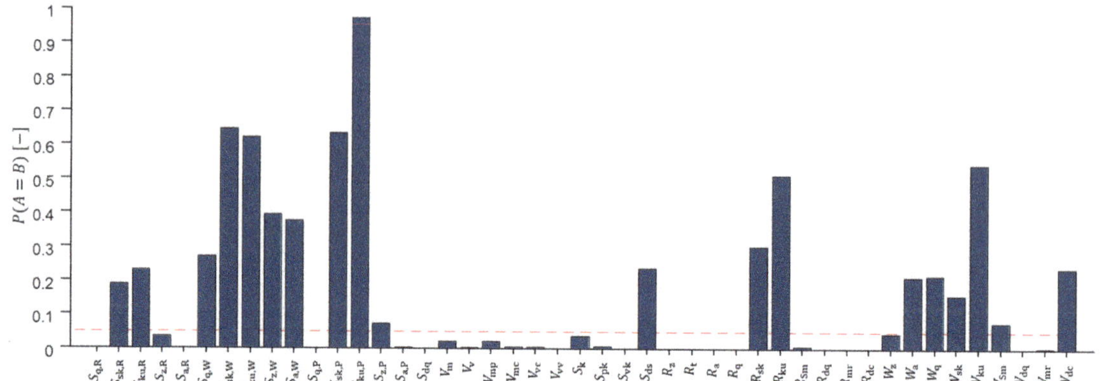

Figure 9. Result of the Wilcoxon rank-sum test for the investigation of the probability of equal sets A and B.

This reference value is the upper limit, above which all measured data points are considered outliers. Thus, the relative variation of the damage grade is calculated from the arithmetic mean of the interquartile range $\overline{Q}_{DG}$,

$$s_{DG} = \frac{0.5 \cdot \overline{Q}_{DG}}{z_{u,DG}}, \tag{2}$$

or for a surface property X from the arithmetic mean of standard deviation $\overline{\sigma}_X$ respectivly,

$$s_X = \frac{0.67 \cdot \overline{\sigma}_X}{z_{u,DG}}. \tag{3}$$

The different factors in Equations (2) and (3) are necessary to normalize the half interquartile range and the standard deviation.

Figure 10 shows the relative variation of the investigated surface properties. As a reference, the variation of the assessed damage grade $s_{DG} = 0.13$ is shown as a dashed line. It is particularly noticeable that all values describing the skewness of the profile show a very high variation. In total, 21 properties fulfill the set requirement.

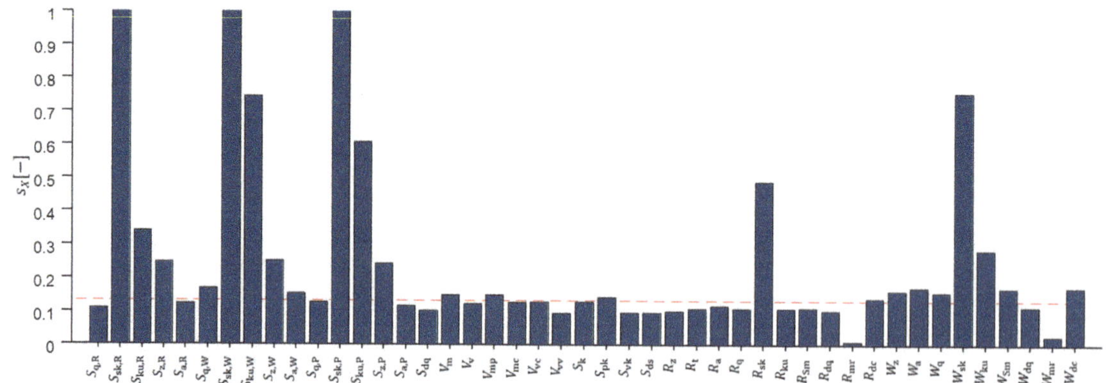

Figure 10. Relative variation of the investigated surface properties in comparison to the variation of the assessed damage grade (dashed line).

3.2.4. Investigation of the Correlation between the Magnitude and Variation of the Surface Property

The final requirement for suitable surface properties is the independence between the magnitude and the standard deviation of a measured property. This correlation can be by quantified by the rank correlation coefficient as shown in Figure 11. In total, 35 surface properties yield a coefficient below the critical value of $|R_{X,\sigma_X}| < 0.8$, and thus, they fulfill the requirement of a low magnitude to medium correlation between the magnitude and the standard deviation of a measurement.

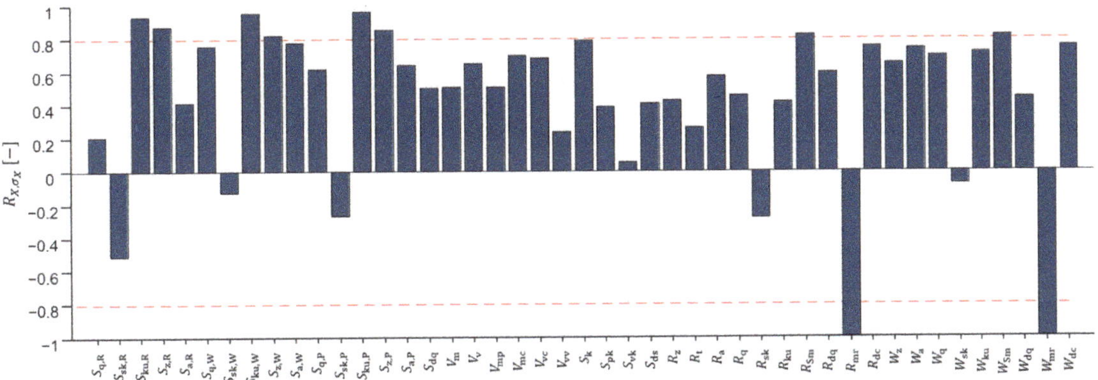

Figure 11. Correlation between mean value and standard deviation of surface properties.

4. Discussion

From the examination of the requirements set, it is recognizable that 12 out of the 45 surface properties investigated meet all the requirements and, therefore, are suitable for qualitative and quantitative evaluation of electrical surface damage. However, especially for surface properties, which fulfill the first two requirements and fail the last two requirements, it is worth investigating whether a larger amount of scanned surface areas enables a reduction of the standard deviation or its correlation to the magnitude of the measured surface property.

A qualitative evaluation of electrical surface damage is possible by defining a threshold value for the respective surface property, which enables a differentiation of surfaces with corrugation damage from surfaces without corrugation damage. This differentiation is defined at the transition between the third quartile of set A $(X_{0.75,A})$ and the first quartile of set B $(X_{0.25,B})$. If the quartiles overlap, the arithmetic mean of $X_{0.75,A}$ and $X_{0.25,B}$ is calculated as the critical value. If $X_{0.25,B}$ is greater than $X_{0.75,A}$, $X_{0.25,B}$ is considered the critical value. Mathematically, this relationship is described as

$$X_{crit} := \begin{cases} \frac{X_{0.25,B}+X_{0.75,A}}{2} & \text{if } X_{0.25,B} < X_{0.75,A} \\ X_{0.25,B} & \text{if } X_{0.25,B} \geq X_{0.75,A} \end{cases}. \quad (4)$$

These calculated critical values of all surface properties meeting the requirements are summarized in Table 4, the boxplots for all surface properties are found in Figure A2 in the Appendix A.

Next, it is checked whether the defined critical values are compatible with measured values from the literature. If test data from the literature can be delimited with the help of the critical values shown in Table 4, this is taken as confirmation of the corresponding threshold value. If a source gives values for surface damage that contradict the specific limit value but are within the 25% above or below the limit value, the data from the corresponding source are not considered to contradict the threshold value. If the published data are outside this range, they are considered a deviation. For the remaining surface properties, no comparable data were found.

Table 4. Summary of critical values for corrugations for the respective surface properties.

Surface Property	Critical Value X_{crit}	Comparison to Literature
$S_{q,R}$	0.110 µm	No comparable data
$S_{a,R}$	0.069 µm	No contradiction
$S_{q,P}$	0.189 µm	Deviation; no contradiction [15,19]
S_{dq}	0.353	No comparable data
V_{vv}	0.030 $\frac{\mu m^3}{\mu m^2}$	No comparable data
S_{vk}	0.289 µm	Deviation; no contradiction [14,15,19]
R_z	0.285 µm	Deviation; no contradiction [17,18]
R_t	0.824 µm	Confirmed [16,19]
R_a	0.043 µm	No contradiction [17–19]
R_q	0.062 µm	No contradiction [16–19]
R_{dq}	4.596°	No comparable data
W_{dq}	0.436°	No comparable data

From comparison with measured data presented in the literature, it can be seen that four properties can either be confirmed or at least the data in the literature do not contradict the defined critical values. For three properties, a deviation is found in the literature, although for all of these properties, additional data were found that show no contradiction to the presented values. In this case, surfaces properties were presented, which are above the here-defined critical values without any corrugation patterns. This can be for several reasons: On the one hand, these publications use different bearings with different surface properties. On the other hand, the above-mentioned sources mainly show a false positive result (α-error). This means that based on the surface properties, corrugation patterns should be present but are not. In contrast, false negative data (β-error) have not been observed in the literature. Thus, it can be assumed that the critical surface property limit values are a statistical threshold below which corrugations rarely occur, and they can therefore be used as a critical damage value. To either confirm or adjust this limit values, larger data sets for the surface properties of electrically damaged bearings without corrugations, and especially for electrically damaged bearings with corrugations, are required.

5. Conclusions and Outlook

This paper presents surface properties suitable for the quantification of voltage-induced bearing damages and defines critical limit values that allow for a distinction between grey frosting and corrugations of the bearing surface. Therefore, the used experimental setup was presented to create electrical damages on thrust ball bearings, and the damage progression was evaluated. The damage grade of these surfaces is afterwards evaluated in a study, and the surface properties are measured. Based on defined requirements for suitable surface properties, the suitability of the investigated properties is evaluated and leads to 12 suitable surface properties. A critical value is derived for each of these properties, which allows for a distinction between crater damages like grey frosting and corrugation patterns. These values are discussed in comparison with measurements presented in other publications.

Further research can aim in different directions. First, the evaluated surface properties and the critical values have to be validated with additional surface data of bearings from test bench experiments as well as from real applications. Based on the surface properties, it is possible to derive changing rates of the surface property and investigate the effect of the operating conditions and the applied electric load on the changing rate of the respective property. Finally, the change of the surface properties has to be connected to changes in the condition monitoring data of the bearing. This can give the opportunity to evaluate the surface properties and allow for statements on the current electrical bearing damage state. If one can quantify the current damage state of a bearing surface via condition monitoring and if one is able to calculate the damage progression rate respective to the mechanical and electrical operating conditions, it will be possible to estimate the time that is left until a bearing reaches a critical surface property level. The time until a critical surface level is reached can be defined as the remaining safe operation time of the bearing. The here-presented surface properties and their critical values can therefore be a step towards a calculation of such a safe operation time until corrugation damages occur.

Author Contributions: Conceptualization, A.H.; methodology, A.H., A.Z. and F.M.B.-D.; investigation, A.H. and F.M.B.-D.; writing—original draft preparation, A.H., A.Z., F.M.B.-D. and E.K.; writing—review and editing, A.H. and S.P.; visualization, A.H. and A.Z.; supervision, E.K.; project administration, A.H.; funding acquisition, E.K. All authors have read and agreed to the published version of the manuscript.

Funding: Funded by the Deutsche Forschungsgemeinschaft (DFG, German Research Foundation), project numbers 401671541, 467849890 and 463357020.

Informed Consent Statement: Informed consent was obtained from all subjects involved in the study.

Data Availability Statement: The data presented in this study are openly available on TUdatalib at https://doi.org/10.48328/tudatalib-932, reference number [24].

Conflicts of Interest: The authors declare no conflict of interest.

Appendix A

Table A1. Operating conditions of all tests.

Test Nr.	Axial Load (kN)	Rotation Speed (rpm)	Voltage (V)	Frequency (f)	Signal Form (-)
V01	3.5	2500	5	20	Sine
V02	1	2500	5	5	Sine
V03	1	500	2.5	5	Sine
V04	3.5	500	2.5	20	Sine
V05	1	2500	5	20	Square
V06	3.5	2500	5	5	Square
V07	3.5	500	2.5	5	Square
V08	1	500	2.5	20	Square
V09	1	2500	2.5	5	Square
V10	3.5	500	5	20	Square
V11	1	500	5	20	Sine
V12	1	500	5	5	Square
V13	3.5	2500	2.5	20	Square
V14	3.5	2500	2.5	5	Sine
V15	1	2500	2.5	20	Sine
V16	3.5	500	5	5	Sine

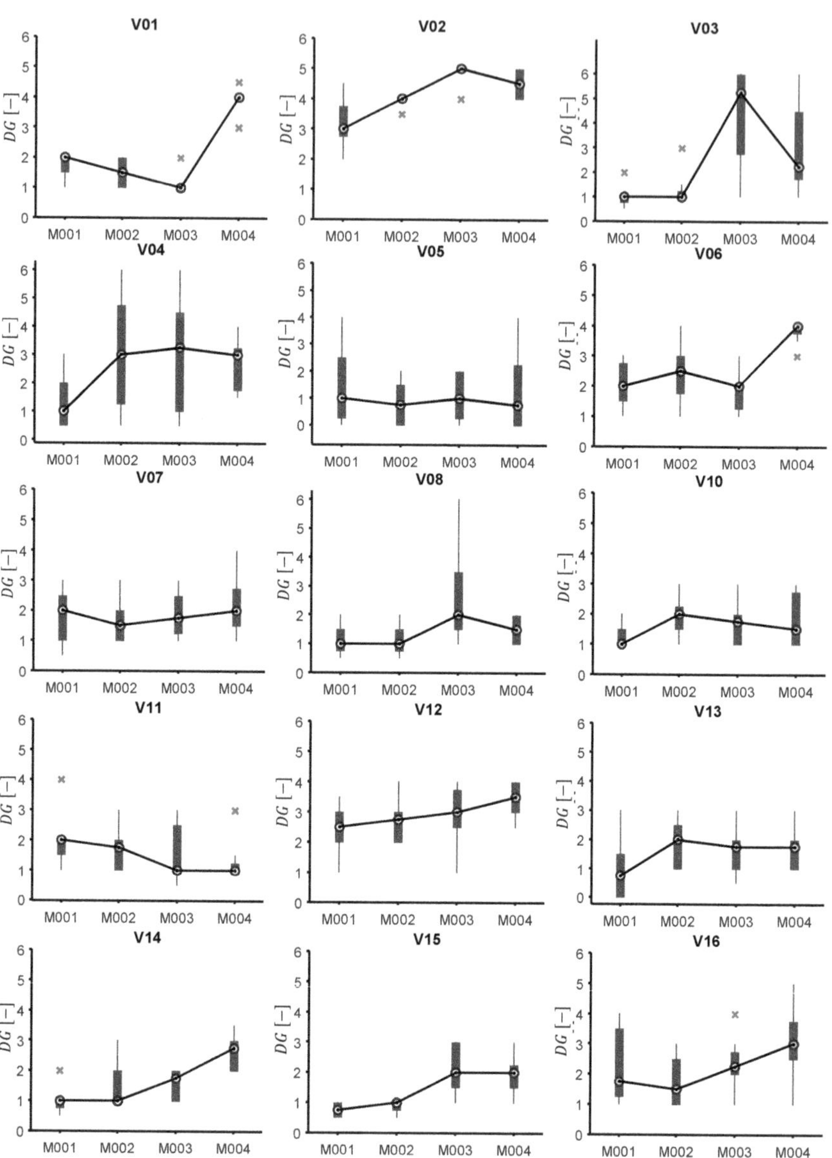

Figure A1. Assessed damage grade (*DG*) of tests V01–V16 of a 51305 axial ball bearing after 3 h (M001), 6 h (M002), 9 h (M003) and 21 h (M004) of electric damaging. Circles mark the median, boxes mark the interquartile range, vertical lines mark the overall range of values and crosses mark outliers.

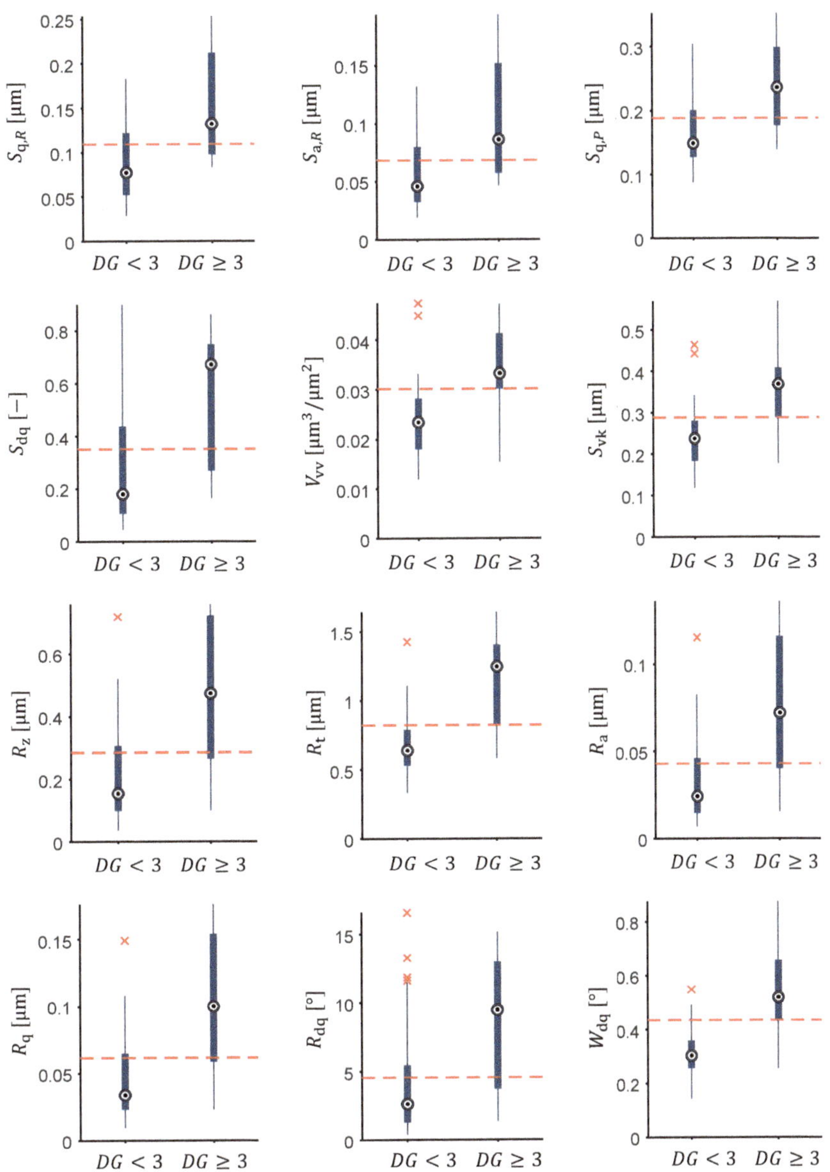

Figure A2. Boxplots of the measured surface properties. The data are separated in two groups for data points with and without corrugations. The red line is the critical value, above which corrugations are likely.

References

1. Muetze, A. Thousands of hits: On inverter-induced bearing currents, related work and the literature. *e & i Elektrotechnik und Informationstechnik* **2011**, *128*, 82–388. [CrossRef]
2. Prashad, H. *Tribology in Electrical Environments*; Elsevier: Amsterdam, The Netherlands, 2006; Volume 49, pp. 1–494.
3. Boyanton, H.E.; Hodges, G. Bearing fluting. *IEEE Ind. Appl. Mag.* **2002**, *8*, 53–57. [CrossRef]
4. Busse, D.; Erdman, J.; Kerkman, R.J.; Schlegel, D.; Skibinski, G. System electrical parameters and their effects on bearing currents. *IEEE Trans. Ind. Appl.* **1997**, *33*, 577–584. [CrossRef]

5. Gennaro, M.D.; Scheuermann, P.; Wellerdieck, T.; Ravello, V.; Pellegrino, G.; Trancho, E. The H2020 project FITGEN: Preliminary results and design guidelines of an integrated e-axle for the third-generation electric vehicles. In Proceedings of the 8th Transport Research Arena TRA 2020, Helsinki, Finland, 27–30 April 2020.
6. Beyer, M.; Brown, G.; Gahagan, M.; Higuchi, T.; Hunt, G.; Huston, M.; Jayne, D.; McFadden, C.; Newcomb, T.; Patterson, S.; et al. Lubricant Concepts for Electrified Vehicle Transmissions and Axles. *Tribol. Online* **2019**, *14*, 428–437. [CrossRef]
7. Graf, S.; Sauer, B. Simulative Untersuchungen des Einflusses elektromechanischer Belastungen an einem Axiallager auf die Glättung der Oberflächen und den Traganteil im Kugel-Laufbahnkontakt. *Forsch. Ing.* **2022**, *86*, 145–159. [CrossRef]
8. Schiferl, R.F.; Melfi, M.J. Bearing current remediation options. *IEEE Ind. Appl. Mag.* **2004**, *10*, 40–50. [CrossRef]
9. Lin, F.; Chau, K.T.; Chan, C.; Liu, C. Fault Diagnosis of Power Components in Electric Vehicles. *J. Asian Electr. Veh.* **2013**, *11*, 1659–1666. [CrossRef]
10. Mütze, A. Bearing Currents in Inverter Fed AC-Motors. Ph.D. Thesis, Technische Universität Darmstadt, Darmstadt, Germany, 2004.
11. He, F.; Xie, G.; Luo, J. Electrical bearing failures in electric vehicles. *Friction* **2020**, *8*, 4–28. [CrossRef]
12. Schneider, V.; Behrendt, C.; Höltje, P.; Cornel, D.; Becker-Dombrowsky, F.M.; Puchtler, S.; Gutiérrez Guzmán, F.; Ponick, B.; Jacobs, G.; Kirchner, E. Electrical Bearing Damage, A Problem in the Nano- and Macro-Range. *Lubricants* **2022**, *10*, 194. [CrossRef]
13. Wittek, E.; Kriese, M.; Tischmacher, H.; Gattermann, S.; Ponick, B.; Poll, G. Capacitances and lubricant film thicknesses of motor bearings under different operating conditions. In Proceedings of the XIX International Conference on Electrical Machines—ICEM, Rome, Italy, 6–8 September 2010; pp. 1–6. [CrossRef]
14. Loos, J.; Bergmann, I.; Goss, M. Influence of Currents from Electrostatic Charges on WEC Formation in Rolling Bearings. *Tribol. Trans.* **2016**, *59*, 865–875. [CrossRef]
15. Schirra, T.; Martin, G.; Vogel, S.; Kirchner, E. Ball Bearings as Sensors for Systematical Combination of Load and Failure Monitoring. In Proceedings of the DESIGN 2018 15th International Design Conference, Dubrovnik, Croatia, 21–24 May 2018; pp. 3011–3022. [CrossRef]
16. Martin, G.; Becker, F.; Kirchner, E. A novel method for diagnosing rolling bearing surface damage by electric impedance analysis. *Tribol. Int.* **2021**, *170*, 107503. [CrossRef]
17. Tischmacher, H. Bearing Wear Condition Identification on Converter-fed Motors. In Proceedings of the 2018 International Symposium on Power Electronics, Electrical Drives, Automation and Motion, Amalfi, Italy, 20–22 June 2018. [CrossRef]
18. Bechev, D.; Gonda, A.; Capan, R.; Sauer, B. Untersuchung der Oberflächenmutationen und der Riffelbildung bei spannungsbeaufschlagten Wälzlagern. In *VDI-Tagung Gleit-und Wälzlagerungen*; VDI Verlag: Düsseldorf, Germany, 2019; pp. 301–312.
19. Graf, S.; Sauer, B. Surface Mutation of the Bearing Raceway caused by electrical Current Passage in Mixed Friction Operation. *Bear. World J.* **2020**, *5*, 137–147.
20. Zika, T.; Gebeshuber, I.C.; Buschbeck, F.; Preisinger, G.; Gröschl, M. Surface analysis on rolling bearings after exposure to defined electric stress. *Proc. Inst. Mech. Eng. Part J J. Eng. Tribol.* **2009**, *223*, 787–797. [CrossRef]
21. Gemeinder, Y. Lagerimpedanz und Lagerschädigung bei Stromdurchgang in Umrichtergespeisten Elektrischen Maschinen. Ph.D. Thesis, TU Darmstadt, Darmstadt, Germany, 2016.
22. Forschungsvereinigungs Antriebstechnik. *Schädlicher Stromdurchgang 1—Untersuchung des Schädigungsmechanismus und der zulässigen Lagerstrombelastung von Wälzlagern in E-Motoren und Generatoren Verursacht Durch Parasitäre Hochfrequente Lagerströme*; Forschungsvorhaben Nr. 650 I Heft 1127; Forschungsvereinigungs Antriebstechnik: Frankfurt am Main, Germany, 2015.
23. Forschungsvereinigungs Antriebstechnik. *Schädlicher Stromdurchgang 2—Methodik zur Praxisnahen Charakterisierung von Elektrischen Schmierstoffeigenschaften zur Verbesserung der Rechnerischen Vorhersage von Lagerströmen*; Forschungsvorhaben Nr. 650 II Heft 1387; Forschungsvereinigungs Antriebstechnik: Frankfurt am Main, Germany, 2020.
24. *DIN EN ISO 4287:2010-07*; Geometrische Produktspezifikation (GPS)—Oberflächenbeschaffenheit: Tastschnittverfahren—Benennungen, Definitionen und Kenngrößen der Oberflächenbeschaffenheit. ISO: Geneva, Switzerland, 2010.
25. *ISO 25178-2:2012*; Geometrical Product Specifications (GPS)—Surface Texture: Areal—Part 2: Terms, Definitions and Surface texture Parameters. ISO: Geneva, Switzerland, 2012.
26. Wilcoxon, F. Individual Comparisons by Ranking Methods. *Biom. Bull.* **1945**, *1*, 80. [CrossRef]
27. Schirra, T.; Martin, G.; Puchtler, S.; Kirchner, E. Electric impedance of rolling bearings—Consideration of unloaded rolling elements. *Tribol. Int.* **2021**, *158*, 106927. [CrossRef]
28. Harder, A.; Kirchner, E. *Voltage Induced Damage Progression on the Raceway Surfaces of Thrust Ball Bearings, Part I: Surface Measurements*; TUdatalib: Darmstadt, Germany, 2022. [CrossRef]
29. Fahrmeir, L. *Statistik: Der Weg zur Datenanalyse*; Springer: Berlin/Heidelberg, Germany, 2016; ISBN 13 978-3662503713.

Article

Degeneration Effects of Thin-Film Sensors after Critical Load Conditions of Machine Components

Rico Ottermann [1,*], Tobias Steppeler [1], Folke Dencker [1] and Marc Christopher Wurz [1,2]

1 Institute of Micro Production Technology (IMPT), Leibniz University Hannover, 30823 Garbsen, Germany
2 DLR Institute for Quantum Technologies, Ulm University, 89069 Ulm, Germany
* Correspondence: ottermann@impt.uni-hannover.de; Tel.: +49-511-762-5747

Abstract: In the context of intelligent components in industrial applications in the automotive, energy or construction sector, sensor monitoring is crucial for security issues and to avoid long and costly downtimes. This article discusses component-inherent thin-film sensors for this purpose, which, in contrast to conventional sensor technology, can be applied inseparably onto the component's surface via sputtering, so that a maximum of information about the component's condition can be generated, especially regarding deformation. This article examines whether the sensors can continue to generate reliable measurement data even after critical component loads have been applied. This extends their field of use concerning plastic deformation behavior. Therefore, any change in sensor properties is necessary for ongoing elastic strain measurements. These novel fundamentals are established for thin-film constantan strain gauges and platinum temperature sensors on steel substrates. In general, a k-factor decrease and an increase in the temperature coefficient of resistance with increasing plastic deformation could be observed until a sensor failure above 0.5% plastic deformation (constantan) occurred (1.3% for platinum). Knowing these values makes it possible to continue measuring elastic strains after critical load conditions on a machine component in terms of plastic deformation. Additionally, a method of sensor-data fusion for the clear determination of plastic deformation and temperature change is presented.

Keywords: thin-film sensor; strain gauge; temperature sensor; sputtering; temperature coefficient of resistance; k-factor; plastic deformation; tribological contact; bearings; sensor data fusion

Citation: Ottermann, R.; Steppeler, T.; Dencker, F.; Wurz, M.C. Degeneration Effects of Thin-Film Sensors after Critical Load Conditions of Machine Components. *Machines* **2022**, *10*, 870. https://doi.org/10.3390/machines10100870

Academic Editors: Sven Matthiesen and Thomas Gwosch

Received: 31 August 2022
Accepted: 24 September 2022
Published: 27 September 2022

1. Introduction

The sensory acquisition of measurement data is necessary in many branches of industry, e.g., to comply with safety, process or application limits. Important measurement variables are strain and temperature, which are usually measured using resistive sensors. In the automotive sector, for example, metallic strain gauges are used to measure the strain and compression condition of cast parts and engine blocks [1]. In the energy sector, strain gauges are used in 3D-woven composite spar caps for structural health monitoring (SHM) of wind-turbine blades and a reduction in measurement costs [2]. In the construction sector, deformations and crack building in buildings are measured with specially encapsulated strain gauges inserted into deep measurement boreholes that thus detect potential damage at an early stage [3]. For higher-level monitoring of machine components such as bearings, different approaches are used. They all aim to prevent damage caused by complex conditions like high speeds and heavy loads over a long period of time [4]. Here, the behavior at the elastohydrodynamic (EHD) rolling contact is of particular interest. Different approaches for the sensory monitoring of bearings are currently carried out by externally mounted sensors outside of the tribologically loaded EHD contact. Gao et al. presented a piezoelectric vibration sensor unit that was attached on the backside of one bearing washer to measure the force of the roller on the washer [5,6]. Marble et al. measured the cage motion and temperature with a sensor that is integrated to the cage to draw conclusions about the condition of the lubrication [7]. Large magnetic stray flux sensors were used to detect bearing faults such as cracks or holes in the outer race or a deformation of the seal [8]. Newer approaches show the use of neural network models using multi-sensor data from two accelerometers used as vibration sensors [9].

One approach to obtain the maximum measurement information directly at the contact point of the bearing washer and rollers is to use strain gauges in thin-film technology, which are produced directly on the bearing washer where the EHD contact is located. They can generate data at previously inaccessible measurement positions [10] due to their small thickness of less than 5 μm [11]. They withstand temperatures up to at least 400 °C [12]—which surpasses the range of conventional polymer film-based sensors [13]—, have high adhesion on steel [12] and can be applied and structured on curved surfaces as well [14].

Different sensor layers have been used in the literature. With a sensor layer based on amorphous diamond-like carbon (DLC) produced in a chemical vapor deposition (CVD) process, high measurement signals with increasing strain were detected. Due to the semiconductor behavior of the DLC layer, a high temperature sensitivity was observed as well [15]. Other investigations show the measurement of the temperature change at the rolling contacts in a two-disk test rig with different machine conditions concerning pressure and friction [16]. Here, a 4.6 μm Al_2O_3 insulation layer and a one-dimensional sensor line of 200 nm chromium was used that revealed a positive resistance change with increasing temperature and a negative resistance change for increasing pressure.

To further extend the application range of component-inherent thin-film sensors to extreme mechanical stresses, knowledge of their behavior under plastic deformation and of their application limits is important. This is due to a possible change in the sensor properties with plastic deformation because of, e.g., geometry effects, crystal defects or crack formation. Only if the property change is known—especially the sensitivity for strain and temperature—is a further use of the sensors with high accuracy possible. Therefore, this article targets the influence of plastic deformation on sensor properties such as resistance, strain and temperature behavior, which is presented after the details of sensor manufacturing. In the end, a sensor-data fusion method is shown to determine the stress condition of a machine component regarding deformation and temperature.

2. Materials and Methods

This section describes the processes for thin-film sensor production on stainless steel substrates in detail using cathode sputtering. As insulation layer, Al_2O_3 was applied. Constantan ($Cu_{54}Ni_{45}Mn_1$) is used for the strain-gauge sensor layer and platinum (Pt) for the temperature sensor. Afterwards, the special tensile specimen geometry of the substrate is shown. Finally, the characterization and test methods are explained concerning elastic and plastic deformation and temperature behavior.

2.1. Sensor Manufacturing

2.1.1. Sensor Design

The sensor design contains three strain gauges (rotated by 45° each, named Sg 0°, Sg 45° and Sg 90°) and a rotationally symmetric temperature sensor (T sensor). The strain gauges build a rosette that can be used in the future for the determination of the two main directions and quantities of mechanical stress [17]. The strain gauges consist of 10 parallel lines with a width of 10 μm connected by meander curves resulting in a total length of 5 mm. The temperature sensor has a total length of 6.54 mm and a width of 13 μm. All sensors cover small areas of approximately 0.18 mm^2 to enable local measurements for the application in the future.

The complete sensor system is shown in Figure 1.

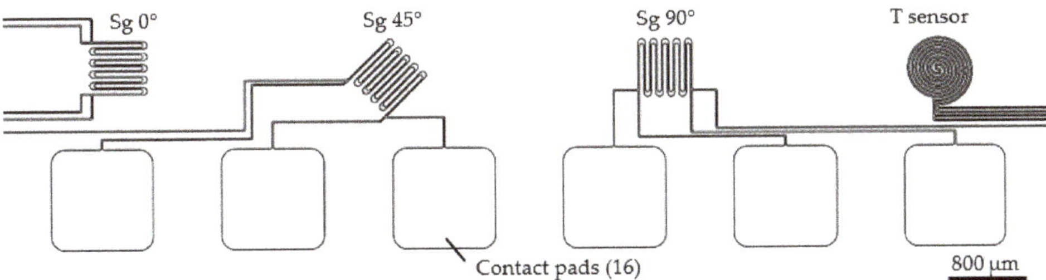

Figure 1. Overview of the sensor layout on the stainless steel tensile specimen, consisting of three constantan strain gauges in three different alignments (Sg 90°, Sg 45°, Sg 0°) and one symmetric platinum temperature sensor (T sensor).

2.1.2. Sputter Deposition

For plastic and elastic investigations, an austenitic stainless steel disc (1.4301 (X5CrNi18-10), diameter: 100 mm, thickness: 0.8 mm) was used as a substrate for the deposition of component-integrated strain gauges and temperature sensors. The arithmetic mean roughness value R_a was 9 nm and the mean roughness depth R_z was 86 nm, measured with a tactile roughness measurement device. At the beginning of the deposition of the insulation layer in a *SenVac* Z550 sputtering system, the disc was subjected to a sputter etching process in order to improve the adhesion strength of the following insulation layer. The etching was performed with a power of 200 W and a DC bias voltage of 108 V for a period of 5 min with a base pressure of $1.1 \cdot 10^{-4}$ mbar and a sputtering pressure of $3.1 \cdot 10^{-3}$ mbar in a pure argon atmosphere. Afterwards, the insulation layer was directly applied to prevent contamination of the surface. The insulation layer, formed out of Al_2O_3, was applied with a power of 400 W (1.88 W/cm^2) at a base pressure below $2.3 \cdot 10^{-5}$ mbar and a process pressure of $3.1 \cdot 10^{-3}$ mbar in a pure argon atmosphere. With a deposition rate of 8.3 nm/min, a 2 μm-thick insulation layer was deposited. The sensors were then processed by microtechnological methods. For deposition and structuring of the strain gauges and the temperature sensor, a lift-off process was used. After the spin coating of the resist AZ® 5214 E, a softbake, the exposure and the development, the constantan strain gauges were sputtered with the *SenVac* Z550 system. A 300 nm-thick layer was produced with a power of 200 W and a sputtering pressure of $4.2 \cdot 10^{-3}$ mbar in a pure argon atmosphere, resulting in a deposition rate of 16.2 nm/min over a period of 18.5 min. After the final lift-off, the platinum temperature sensor was manufactured in the same lift-off process. This time, a *Kenotec* MRC sputtering system was used for platinum sputter deposition in a pure argon atmosphere with a base pressure of $2.7 \cdot 10^{-7}$ mbar, a sputtering pressure of $9.1 \cdot 10^{-3}$ mbar and a power of 200 W (0.94 W/cm^2). With a deposition rate of 14.4 nm/min, a 260 nm-thick layer was deposited and the final lift-off took place.

2.1.3. Sample Preparation

For the sensor evaluation, tensile specimens were milled from the disc. The layout was created based on the standard DIN 50125 [18] with a total length of 68 mm, a width of 3.6 mm and a thickness of 0.8 mm. The test length was 22 mm. The width was chosen to enable high strains in the tensile specimen with a force of up to 2500 N, which is the maximum load of the tensile testing machine used. As a special feature, a steel bar was left at one edge where a circuit board was attached using high-temperature silicone. A conductive two-component silver adhesive was used for the electrical contacting of the sensor's contact pads to the soldering pads of the circuit board with thin wires. Due to the possibility of high plastic deformation of the specimen, they were used in an appropriate loop. A final tensile specimen is shown in Figure 2. Two such samples were manufactured and characterized.

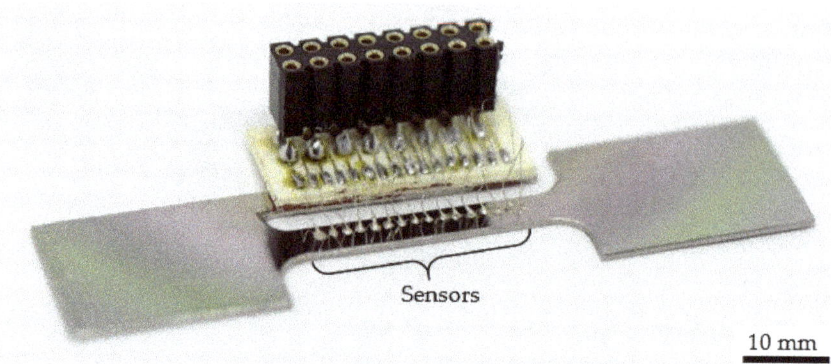

Figure 2. Overview of the sensor layout consisting of three constantan strain gauges in three different alignments (Sg 90°, Sg 45°, Sg 0°) and one symmetric platinum temperature sensor.

2.2. Measurement Details

The basic characteristics for resistance-based thin-film sensors are the initial resistance R_0 at room temperature and the insulation resistance R_{ins}. Additional important values for strain gauges are the k-factor (k), which describes the strain sensitivity, and the temperature coefficient of resistance (TCR), which qualifies the resistance change due to a change in temperature, given by Equations (1) and (2). Here, ΔR represents the resistance change in the initial resistance value R_0 caused by either a variation in strain $\Delta\varepsilon$ or in temperature ΔT. These four characteristics (R_0, R_{ins}, k-factor and TCR) will be examined for their dependence on plastic deformation.

$$k = (\Delta R/R_0)/\Delta\varepsilon \tag{1}$$

$$TCR = (\Delta R/R_0)/\Delta T \tag{2}$$

The resistance measurements were performed using the multimeter *Keithley* DAQ6510 in four-wire technology. For the insulation resistance, a tera-ohm meter *Fischer* TO3 was used with a measurement voltage of 10 V. The two electrodes were connected to one contact pad of each sensor and to the metal substrate.

The strains in the elastic range for the evaluation of the k-factor and in the plastic range for the irreversible plastic deformation were initiated with the tensile testing machine *Mecmesin* MultiTest 2.5-xt. For the elastic strain, a preload F_{pre} of 100 N and a maximum load F_{max} of 350 N were chosen. This results in strain values of 172 µm/m and 602 µm/m, which finally led to an elastic strain difference $\Delta\varepsilon$ of about 430 µm/m for the characterization of the k-factor. The strain was calculated according to Equations (3), (4) and finally Equation (5) [12]. With the strain ε, the mechanical stress σ, the Young's modulus E, and the cross-sectional area A—which is the product of the sample width w and the thickness t—, the strain difference $\Delta\varepsilon$ is calculated. For the plastic deformation, values of 0.01%, 0.19%, 0.54%, 0.92% and 1.30% were achieved with force values of 616 N, 800 N, 900 N, 950 N and 1250 N.

$$\varepsilon = \sigma/E \tag{3}$$

$$\sigma = F/A \tag{4}$$

$$\Delta\varepsilon = (F_{max} - F_{pre})/(E\cdot w\cdot t) \tag{5}$$

For determination of the TCR, a hot plate was used in the temperature range from 30 °C to 100 °C. Here, only the cooling curves were used for the evaluation since the samples were removed from the hot plate to accelerate the cooling process which results in a homogenous temperature decrease. The temperature measurement took place with a commercial temperature sensor (Pt100, type PTFM101B1A0) from *TE Connectivity*. This was attached to the tensile specimen with high-temperature silicone from *Pattex* to enable live measurements of the temperature.

The characterization procedure started with the measurement of the initial resistance and the insulation resistivity, followed by the k-factor and the *TCR*. A plastic deformation in steps of approximately 0.35% beginning with the yield strength (plastic deformation of 0.2%) was then carried out until the irreversible destruction of the sensors. The resulting plastic deformation was determined using a reflected light microscope from *Nikon* with a lens with a magnification of 10. The change in distance between the left contact pad of the strain gauge Sg 0° to the right contact pad of the temperature sensor was measured. Since these structures exhibit the longest distance for measurement, the accuracy increases when normalized on the initial length, which was 18.5 mm. It could be measured with a standard variation of ±1 μm. A plastic deformation of, for example, 1% (185 μm), would lead to a percentage error of ±0.54%, which is negligible. For the optical analysis of the sensor failure, lenses with a magnification of 10 and 100 were used.

For all measurement parameters, at least three values were taken. The following figures include the resulting standard deviations, which are basically not visible due to their small values.

The surface roughness values were measured with the tactile measurement device HOMMEL-ETAMIC W5 from *Jenoptik*. Here, the length of the measurement was set to 4.8 mm and the velocity of the diamond tip was 0.5 mm/s. For the analysis, a filter according to ISO 11562 [19] was used.

3. Results

With the fabricated sensors, the influence of plastic deformation on the insulation resistance, initial resistance, the k-factor and the temperature coefficient of resistance (TCR) can be determined. Two samples were characterized. Since the thin-film manufacturing happened with the same deposition processes, similar behavior of both tensile specimens was measured resulting in maximum standard deviations of 5.6%.

3.1. Influence of Plastic Deformation on the Insulation Resistivity

The insulation resistance and therefore its resistivity of the Al_2O_3 layer stays constant, as Figure 3 proves. The resistivity—calculated according to Equation (6)—does not change significantly up to a plastic deformation of at least 0.9% with a mean value of $2.2{\cdot}10^{14} \pm 1.0{\cdot}10^{14}$ Ωcm, which is comparable to the literature [20,21]. The mean resistance value was 1.9 ± 0.8 TΩ. For simplification, only the values of one constantan strain gauge and the platinum temperature sensor are shown.

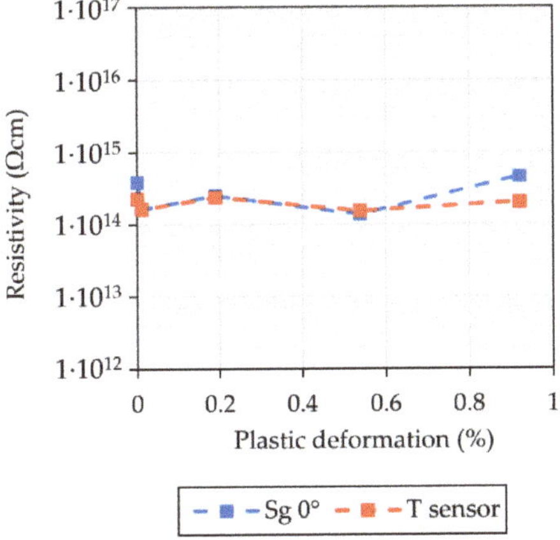

Figure 3. Influence of the plastic deformation on the insulation resistivity.

3.2. Influence of Plastic Deformation on the Initial Resistance

The initial resistance values for the three strain-gauge sensors and the temperature sensor were first measured at room temperature without applying any elastic or plastic strain. The result and a comparison with values from the literature is given in Table 1.

Table 1. Resistivity values of the manufactured thin-film sensors and comparison with bulk and thin-film values from the literature.

Material	T Sensor [10^{-4} Ωcm]	Sg 0° [10^{-4} Ωcm]	Sg 45° [10^{-4} Ωcm]	Sg 90° [10^{-4} Ωcm]	Average [10^{-4} Ωcm]	Std. dev. [10^{-4} Ωcm]	Bulk [10^{-4} Ωcm]	Thin-Film [10^{-4} Ωcm]
Constantan	-	1.34	1.36	1.33	1.34	0.013	0.49 [22]	1.10 [11]
Platinum	0.30	-	-	-	-	-	0.11 [22]	0.18 [23]

The initial resistance of the individual sensors changed with increasing plastic deformation. Different behaviors could be observed for the absolute values, as Figure 4a reveals. For comparing without the influence of the different initial resistance values before the plastic deformation, Figure 4b shows the normalized resistance change in the individual sensors. Regarding the three strain gauges, the strain gauge aligned in the direction of elongation (Sg 0°) experienced the highest resistance change of 0.85% after a plastic deformation of approximately 0.5%. An increase of 0.4% can be observed for the strain gauge aligned at 45° (Sg 45°). In addition, it can be seen that the strain gauge arranged perpendicularly to the direction of elongation (Sg 90°) has a negative change in resistance of up to −0.2%. After a plastic deformation of 0.9%, the strain-gauge resistance values could not be measured anymore. Only the temperature sensor still showed reasonable resistance values at 0.9% plastic deformation. After additional plastic deformation up to a value of 1.3%, the platinum temperature sensor was no longer measurable, and thus no longer functional as well. An increase of 1.4‰ per 1‰ plastic deformation was detected.

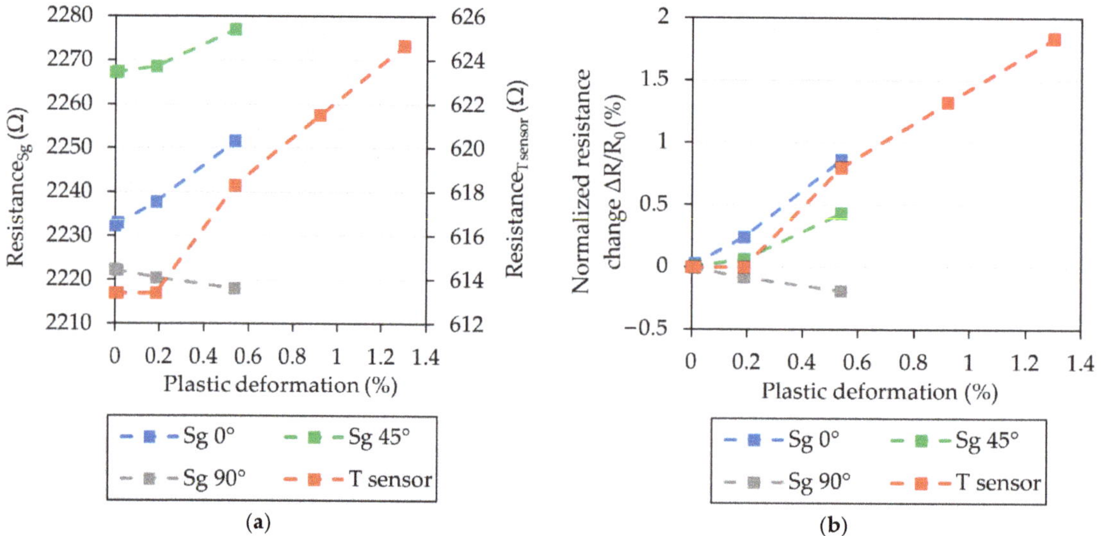

Figure 4. Influence of the plastic deformation on (**a**) the absolute initial resistance values and (**b**) the normalized resistance change. Sg 0° = Strain gauge aligned parallel to the elongation direction, Sg 45° = Strain gauge aligned in an angle of 45°, Sg 90° = Strain gauge aligned perpendicular to the elongation direction, T sensor = Temperature sensor.

3.3. Influence of Plastic Deformation on the Strain Sensitivity (k-Factor)

The most interesting characteristic of a strain gauge is the k-factor k, which describes the strain sensitivity. The k-factor decreases continuously with increasing plastic strain in Figure 5a for all four sensors. The constantan strain gauge (Sg 0°) has a value of 2.15 ± 0.02 in the beginning and a value of 1.96 ± 0.02 after a plastic deformation of 0.5%.

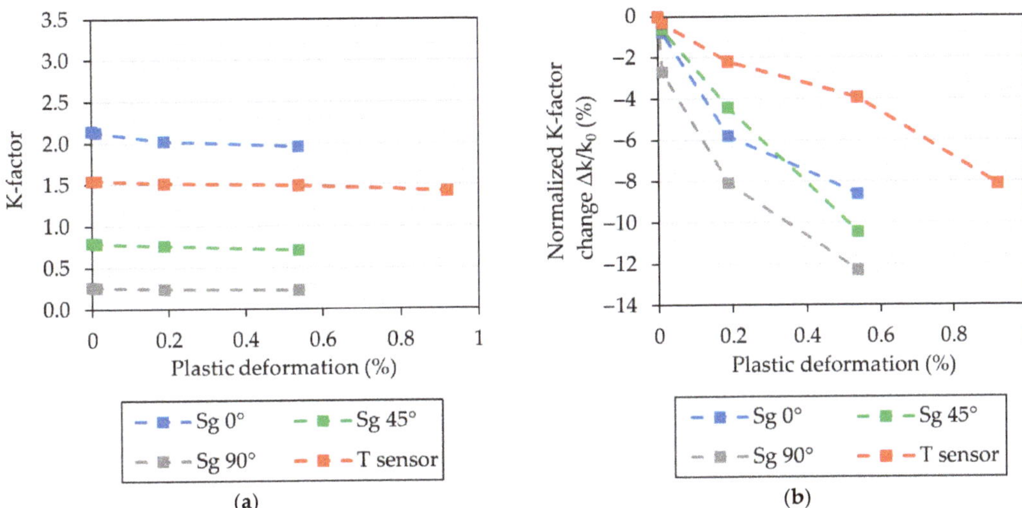

Figure 5. Influence of the plastic deformation on (**a**) the k-factor and (**b**) the normalized k-factor change.

In the studied plastic region zone, the behavior can be considered approximately linear, which leads to a reduction of 17‰ per 1‰ plastic deformation. The strain gauge Sg 45° showed values from 0.79 ± 0.02 to 0.71 ± 0.01 (19‰ reduction per 1‰ plastic deformation). For the strain gauge Sg 90°, an initial k-factor of 0.26 ± 0.02 was measured, which decreased to a value of 0.23 ± 0.01 (22‰ reduction per 1‰ plastic deformation).

The k-factor of the temperature sensor decreased from 1.55 ± 0.02 to 1.42 ± 0.02 after a plastic strain of 0.9%, which is a reduction of 9‰ per 1‰ plastic deformation.

Once again, Figure 5b reveals comparable percentual decreases for the k-factor of the strain gauges and the significantly different behavior of the symmetric temperature sensor.

3.4. Influence of Plastic Deformation on the Temperature Coefficient of Resistance (TCR)

A change in the *TCR* could be observed as well. For each resistive sensor, a linear behavior of the resistance with increasing and decreasing temperature was assumed and observed. This enables a comparison of the single curves. As an example, the resistance change with temperature for strain gauge Sg 0° is shown in Figure 6 at different plastic deformation states. The linear regression showed coefficients of determination (R^2) beginning with 0.9985 before any plastic deformation, decreasing to 0.9933 for maximal plastic deformation.

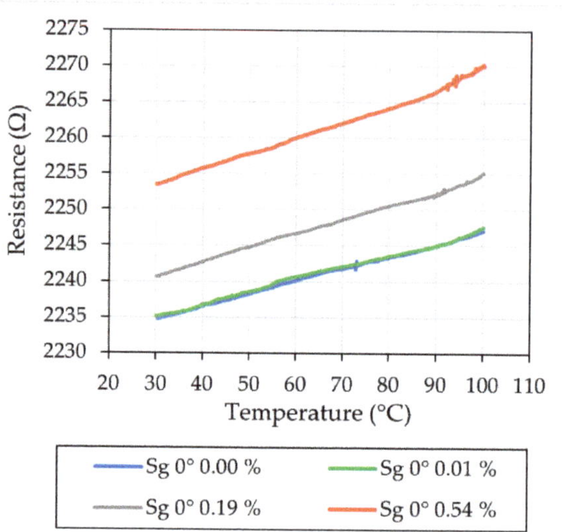

Figure 6. Resistance changes with temperature at different plastic deformation states.

With increasing plastic deformation, the gradient of the resistance change increases, leading to increased *TCR* values for all sensors, shown in Figure 7a. The *TCR* of the constantan strain gauge Sg 0° increases from 89 ± 5 ppm/°C to a value of 119 ± 5 ppm/°C after a plastic strain of 0.5%, which is an increase of 62‰ per 1‰ plastic deformation. The strain gauges Sg 45° and Sg 90° showed values from 93 ± 3 ppm/°C to 112 ± 2 ppm/°C (39‰ per 1‰ plastic deformation) and 95 ± 4 ppm/°C to 105 ± 1 ppm/°C (20‰ per 1‰ plastic deformation).

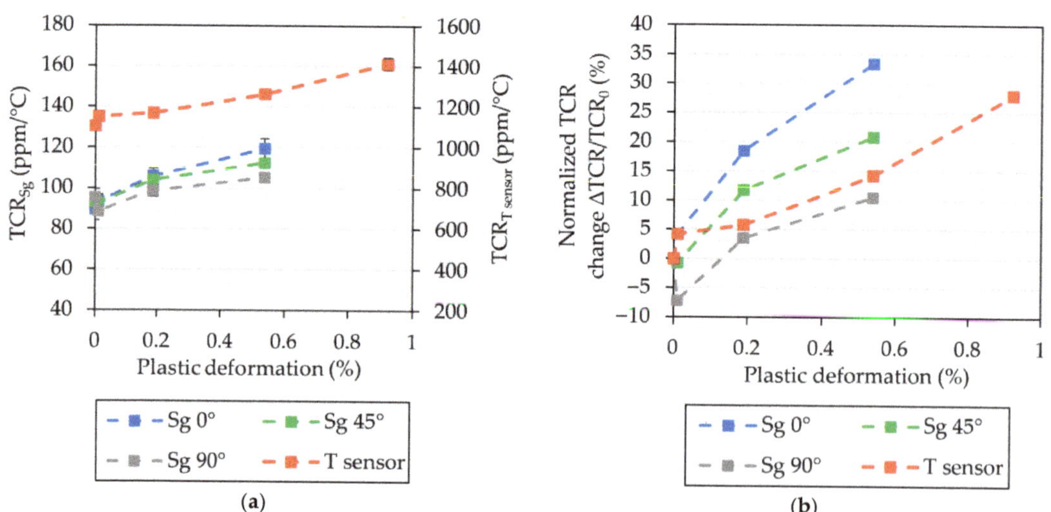

Figure 7. Influence of the plastic deformation on (**a**) the *TCR* and (**b**) the normalized *TCR* change.

The platinum temperature sensor shows an increase of 30‰ per 1‰ plastic deformation with values from 1103 ± 12 ppm/°C to 1,412 ± 32 ppm/°C.

4. Discussion

This section interprets the results based on the detailed description of the influence of plastic deformation on the different thin-film sensors and finally shows a method of sensor-data fusion to determine the stress condition of a machine component regarding deformation and temperature.

4.1. Insulation Resistivity

The comparison of the mean resistivity value of $2.2 \cdot 10^{14}$ Ωcm (mean resistance of 1.9 TΩ) with thin-film values from the literature shows good agreement (e.g., $1.2 \cdot 10^{14}$ Ωcm [10]). As expected, there is no difference between strain gauge and temperature sensor. Since the insulation resistance values stay constant over the investigated plastic deformation region up to at least 0.9%, no significant damage can be measured at first. Nevertheless, after optical microscopy, thin cracks in the Al_2O_3 layer could be observed after a plastic deformation of 0.9%, as shown in Figure 8. As expected, these defects are aligned perpendicularly to the strain direction of the tensile specimen.

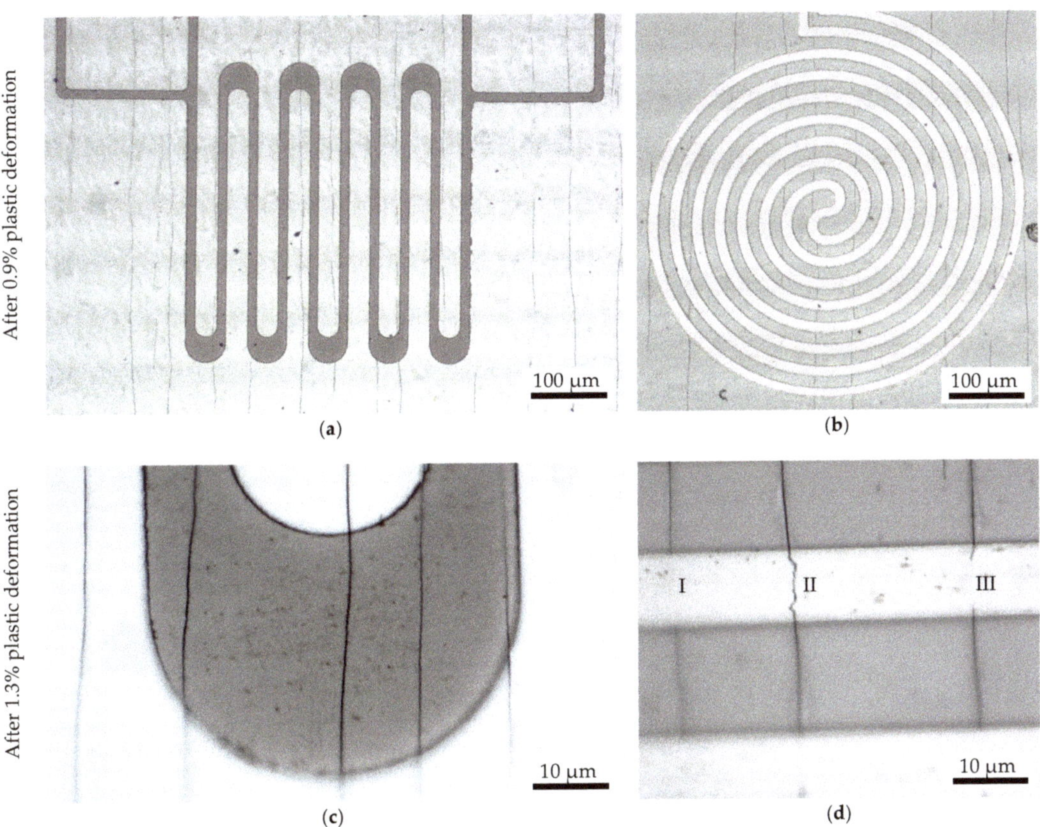

Figure 8. Optical images of (**a**) a constantan strain gauge (Sg 90°) after a plastic deformation of 0.9%, (**b**) a platinum temperature sensor after a plastic deformation of 0.9%, (**c**) a constantan strain gauge (Sg 90°) after a plastic deformation of 1.3%, and (**d**) a platinum temperature sensor after a plastic deformation of 1.3%. Cracks in the Al_2O_3 insulation layer perpendicular to the strain direction are visible in (**a**,**c**,**d**). I shows a location where a crack is not transferred into the platinum layer, II shows a complete crack in platinum layer and III shows the beginning of the formation of a crack in the platinum layer.

4.2. Initial Resistance

The constantan resistivity shows a mean value of $1.34 \cdot 10^{-4}$ Ωcm. A small standard deviation below 1% could be measured for the three differently aligned constantan strain gauges which proves the existence of an isotropic and homogeneous sputter coating. Compared to the bulk value of $0.49 \cdot 10^{-4}$ Ωcm, an increase of factor 2.7 exists. For platinum, the same factor can be calculated even though the sensors were deposited with two different coating systems. One of the explanations for the increased resistivity values compared to values of bulk material can be the influence of scattering at grain boundaries and at interfaces with other materials, which play a major role for the resistivity value [24]. In the literature, a general decrease in the resistivity with increasing layer thickness can be found for different materials [25,26], which supports the explanation. Another aspect is the impact of surface roughness, which would influence both sensor types in the same way. A tactile roughness measurement resulted in an arithmetic mean roughness value R_a of 9 nm and a mean roughness depth R_z of 86 nm. The detailed influence of different roughness values on the thin-film resistivity has to be investigated in the future. A thin-film resistivity value for constantan from the literature is $1.10 \cdot 10^{-4}$ Ωcm [11], which is in the region of the measured value but still shows a deviation of 18%. Slightly different alloy compositions could be the reason here.

The normalized resistance changes show values of 1.7‰, 0.8‰ and −0.4‰ per 1‰ plastic deformation for the strain gauges Sg 0°, Sg 45° and Sg 90°. This is because the geometry of the strain gauges changes with increasing plastic deformation, which results in a change in resistance, according to Equation (6), where ρ, l and A represent the resistivity, the length and the cross-sectional area.

$$R = \rho \cdot l/A \tag{6}$$

The highest change could be observed for the strain gauge aligned in strain direction (Sg 0°) because the length of the conductive tracks increases the most due to their alignment. The strain gauge aligned perpendicularly to the strain direction (Sg 90°) reveals a resistance decrease because the plastic deformation increases the width—and therefore the cross-sectional area A—and decreases the length l of the sensor's conductive tracks (Equation (6)).

Since the insulation resistance was still present up to at least 0.9% plastic deformation (compare Figure 3), it can be assumed that cracks in the constantan layer are the reason for this behavior, which occurs at a plastic deformation between 0.5% and 0.9%. Optical images confirm this assumption, as Figure 8c proves. In contrast, the resistance of the platinum temperature sensor increases up to 1.8% at 1.3% plastic deformation when the failure takes place. The higher resilience of platinum against crack building can be seen in Figure 8d. This image shows a platinum conductor track after 1.3% plastic deformation. Several things have become obvious: First, on the left side, a crack of the Al_2O_3 layer is present which did not exist in the platinum layer (I). On the right side, a crack of the Al_2O_3 layer leads to the beginning of the formation of a crack in the platinum layer (III). Finally, in the middle, a complete crack has built that is responsible for the failure of the sensor (II). Due to its metal properties, platinum is more ductile than the alloy constantan and can withstand higher strains, which would correspond to the literature: The elongation at break is 25% [27] for constantan and 35% [28] for platinum. Since these values are for bulk material, further investigations have to be performed. Nevertheless, the results give important information about the maximum permitted plastic deformation. With this knowledge, on the one hand, the sensors can be applied at specific positions on machine components based on, for example, mechanical FEM simulations, where strain values appear that are below the critical plastic deformation. On the other hand, it can be assumed that a plastic deformation above 0.9% (1.3%) occurs if a failure of constantan (platinum) sensors is detected.

4.3. K-Factor

The k-factor value 2.15 for the constantan ($Cu_{54}Ni_{45}Mn_1$) strain gauge is in agreement with values from the literature for bulk material (2.15 for $Cu_{60}Ni_{40}$ and 2.0 for $Cu_{56}Ni_{44}$ [13]) and thin-film sensors (2.07 for $Cu_{54}Ni_{45}Mn_1$ [11]). The k-factor for the strain gauge perpendicular to the elongation direction shows a value of 0.26. According to Equation (7) [13], this corresponds to a cross-sensitivity of 12%. Polymer-foil based sensors are optimized for a low cross-sensitivity. Manufacturer specify values between 0.2 and 0.6% [29]. The comparatively high cross-sensitivity of 12% has to be reduced mainly through the adaption of the sensor layout. In any case, it has to be considered for further measurements on machine components. As expected, the k-factor of the strain gauge Sg 45° lies between the others.

$$q = k_{0^\circ} / k_{90^\circ} \tag{7}$$

For platinum, a k-factor of 1.55 was the result. It is remarkable that the resulting k-factor could be reduced with the developed symmetric sensor design by at least factor 2 compared to values from the literature of 3.8 [24]. This is an advantage concerning its usage as a temperature sensor with low strain impact.

The plastic deformation has a significant impact on the k-factor. The nearly linear decrease amounts to 17‰, 19‰ and 22‰ per 1‰ plastic deformation for the strain gauges Sg 0°, Sg 45° and Sg 90°. The k-factor decrease in the platinum temperature is 9‰ per 1‰ plastic deformation. These values show that strain measurements are still possible after critical loads of machine components with plastic deformations below 0.5%, even though the strain sensitivity will be reduced. Since the normalized resistance change is under 1.5% for all sensors, this behavior cannot be the reason for the maximum absolute k-factor decrease of 12% (compare Equation (1)). In general, mechanical strain influences the charge carrier transport mechanism. In this case, it is most likely that the plastic deformation creates different types of crystal defects in the material, which reduce the charge carrier mobility. It can be assumed that these defects influence the grain boundaries, whose condition plays a major role for the electrical resistance [24]. Further investigations have to be made to understand this behavior in detail.

4.4. Temperature Coefficient of Resistance (TCR)

The *TCR* describes the temperature sensitivity of the sensors. As the temperature influences the strain signal, low values close to zero are desired. For temperature sensors, the value should be high and the strain sensitivity low. That is the reason why constantan and platinum are chosen for the strain and temperature sensors.

Since the *TCR* should not depend on the alignment of the strain gauges, the results show an expected behavior of constant *TCR* values of the three strain gauges aligned in three different directions with a mean value of 92 ± 3 ppm/°C. It is in good agreement with the alignment-independent resistivity, as explained before. The literature for constantan thin films shows values close to zero as well (−52 ppm/°C [11], +75 ppm/°C [14]) even though bulk values seem to be even lower (±10 ppm/°C [30]). Differences can be found due to different substrate conditions or slightly different alloy compositions.

For the normalized *TCR* change with plastic deformation, differences were detected. This is because of the different alignments, since the plastic deformation only works in one direction, so the isotropic *TCR* behavior interferes with the anisotropic strain impact. This is the conclusion of the decreasing impact of the plastic strain on the *TCR* change with changing alignment of the strain gauge towards the perpendicular direction. The change decreased from 62‰ (Sg 0°) over 39‰ (Sg 45°) to 20‰ (Sg 90°) per 1‰ plastic deformation.

As expected, the *TCR* of the platinum temperature sensor is significantly higher than the value of the constantan strain gauges. Nevertheless, compared to bulk values (3850 ppm/°C [31]) and to other thin-film platinum sensors (1937 ppm/°C [10], 2600 ppm/°C [24]), the *TCR* is low with a value of 1103 ppm/°C. Reasons for this behavior might be the low thickness of 200 nm compared to 1 μm [24] for the value of 2600 ppm/°C so that the influence of scattering at defects, grain boundaries and interfaces increases [24].

As for the k-factor, the increase in the initial resistance of up to approximately 1.5% cannot be the main reason for the absolute *TCR* increase of up to 33% (compare Equation (2)).

4.5. Condition Monitoring

In summary, the plastic deformation results in a degradation of the thin-film sensors. These means a negative influence on the characteristic properties of strain gauges and a positive influence on the temperature sensor. Nevertheless, the knowledge about the behavior of the sensors properties enables the further use of the sensors after crucial machine loads, resulting in plastic deformations up to at least 0.5% without the need for generating a new characteristic curve of the sensor. With the combination of all this information, it is possible to determine the elastic strain ε_{el}, the plastic strain ε_{pl} and the temperature difference ΔT. In a real application on a machine component, the strains can occur in x-, y- and z-directions. To simplify the procedure, Equation (8) shows the dependency of a resistive sensor in only one direction.

$$R\ (\Delta T,\varepsilon_{el},\varepsilon_{pl}) = R_0 \cdot [1 + a_R \cdot \varepsilon_{pl} + TCR_0 \cdot \Delta T \cdot (1 + a_{TCR} \cdot \varepsilon_{pl}) + k_0 \cdot \varepsilon_{el} \cdot (1 + a_k \cdot \varepsilon_{pl})] \quad (8)$$

Since it is only the information of one sensor, no final statement can be made for the origin of the resistance value. Without the elastic strain, Equation (8) simplifies and for each sensor, Equation (9) results, showing the dependency of the measurable resistance value R and the unknown plastic deformation ε_{pl} on the temperature change ΔT.

$$\Delta T\ (R,\varepsilon_{pl}) = (R/R_0 - 1 - a_R \cdot \varepsilon_{pl})/(TCR_0 \cdot (1 + a_{TCR} \cdot \varepsilon_{pl})) \quad (9)$$

With its application on all sensors, the dependency of the temperature change can be illustrated in the dependency of the plastic deformation when a resistance value R is measured for each sensor. Due to the fact that this resistance value results out of both a possible temperature change and a possible plastic deformation, the values of these two physical values are not clear when measuring only one sensor. Sensor-data fusion has to be applied, as shown in Figure 9b as an example for one stress condition. Here, a temperature change of 20 °C and a plastic deformation of 0.5% are present, clearly displayed by the intersection of the single curves. Since there are only two unknown values, two sensors would have been enough for a clear assignment.

With this procedure, it is possible to determine the unknown values for the temperature change and the plastic deformation if the parameters of Equations (8) and (9) are known. Additionally, the deformation only takes place in one direction. Here, the direction parallel to strain gauge Sg 0° was chosen. If the direction of the plastic deformation on a real machine component is completely unknown, then seven unknown values appear in total: ΔT, ε_{pl_x}, ε_{pl_y}, ε_{pl_z}, ε_{el_x}, ε_{el_y}, ε_{el_z}. For there to be clear determination, all in all seven sensors would be needed. The solution of the emerging nonlinear system of equations would need the use of numerical algorithms such as Newton's method [32], as mentioned above. In this way, and with the sensors applied on a real machine component, a measurement of the unknown values would be possible without interrupting the operation of the machine. For example, the thin-film sensors could be used on the racetracks of roller bearings by coating one of the bearing washers. In this use case, high loads can occur and a plastic deformation of the bearing can negatively influence the machine component's behavior. Here, the sensor application is possible due to the advantage of the component-inherent sensor thickness, which can be below 5 µm [11], even when the sensor-layer system contains a necessary additional protection layer.

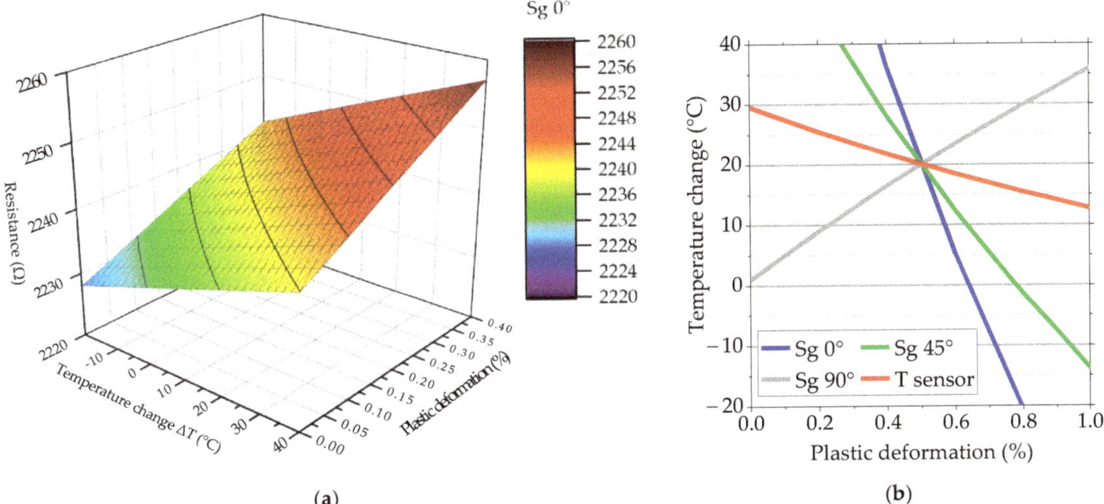

Figure 9. (**a**) Exemplary resistance plot for strain gauge Sg 0° in dependency of the temperature change and the plastic deformation according to Equation (8). Any elastic strain is set to zero. (**b**) Correlation of the temperature change with the plastic deformation for exemplary visualization with measured resistance values in a stress case with a temperature change of 20 °C and a plastic deformation of 0.5% (see Equation (9)).

All in all, these investigations extend the state of the art concerning thin-film sensors with regard to the influence of plastic deformation. Through sensor-data fusion and the solution of linear (nonlinear) systems of equations, a statement can be made about both the plastic deformation and temperature change (and elastic deformation) at the same time.

5. Conclusions

This article shows first-time results of degeneration effects concerning the influence of plastic deformation on the properties of thin-film constantan strain gauges and platinum temperature sensors. Due to changing sensor properties, such as initial resistance, k-factor and temperature coefficient of resistance (*TCR*), the knowledge of their behavior is mandatory to enable elastic strain measurements after plastic deformation, which can occur in machine components due to critical loads. For the initial resistance, an increase in the range of 1% was detected, depending on the alignment of the strain gauges. The resistivity of the Al_2O_3 insulation layer showed a constant value of $2.2 \cdot 10^{14}$ Ωcm up to a plastic deformation of at least 0.9%. This results in a mean insulation resistance of 1.9 TΩ which is sufficient for the application. For constantan, a nearly linear k-factor decrease of 17‰ (9‰ for platinum) and an increase in the temperature coefficient of resistance of 62‰ (30‰ for platinum) per 1‰ plastic deformation could be observed until a sensor failure above 0.5% plastic deformation (about 1.3% for platinum). Optical analysis revealed that micro cracks are the reason. This new fundamental knowledge offers the potential to operate the sensors outside of the elastic deformation conditions and to draw conclusions about plastic processes in the work piece. In fact, intelligent sensor-data fusion enables the clear interpretation of the measured resistance values so that the elastic deformation, the plastic deformation and the temperature change can be determined precisely at the same time. As shown in this article for only one direction of deformation and with the elastic deformation set to zero, a linear system of equations has to be solved.

Based on the results, the position of the sensors can be chosen with respect to the maximum expected plastic deformation based on simulations. Thereby, the conflict of goals can be addressed regarding a measurement signal and an as high as possible resolution on the one hand and a long sensor lifetime at the other hand. Additionally, further elastic strain measurements after a critical load in terms of plastic deformation are enabled.

In the future, the application of the developed sensors will take place on large-diameter bearings coated with a unique sputtering system [33]. Therefore, a resilient protection layer has to be used. Additionally, an algorithm for the solution of the nonlinear system of equations has to be implemented. Based on the measured strains and normal and tangential forces, the slippage and the temperature are to be measured.

Author Contributions: Conceptualization, R.O., F.D. and M.C.W.; methodology, R.O. and T.S.; software, R.O. and T.S.; validation, R.O. and T.S.; formal analysis, R.O. and T.S.; investigation, T.S.; data curation, T.S.; writing—original draft preparation, R.O.; writing—review and editing, R.O., F.D. and M.C.W.; visualization, T.S. and R.O.; supervision, F.D. and M.C.W.; project administration, F.D. and M.C.W.; funding acquisition, M.C.W., F.D. and R.O. All authors have read and agreed to the published version of the manuscript.

Funding: The authors thank the German Research Foundation (DFG) that funded this work within the research project "Integrated sensors for intelligent large-diameter bearings" (WU 558/41-1) as part of the Priority Program 2305 "Sensor-integrating machine elements".

Institutional Review Board Statement: Not applicable.

Informed Consent Statement: Not applicable.

Data Availability Statement: The data presented in this study are available on request from the corresponding author.

Conflicts of Interest: The authors declare no conflict of interest.

References

1. Sediako, D.; Stroh, J.; Kianfar, S. Residual Stress in Automotive Powertrains: Methods and Analyses. *Mater. Sci. Forum* **2021**, *1016*, 1291–1298. [CrossRef]
2. Zhao, D.; Rasool, S.; Forde, M.; Weafer, B.; Archer, E.; McIlhagger, A.; McLaughlin, J. Development of an embedded thin-film strain-gauge-based SHM network into 3D-woven composite structure for wind turbine blades. In Proceedings of the SPIE Smart Structures and Materials + Nondestructive Evaluation and Health Monitoring, Portland, OR, USA, 25–29 March 2017; Volume 10171, pp. 1–9. [CrossRef]
3. Tegtmeier, F.L. Strain gauge based microsensor for stress analysis in building structures. *Measurement* **2008**, *41*, 1144–1151. [CrossRef]
4. Chen, X.F.; Wang, S.B.; Qiao, B.J.; Chen, Q. Basic research on machinery fault diagnostics: Past, present, and future trends. *Front. Mech. Eng.* **2018**, *13*, 264–291. [CrossRef]
5. Gao, R.X.; Holm-Hansen, B.T.; Wang, C. Design of a mechatronic bearing through sensor integration. In Proceedings of the Photonics East, Sensors and Controls for Intelligent Machining, Agile Manufacturing and Mechatronics, Boston, MA, USA, 1–6 November 1998; Volume 3518, pp. 244–250. [CrossRef]
6. Holm-Hansen, R.X. Vibration Analysis of a Sensor-Integrated Ball Bearing. *J. Vib. Acoust.* **2000**, *122*, 384–392. [CrossRef]
7. Marble, S.; Tow, D. Bearing health monitoring and life extension in satellite momentum/reaction wheels. In Proceedings of the 2006 IEEE Aerospace Conference, Big Sky, MT, USA, 4–11 March 2006; pp. 1–7. [CrossRef]
8. Frosini, L.; Harlisca, C.; Szabo, L. Induction Machine Bearing Fault Detection by Means of Statistical Processing of the Stray Flux Measurement. *IEEE Trans. Ind. Electron.* **2014**, *62*, 1846–1854. [CrossRef]
9. Liu, Y.; Yan, X.; Zhang, C.-A.; Liu, W. An Ensemble Convolutional Neural Networks for Bearing Fault Diagnosis Using Multi-Sensor Data. *Sensors* **2019**, *19*, 5300. [CrossRef] [PubMed]
10. Heikebrügge, S.; Ottermann, R.; Breidenstein, B.; Dencker, F.; Wurz, M.C. Residual stresses from incremental hole drilling using directly deposited thin film strain gauges. *Exp. Mech.* **2022**, *62*, 701–713. [CrossRef]
11. Ottermann, R.; Klaas, D.; Dencker, F.; Hoheisel, D.; Rottengatter, P.; Kruspe, T.; Wurz, M.C. Direct Deposition of Thin-Film Strain Gauges with a New Coating System for Elevated Temperatures. In Proceedings of the IEEE Sensors, Rotterdam, The Netherlands, 25–28 October 2020; pp. 1–4. [CrossRef]
12. Klaas, D.; Ottermann, R.; Dencker, F.; Wurz, M.C. Development, Characterisation and High-Temperature Suitability of Thin-Film Strain Gauges Directly Deposited with a New Sputter Coating System. *Sensors* **2020**, *20*, 3294. [CrossRef] [PubMed]
13. Keil, S. *Dehnungsmessstreifen*, 2nd ed.; Springer Verlag GmbH: Wiesbaden, Germany, 2017; ISBN 9783658136123.
14. Ottermann, R.; Klaas, D.; Dencker, F.; Hoheisel, D.; Jung, S.; Wienke, A.; Duesing, J.F.; Koch, J.; Wurz, M.C. Directly Deposited Thin-Film Strain Gauges on Curved Metallic Surfaces. In Proceedings of the IEEE Sensors, Sydney, Australia, 31 October–4 November 2021; pp. 1–4. [CrossRef]

15. Biehl, S.; Luethje, H.; Bandorf, R.; Sick, J.-H. Multifunctional thin film sensors based on amorphous diamond-like carbon for use in tribological applications. *Thin Solid Films* **2006**, *515*, 1171–1175. [CrossRef]
16. Emmrich, S.; Plogmeyer, M.; Bartel, D.; Herrmann, C. Development of a Thin-Film Sensor for In Situ Measurement of the Temperature Rise in Rolling Contacts with Fluid Film and Mixed Lubrication. *Sensors* **2021**, *21*, 6787. [CrossRef] [PubMed]
17. ASTM International. *ASTM E837-20: Standard Test Method for Determining Residual Stresses by the Hole-Drilling Strain-Gage Method*; ASTM: West Conshohocken, PA, USA, 2020. Available online: https://www.astm.org/e0837-20.html (accessed on 20 August 2022).
18. *ICS: 77.040.10*; Deutsches Institut für Normung e.V. Standard DIN 50125: Testing of Metallic Materials-Tensile Test Pieces. Beuth Verlag GmbH: Düsseldorf, Germany, 2004. [CrossRef]
19. *ICS 17.040.40*; Deutsches Institut für Normung e.V. Standard Geometrical product specifications (GPS)-Filtration-Part 21: Linear profile filters: Gaussian filters (ISO 16610-21:2011). Beuth Verlag GmbH: Düsseldorf, Germany, 2013. [CrossRef]
20. Bartzsch, H.; Gloeß, D.; Boecher, B.; Frach, P.; Goedicke, K. Properties of SiO_2 and Al_2O_3 films for electrical insulation applications deposited by reactive pulse magnetron sputtering. *Surf. Coat. Technol.* **2003**, *174–175*, 774–778. [CrossRef]
21. Voigt, M.; Sokolowski, M. Electrical properties of thin rf sputtered aluminum oxide films. *Mater. Sci. Eng. B* **2004**, *109*, 99–103. [CrossRef]
22. Waechter, M. *Tabellenbuch der Chemie*, 1st ed.; Wiley-VCH: Weinheim, Germany, 2012; ISBN 978-3-527-32960-1.
23. Schmid, U. The impact of thermal annealing and adhesion film thickness on the resistivity and the agglomeration behavior of titanium/platinum thin films. *J. Appl. Phys.* **2008**, *103*, 054902. [CrossRef]
24. Fricke, S.; Friedberger, A.; Mueller, G.; Seidel, H.; Schmid, U. Strain gauge factor and TCR of sputter deposited Pt thin film up to 850 °C. In Proceedings of the IEEE Sensors, Lecce, Italy, 26–29 October 2008; pp. 1531–1535. [CrossRef]
25. Gordillo, G.; Mesa, F.; Calderon, C. Electrical and Morphological Properties of Low Resistivity Mo thin Films Prepared by Magnetron Sputtering. *Braz. J. Phys.* **2006**, *36*, 982–985. [CrossRef]
26. Barbalas, D.; Legros, A.; Rimal, G.; Oh, S.; Armitage, N.P. Disorder-enhanced effective masses and deviations from Matthiessen's rule in $PdCoO_2$ thin films. *arXiv* **2022**, arXiv:2205.05006, 1–10. [CrossRef]
27. Isabellenhuette. Isotan Cu55Ni44Mn1 Data Sheet. 2021, pp. 1–4. Available online: https://www.isabellenhuette.de/fileadmin/Daten/Praezisionslegierungen/Datenblaetter_Widerstand/ISOTAN.pdf (accessed on 19 August 2022).
28. AZO Materials. Platinum (Pt)–Properties, Applications–Data Sheet. 2013, pp. 1–3. Available online: https://www.azom.com/article.aspx?ArticleID=9235 (accessed on 19 August 2022).
29. Hottinger Baldwin Messtechnik GmbH. Strain gages–Datasheet–Order No. 1-XY71-3/350, Foil Lot A421/17, Production Batch 812088898. Available online: https://www.hbm.com/en/2122/strain-gauge-datasheets/ (accessed on 20 August 2022).
30. Duemke, A.; Lampe, K.; Machon, W.; Milde, H.; Moussaoui, M.; Scheurmann, M.; Vehreschild, K.; Zantis, F.-P. *Friedrich-Tabellenbuch Elektrotechnik/Elektronik*, 10th ed.; Bildungsverlag EINS: Koeln, Germany, 2007; ISBN 978-3-427-53030-5.
31. *ICS: 17.200.20*; Deutsches Institut für Normung e. V. Standard DIN EN 60751:2009-05: Industrial platinum resistance thermometers and platinum temperature sensors (IEC 60751:2008). Beuth Verlag GmbH: Düsseldorf, Germany, 2009. [CrossRef]
32. Dembo, R.S.; Eisenstat, S.C.; Steihaug, T. Inexact Newton Methods. *SIAM J. Numer. Anal.* **1982**, *19*, 400–408. [CrossRef]
33. Klaas, D.; Becker, J.; Wurz, M.C.; Schlosser, J.; Kunze, M. New coating system for direct-deposition of sensors on components of arbitrary size. In Proceedings of the IEEE Sensors, Orlando, FA, USA, 30 October–3 November 2016; pp. 1–3. [CrossRef]

machines

MDPI

Article

The Effect of Sensor Integration on the Load Carrying Capacity of Gears

Luca Bonaiti [1,2,*], Erich Knoll [2], Michael Otto [2], Carlo Gorla [1] and Karsten Stahl [2]

[1] Department of Mechanical Engineering, Politecnico di Milano, 20133 Milan, Italy
[2] Institute of Machine Elements, Gear Research Centre (FZG), Technical University of Munich (TUM), 85748 Garching, Germany
* Correspondence: luca.bonaiti@polimi.it

Abstract: Classical machine elements have been around for centuries, even millennia. However, the current advancement in Structural Health Monitoring (SHM), together with Condition Monitoring (CM), requires that machine elements should be upgraded from a not-simple object to an intelligent object, able to provide information about its working conditions to its surroundings, especially its health. However, the integration of electronics in a mechanical component may lead to a reduction in its load capacity since the component may need to be modified in order to accommodate them. This paper describes a case study, where, differently from other cases present in the literature, sensor integration has been developed under the gear teeth of an actual case-hardened helical gear pair to be used within an actual gearbox. This article has two different purposes. On the one hand, it aims to investigate the effect that component-level SHM/CM has on the gear load carrying capacity. On the other hand, it also aims to be of inspiration to the reader who wants to undertake the challenges of designing a sensor-integrated gear.

Keywords: gear; gearboxes; sensor integration; sensor integrated gears; gear SHM and CM

Citation: Bonaiti, L.; Knoll, E.; Otto, M.; Gorla, C.; Stahl, K. The Effect of Sensor Integration on the Load Carrying Capacity of Gears. *Machines* **2022**, *10*, 888. https://doi.org/10.3390/machines10100888

Academic Editors: Sven Matthiesen and Thomas Gwosch

Received: 29 August 2022
Accepted: 27 September 2022
Published: 2 October 2022

1. Introduction

Classical machine elements have been around for centuries, even millennia. For instance, mechanical gears can be traced back to the world-famous Antikythera mechanism [1]; one of the very first founders of modern gear theory can be identified in Euler, who in 1781, developed the mathematical formulation of the involute that later became the basis of the modern gear [2]. A similar case is that of bearings, which were studied by Leonardo da Vinci in his friction investigation [3]. However, despite being known and used for centuries, the increasing day-to-day industrial demand for a safe, reliable and lightweight mechanical system requires continuous research as a means to provide industry with a proper solution. Within this context, as a means to increase the system reliability (or at least reduce the occurrence of failure), the idea behind Structural Health Monitoring (SHM) and Condition Monitoring (CM) plays an important role [4], especially for rotating machinery [5].

Focusing on gearboxes, the current state of the art is to monitor the whole system using additional sensor equipment which has to be attached to the outside (e.g., [6–17]) rather than on a single mechanical component, since sensors are not located directly on the component. However, the advancements of SHM and CM are now moving towards the idea of a smart mechanical element that, apart from performing the task it has been designed for, is able to communicate information to the outside world about its working conditions, especially its health.

Within this context, the SPP2305 German priority program has been established in order to evaluate the development of smart machine elements [18] within 10 projects focused on different machine elements. From a mechanical point of view, the basic requirement for all these projects is that all the electronics have to be located inside the component, without requiring additional space. On the other hand, from an electrical point of view, the components have to communicate with the outside world without cabling, implying that each electronic system has to be self-powered and with a wireless data connection. Examples of smart components being investigated and developed within the program are screw, feather key, radial seals and, obviously, gears. Indeed, this paper presents the first results of the SIZA (Sensor-Integrated Gear, in German *Sensorintegrierendes Zahnrad*) project, in which prototypes of smart gears will be developed.

Indeed, current studies are now directing the field of condition monitoring to the mechanical component. Furthermore, it is possible to find applications of measurement systems for the SHM/CM of single machine elements. One of the early examples is the work of Holm-Hansen, B. and Gao, R. X [19–27], where the feasibility of smart bearings with micro-sensors is discussed from the point of view of both the bearing load carrying capacity and SHM/CM. The concepts of a smart bearing have also been discussed recently by Schirra, T. et al. [28,29]. Other examples for sensor integration in mechanical components include feather key [30], fasteners [31,32], ball screw (e.g., [33,34]) and die (seen as part of the overall stamping system) [35].

As a matter of fact, the idea of a smart gear has recently been proven to be a promising and feasible idea by Peters, J. [36–38] and by Sridhar, V. [39]. However, their studies were based on a spur gear geometry in an "open" gearbox (i.e., without housing), and the sensor board was directly attached to the gear body. Therefore, in order to overcome the problem of sensor integration, Binder, M. [40], focusing more on the sensor integration itself, proposed to use additive technologies as a means to embed sensors and electronics inside the gear body. Nevertheless, despite being advantageous for the sensor integration per se, additive-manufactured sensor-integrated gears present some technological difficulties. Firstly, very few data on gear load carrying capacity that can be used for additive-manufactured gear design are available in the literature (e.g., [41–43]). Secondly, the idea of embedding the sensor in the initial manufacturing phase implies that all electronic components must survive the mandatory heat treatment temperatures for high performance gears. As an alternative, the gear would not be heat treated, adopting a material with a low-load capacity with a detrimental effect on the gearbox dimensions.

Within this context, the implementation of SHM/CM within gears made of classical materials (e.g., case hardened steel) seems to be a rational idea. Indeed, those kinds of materials are well known for both their technological and load capacity points of view. Moreover, as the electronics will be fitted only after manufacturing, the aforementioned temperature problem is no longer an issue. However, this idea also implies that the gear body must undergo important modifications in order to accommodate all the electronics, and the effect that such modification will have on the gear load carrying capacity is unknown.

Here, in order to discuss the implementation of in-situ gear SHM/CM and the effect that sensor integration has on gears, an actual gearbox is adapted as a reference case. The aforementioned gearbox, and the related case-hardened helical gears, have already been developed and used in previous studies (e.g., [44–48]). Figure 1 shows the modification that the original gear geometry will undergo in order to accommodate all sensor boards inside the gear, underneath the teeth. Indeed, the gear body will be modified by creating the various ports adopting classical machine tools (e.g., lathe and milling machine). However, as the gear body is strongly changed by the sensor-integration requirements, the stress acting on the component also varies. Therefore, the load carrying capacity of the component is affected as well. Nevertheless, the said variation is not discussed in the literature.

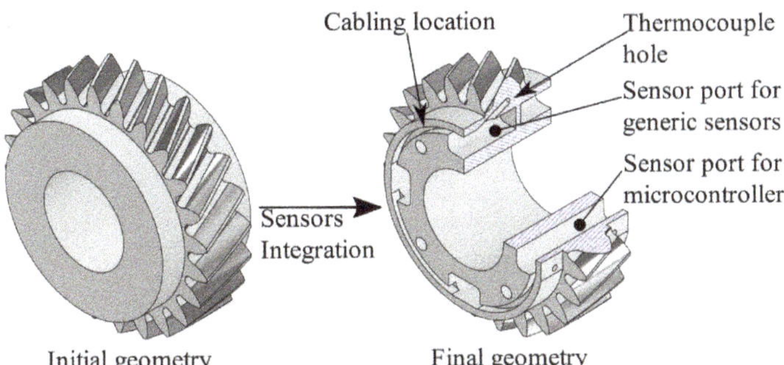

Figure 1. The proposed final gear body geometry (**left**) that has been defined by modifying the original one (**right**).

Therefore, this article has two different purposes. On the one hand, it aims to investigate the effect that component-level SHM/CM will have on the gear load carrying capacity. On the other hand, thanks to the critical discussion about the design choices, as well as the adopted evaluation tools, it also aims to be of inspiration for the reader who wants to undertake the challenges of designing a sensor-integrated gear.

2. Definition of the Components

The gearbox that has been chosen in order to implement the component-level SHM/CM system is part of a back-to-back test rig with a center distance of 112.5 mm, whose typical maximum working conditions are 1000 [Nm] at 3000 [rpm]. The working principle of the 112.5 test rig is completely similar to the one of the more famous 91.5 test rig (shown in Figure 2), which is a part of the ISO 14635 [49]. The test rig is composed of two gearboxes with identical gear ratios; one of which is over-dimensioned (i.e., the slave gearbox) in order to localize the failure on the other one (i.e., the test gearbox). Interested readers are directed to ISO 14635-1 [49], where power-recirculating gearboxes are described. The test gearbox features a 22/24 gear pair, mounted on the shaft using a tapered interference fit. Table 1 lists the gear pair macro geometry data. The gears are made of 18CrNiMo7-6 and have been subject to classical industrial-level case hardening. More detail about the gears can be found in [44–48]).

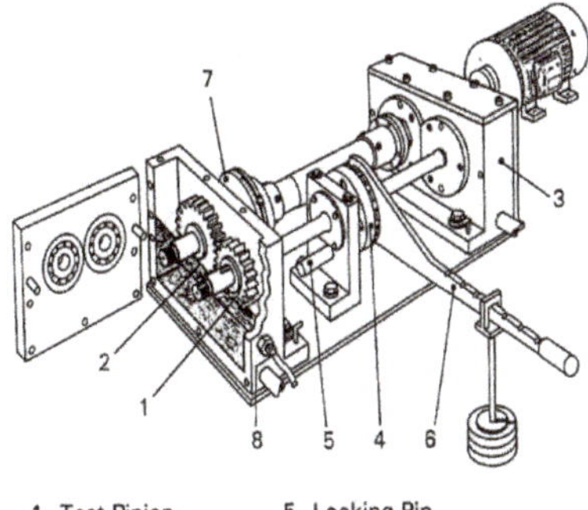

Figure 2. Reference scheme of the 91.5 power-recirculating test rig [50].

Table 1. Main geometrical data and calculated safety factors acc. to ISO 6336 [51–55].

Name	Symbol	Pinion	Wheel	U.M.
Number of teeth	z	22	24	-
Normal modulus	m_n	4.25		Mm
Normal pressure angle	α_n	20		°
Helix angle	β	29		°
Profile shift coefficient	X	0.100	0.077	-
Centre distance	a	112.5		Mm
Tip diameter	d_a	117.70	127.30	Mm
Addendum factor of the basic rack	h^*_{ap}	1.55		-
Tooth root radius factor	ρ^*_{fP}	0.25		-
Facewidth	b	27.6		Mm
Torque	T	1000		Nm
Rotational velocity	n	3000		Rpm
Working hours	H	2000		H
Contact stress	σ_H	1528	1527	MPa
Pitting stress limit	σ_{HG}	1536		MPa
Safety factor, pitting	S_H	1.01	1.01	-
Nominal tooth root stress	σ_F	432	433	MPa
Allowable stress number for bending	σ_{FE}	950		MPa
Safety factor, tooth root bending fatigue	S_F	2.10	2.09	-

Table 1 reports the safety factor for tooth root bending fatigue and macro pitting calculated according to the ISO 6336 framework (i.e., [51–55]). As the adopted geometry was part of a research project regarding contact fatigue, the gear geometry has been defined in order to have a high level of tooth root and flank safety, while respecting the test rig constraints (e.g., the center distance of 112.5 mm).

Moving into the electronic aspect, the implemented SHM/CM system features several sensors that have been defined based on the current state of the art, as well as on the author's experience. Firstly, two accelerometers are adopted as a means to detect gear torsional vibration. Indeed, this peculiar system has been proven to be a valid methodology for evaluating gear dynamic behavior [56,57]. Secondly, a thermocouple is included as tooth temperature (and its sudden variation) can be related to tribological damages (e.g., [58,59]); in order to enhance this aspect, the sensor is placed as close as possible to the contact area. Finally, a Hall-effect sensor is included as a means to obtain information about the rotational speed as well as a microphone, since it can be a cost-effective alternative to an accelerometer. The sensor system is managed by a microcontroller which elaborates sensor data and then provides information about the health status of the component. During the project, the connection of the boards with the outside world will evolve gradually from a cabled connection (via slip ring) to a wireless connection. A similar approach will be followed for the power source of the sensor board, moving from an external board to an integrated board (batteries using an energy-harvesting technique). The selected sensors and microcontroller are deliberately standard-grade electronic components. All aspects of the measurement technique, SHM/CM software, data and energy management is the subject of current research and future publications.

3. Definition of the Gear Body Shape

The development of a smart machine element poses the problem of the conflict of interest between the load carrying capacity of the components and the SHM/CM system reliability. The aim is to have the best possible SHM/CM system without significantly affecting component strength.

In the case of the gears, from the SHM/CM point of view, one aim should be to fit as many sensors as possible. Then, because of the shorter signal path and the related lower signal-to-noise ratio, the sensors should be placed as close as possible to the gear meshing. However, this would lead to a significantly reduced load carrying capacity. Furthermore, the working conditions would be very demanding for the electronic components due to high temperatures and vibrations associated with that region, posing the problem of electrical component durability.

Another challenge when implementing electronics is the board dimensions. Generally, the sensor boards should be as small as possible. Nevertheless, electronics also imply some limitations from a manufacturing standpoint. Indeed, the sensor boards include not only the sensors, but all the necessary internal connections between electronic components. Therefore, their minimum dimensions are limited.

Therefore, the gear body has been modified in order to have the smallest boards, as close as possible to the gear contact, without unreasonably affecting the load carrying capacity.

A first attempt to define a new gear body geometry has been to adopt the annular board, as proposed in [36–39]. Figure 3 shows two geometries which have been designed considering an annular sensor with a 10 mm width. The geometry in Figure 3a also includes 5 mm of radial play in order to leave gaps for the spacer.

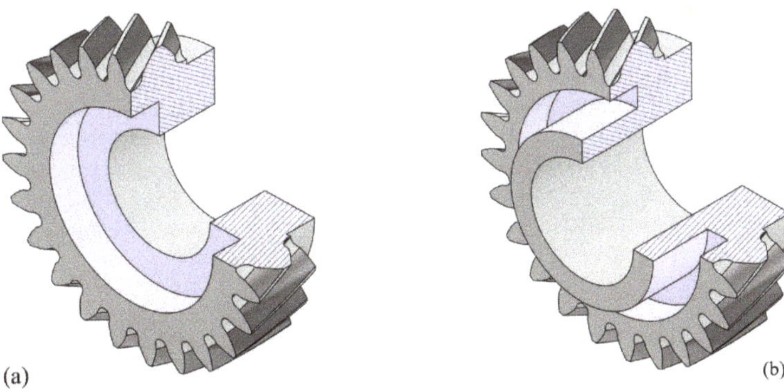

Figure 3. Examples of not-feasible gear body geometries. (**a**,**b**) are two possible geometries with a different position and size of the sensor board.

However, despite being relatively simple, this presents several problems. Firstly, the tapered interference fit is not symmetrical with respect to the gear. If the power flow enters from behind (in respect to Figure 3 view), the problem of fretting may occur at the shaft/hub connection due to the unfavorable power flow [60]. Secondly, half of the teeth are unsupported, with detrimental effects on the tooth root bending fatigue resistance (e.g., [52,61–66]). Finally, this geometry also implies difficulties when transmitting the axial load generated by the helical gear meshing.

Therefore, taking inspiration from the concept of lightweight gears (e.g., [67–69]), an alternative sensor configuration has been studied. This configuration is based on the idea of having several boards placed in different positions and connected by cables, instead of having one single board. The final geometry is shown in Figure 1. The gear body, as well as the related port dimensions, have been defined after several iterations in which both boards and port locations have been defined in order to reduce their effect on the gear carrying capacity. One challenge was the imbalance of the resulting gear body due to the presence of different boards, as well as the presence of some internal walls within the ports. However, such unbalancing, if not prevented, may lead to a higher load acting on the gear than the nominal load. This aspect will be addressed by adopting the industrial-level balancing technique after the gear and boards are manufactured. Figure 1 geometry also features threaded holes which are necessary for the connection with other experimental equipment.

Nevertheless, the load carrying capacity of this new geometry must also be evaluated. Considering this specific case, the new geometry will be assessed by looking at the tapered interference fit and the tooth root bending strength.

4. Evaluation of the Sensor Port Effect on the Interference Fit

Starting from the interference in the contact area, several analytical calculation methods have been proposed as means to evaluate the phenomena occurring within interference fits (e.g., [60,70,71]). Those calculation methods are based on the evaluation of the radial stress generated in the shaft-hub region for a full body gear, that is not the case with the proposed gear body. Furthermore, as observed in experimental campaigns on compound gears [72], variation in the gear body stiffness can lead to a reduction in the contact pressure, leading to the presence of undesired relative movement. As the proposed gear body geometry features important modifications, considering that numerical methods have proven their validity in evaluating the shaft-hub connection with an interference fit specifically (e.g., [70,73,74]), FE models have been adopted as a means to estimate the component behavior. All FE simulations discussed here have been performed using ABAQUS 2020.

For all the evaluated geometries, two cases with different shaft/hub interference have been simulated, aiming to evaluate two different aspects. On the one hand, the models that adopt the minimum interference estimate the contact pressure p acting on the contact area A. This quantity is related to the maximum torque that can be achieved (e.g., [60,70,71]). On the other hand, the models with the maximum interference allow the estimation of the overall stress acting on the component.

Interference fit simulations were carried out adopting quadratic hexahedral, establishing an interference fit between the contact surfaces. Whenever possible, symmetricity has been adopted in order to reduce the computational time, without including the teeth, limiting the external diameter to the tooth root. Figure 4 illustrates the principle behind such simulations, with a particular focus on the mesh. Both the baseline case and the case with the sensor port have been simulated.

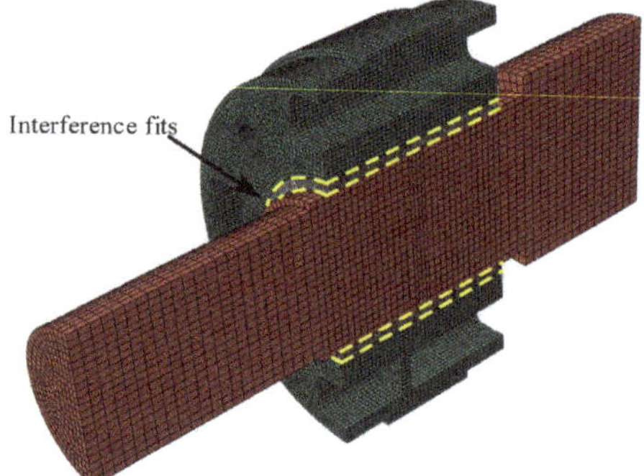

Figure 4. FE simulation procedure for the evaluation of the port effect of the interference fit.

Concerning the specific case, the interference fit must be able to transmit, by means of friction, both the axial load generated by the meshing of helical gears and the torque itself.

In the contact area, they are transferred by means of two different frictional stresses, one in the circumferential (i.e., torque) and one in the axial direction (i.e., the normal force) [71],

$$\begin{cases} \tau_\theta = \frac{T}{\pi d l \frac{d}{2}} \\ \tau_z = \frac{F_a}{\pi d l} = \frac{T}{m_n z/2} \frac{\tan(\alpha_n)}{\cos\beta} \frac{1}{\pi d l} \end{cases} \quad (1)$$

where τ_θ is the circumferential radial shear related to the applied torque T, while τ_z is the axial shear due to the axial force F_a. d and l are the hub diameter (average) and length, respectively. This approach averages the acting load over the total contact surface, neglecting any local effects due to the acting concentrated load in the gear mesh. Nevertheless, considering the general proportionality of friction force and total acting force, it seems reasonable for the intended application. As τ_θ and τ_z are in perpendicular directions, they are summed by means of Pythagoras theorem in order to find the total friction shear stress τ_{tot}.

In order to withstand the applied forces/moments, it is necessary that the interference fit assures a contact pressure p_{min}, which is related to τ_{tot} by means of the friction coefficient f:

$$p_{min} = \frac{\tau_{tot}}{f} \quad (2)$$

In order to be conservative, f equal to 0.1 has been chosen.

Here, as the modified case presents a nonuniform contact pressure, p_{min} will be then compared with $p_{avg,FEM}$, that is the average contact pressure estimated by means of FEM. Therefore, a safety factor for the interference fit S_{IF} is defined accordingly. Table 2 summarizes the results. This implies that the stiffness modification generated by the sensor integration requirements is reducing the safety factor maximum transmissible torque by approximately 16%. Considering that the maximum transmittable torque (not calculated here) is linearly related to the contact pressure in the interference fit region, the maximum transmissible torque in the shaft/hub region is reduced by 16%.

Table 2. Input data and calculated safety factors for the interference fit.

	d	l	T	τ_θ	τ_z	p_{min}	$p_{avg,FEM}$	S_{IF}
	mm	mm	Nm	MPa	MPa	MPa	MPa	-
Baseline	50	55	1000	4.63	1.372	48.3	220	4.56
Modified							186	3.85

Figures 5 and 6 show the von Mises stress distribution for the case with the sensor port. Neglecting the edges, where the high constant stress is due to the fact that chamfers have not been modelled, it is possible to notice that the most stressed area is the port. This peculiar situation can be related to how the component stiffness is distributed. Indeed, the more rigid parts (i.e., the one without the sensor port) try to modify their position, moving away from their undeformed position. This movement is impeded by gear sections with the port that act as a spring, keeping the gear body joined. Therefore, this section results in being highly stressed. As can be seen in Figure 6, the highest stress in that location is around 580 [MPa]. It is worth mentioning that while evaluating the feasibility of the current geometry, this aspect resulted to be the stricter one.

Furthermore, it is possible to notice that the external diameter region corresponding to the tooth root is subject to non-negligible stress. This is also confirmed by the literature, where the effect of the interference fit on the tooth root stress is investigated (e.g., [73–76]).

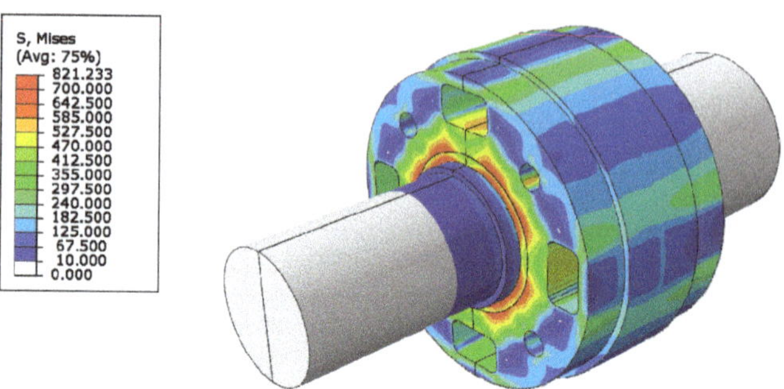

Figure 5. FE results of the interference fit.

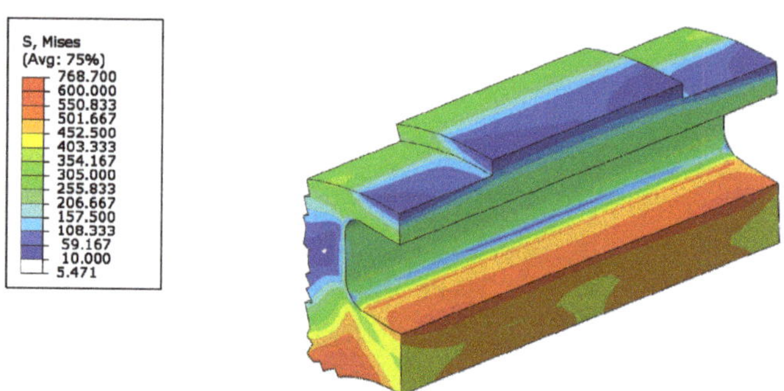

Figure 6. FE results of the interference fit, with a focus on the stress inside the sensor port.

5. Evaluation of the Sensor Port Effect on the Tooth Root Stress

A first attempt to estimate the effect that sensor integration will have on the tooth load carrying capacity can be carried out following the classical gear standards, such as ISO 6336-1 [50] and ISO 6336-3 [52]. On the one hand, by means of the gear blank factor C_R, which modifies the gear stiffness, ISO 6336-1 [50] allows designers to include the effect of the gear body shape within the classical gear assessment methods. On the other hand, the effect of the rim thickness on the tooth root stress is also included in ISO 6336-3 [52] by means of the rim thickness factor Y_B. It is worth mentioning that according to ISO 6336-2 [51], the rim thickness has no direct effect on the contact stress.

However, [50,52] consider a simple gear geometry, in which the whole gear body presents a thin rim. That is, while it seems reasonable to evaluate the possibility of a purely standards-based assessment for a gear body with a shape similar to those shown in Figure 3, the same cannot be applied to the proposed geometry because the result outcome is quite different to the standard. Hence, the necessity of adopting a generalized evaluation method. Following the suggestion of [61–66], numerical models have been implemented.

Here, static meshing gear simulation has been evaluated by following the procedure described in previous papers [77,78], where the simulation is carried out in separate rotational steps. Two separate reference points, located at the center of each gear, are connected to the gear hub by a multi-point constraint. Tooth flank contact is defined by surface-to-surface contact. The pinion is fixed while the wheel is free to rotate. When torque is applied on the wheel, it rotates by a small quantity in order to establish the loaded contact between teeth flanks, and the load is then transmitted from one gear to the other. Figure 7 summarizes the FE procedure. All the FE simulations discussed here have been performed using ABAQUS 2020.

Figure 7. FE simulation procedure for the estimation of tooth root bending stress. The "⏚" symbol rep-resent an encastre while the "✲" one represent a locating-type constraint.

Several rotational steps are then simulated thanks to an external phyton code which rotates each gear on its axis by a rotational step that depends on the wheel/pinion number of teeth and the number of the simulation step. The comparison was been made by looking at the meshing position that results in a higher tooth root stress. Following the indication of Conrado, E. and Davoli, P. [65], the maximum principal component of the stress tensor is adopted as a reference stress.

Here, it is worth mentioning that the FE tooth root stress result and the result estimated according to the ISO 6336 series framework [50–53] cannot be directly compared. Indeed, even though they are both estimating a stress in the tooth root region, the estimated stress values are calculated with different assumptions and methodologies, and cannot be directly compared. The stress number predicted by ISO 6336-3 [52] is the maximum local principal stress, calculated by considering only the tangential force and evaluated at 30° tangents to the contact of the root fillets, while the stress predicted by FEM is a stress tensor, defined all over the model, which considers all the forces exchanged by the teeth.

In order to reduce the computational time, the calculation procedure was based on two models. The first, based on tetrahedral elements, estimated the load sharing between teeth. Then, a second model based on hexahedral elements exploited the sub-modeling technique in order to estimate the actual tooth root stress.

Both the original geometry, the modified geometry (i.e., with the port), as well as the geometry with the hole for the thermocouple were simulated. However, due to the complex geometry, it was not possible to proceed with the sub-modelling for the case with the hole for the thermocouple. Figures 8 and 9 show an example of results for the main model with port and its sub-model, respectively. Figure 10 highlights the contact area estimated for the case in Figure 9. The calculated contact area was in agreement with the general assumption that the contact line ending at the top edge of a helical gear is relevant for maximum tooth root stress levels [79]. The calculated equivalent stresses are summarized in Table 3.

Table 3. Maximum tooth root stress estimated by the FE models.

		σ_I [MPa]	σ_{VM} [MPa]
Original Geometry	Main model	396	337
	Sub model	336	290
Modified Geometry	Main model	407	353
	Sub model	349	302
Modified Geometry with thermocouple hole	Main model	440	380

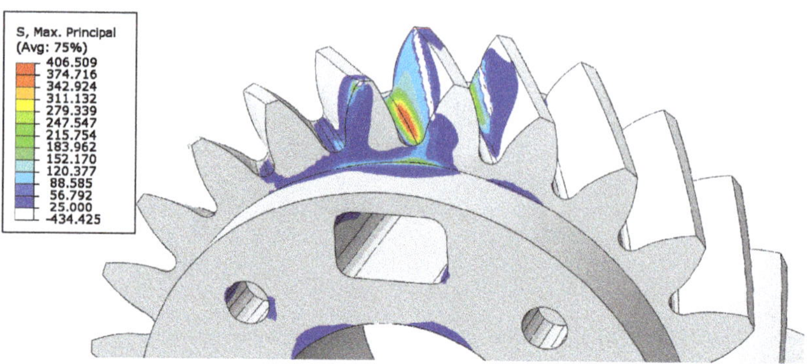

Figure 8. Example of main model FE results showing the maximum principal stress (Modified Geometry).

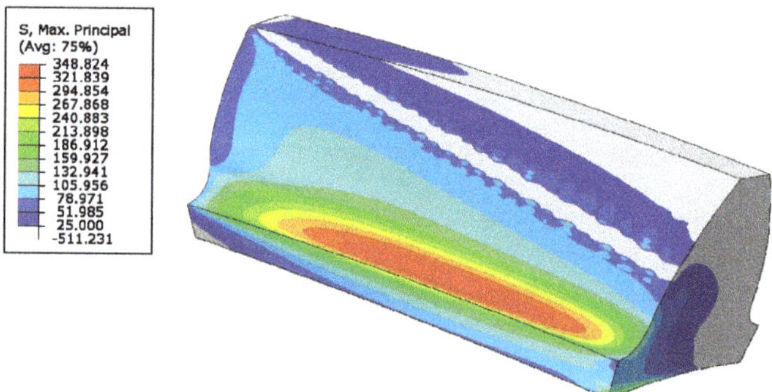

Figure 9. Example of sub-model FE results showing the maximum principal stress (Modified Geometry).

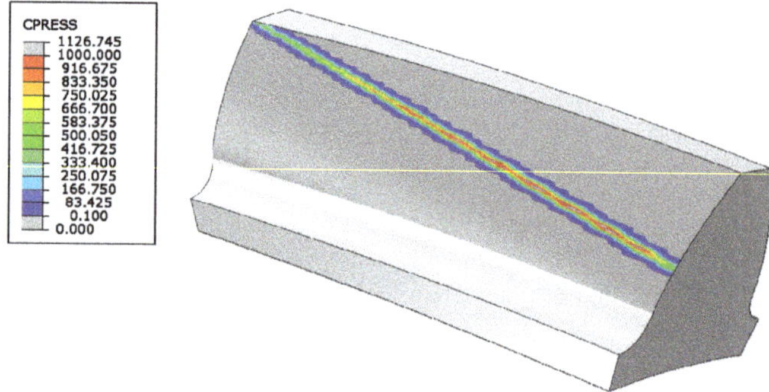

Figure 10. Example of sub-model FE results showing the contact pressure (Modified Geometry).

Nevertheless, in order to properly estimate the effect of sensor integration within the tooth root, it is necessary to consider that the interference fit is loading the root region, as shown in Figure 5. However, while the meshing stress can be schematized as a pulsating stress, the same cannot be done for the interference fit because it is a constant stress. Therefore, a simple arithmetic summation of the stress is not correct from the point of fatigue, but considering the multiaxial nature of the stress, high-cycle multiaxial fatigue criteria were adopted. Among the numerous criteria available in the literature (e.g., [80]), considering the simple load case, the Sines criteria was adopted.

Hence, the stress tensors estimated by both the two FE models were extracted and combined. The equivalent stress was compared for the cases as a means to estimate the effect of the gear body modification. The result of such comparison was the ratio between the various representative stress levels. Both cases with and without the interference fit were evaluated. Table 4 summarizes the results. As can be seen, the effect of the interference fit was predominant. Therefore, its neglection would have led to an underestimation of the sensor port effect. In any case, if the ratios Table 4 were considered as factors multiplying the tooth root stress, thanks to proper gear design, S_F always remains higher than 1.5.

Table 4. Estimated effect of the sensor integration in the tooth root bending stress.

		Without Interference Fit	With Interference Fit
Modified Geometry	Main model	1.05	1.12
	Sub model	1.04	1.12
Modified Geometry with thermocouple hole	Main model	1.13	1.22

6. Results and Conclusions

Within the context of the development of smart mechanical elements, the effect of sensor integration within a gear has been discussed. The focus of this article was an actual case-hardened helical gear pair to be used within an actual gearbox. This article shows that sensor integration in an industrial-grade gear is a feasible idea, however, the designer should expect a reduction in the gear load carrying capacity.

After introducing the proposed gear body modifications and their rationale, a procedure for estimating the related load carrying capacity reductions was presented. Due to the limitations of the classical analytical approach, a FE procedure was adopted. The performed simulations aimed to evaluate the influence that the addition of the sensor ports had on the load carrying capacity of the gear.

Interference fit simulations were performed in order to estimate the effect of the sensor port on the tapered interference fit. Meshing gear simulations were also conducted as a means to evaluate the impact of the sensor port on tooth root stress.

The effect of the geometry modification was estimated by comparing the results of both the modified and the unmodified gear pair, adopting the model with the same parameters (e.g., element type and dimensions). Furthermore, in order to consider the interaction of gear meshing and interference fit, the Sines criteria was adopted as a means to determine the effect of the sensor integration on the tooth root from the point of view of fatigue. The geometry proposed here is indeed the product of all the aforementioned considerations. During its development, thanks to robust gear design (i.e., see Table 1 safety factor), the real bottle neck was found to be in the interference fit rather than in the tooth root bending strength.

As a general result, it holds that the proper design of a sensor-integrated gear needs to trade part of its load carrying capacity for the additional intelligence in the form of space for sensors.

Only static behavior has been considered. Since gear stiffness is modified by the sensor integration and load sharing between the teeth, the Noise Vibration Harshness (NVH) behavior will also subsequently change. This aspect will be a part of future numerical and experimental analysis.

Author Contributions: Conceptualization, L.B. and E.K.; methodology, L.B., E.K. and M.O.; software, L.B.; supervision, M.O., C.G. and K.S.; Writing, L.B. and E.K. All authors have read and agreed to the published version of the manuscript.

Funding: This research was funded by founded by the German Research Foundation (DFG, Deutsche Forschungsgemeinschaft) through the SIZA (Sensor integrated gear, Sensor-Integrierendes Zahnrad) project (grant number 466653706).

Data Availability Statement: Not applicable.

Acknowledgments: The authors would like to express their gratitude to Ermal Fierza and Ralf Brederlow for the development of the electronic boards as well their collaboration in the integration process of the electrical components within the SIZA project.

Conflicts of Interest: The authors declare no conflict of interest.

References

1. De Solla Price, D. Gears from the Greeks. The Antikythera Mechanism: A Calendar Computer from ca. 80 B. C. *Trans. Am. Philos. Soc.* **1974**, *64*, 1. [CrossRef]
2. Litvin, F.L. *Development of Gear Technology and Theory of Gearing*; National Aeronautics and Space Administration, Lewis Research Center: Cleveland, OH, USA, 1997; Volume 1406.
3. Reti, L. Leonardo on Bearings and Gears. *Sci. Am.* **1971**, *224*, 100–111. [CrossRef]
4. Ziegler, L.; Gonzalez, E.; Rubert, T.; Smolka, U.; Melero, J.J. Lifetime Extension of Onshore Wind Turbines: A Review Covering Germany, Spain, Denmark, and the UK. *Renew. Sustain. Energy Rev.* **2018**, *82*, 1261–1271. [CrossRef]
5. Farrar, C.R.; Worden, K. An Introduction to Structural Health Monitoring. *Philos. Trans. R. Soc. A Math. Phys. Eng. Sci.* **2006**, *365*, 303–315. [CrossRef] [PubMed]
6. Sait, A.S.; Sharaf-Eldeen, Y.I. A Review of Gearbox Condition Monitoring Based on Vibration Analysis Techniques Diagnostics and Prognostics. In *Rotating Machinery, Structural Health Monitoring, Shock and Vibration, Volume 5*; Conference Proceedings of the Society for Experimental Mechanics Series; Springer: New York, NY, USA, 2011; Volume 5, pp. 307–324. [CrossRef]
7. Wang, T.; Han, Q.; Chu, F.; Feng, Z. Vibration Based Condition Monitoring and Fault Diagnosis of Wind Turbine Planetary Gearbox: A Review. *Mech. Syst. Signal Process.* **2019**, *126*, 662–685. [CrossRef]
8. Nie, M.; Wang, L. Review of Condition Monitoring and Fault Diagnosis Technologies for Wind Turbine Gearbox. *Procedia CIRP* **2013**, *11*, 287–290. [CrossRef]
9. Salameh, J.P.; Cauet, S.; Etien, E.; Sakout, A.; Rambault, L. Gearbox Condition Monitoring in Wind Turbines: A Review. *Mech. Syst. Signal Process.* **2018**, *111*, 251–264. [CrossRef]
10. Sendlbeck, S.; Fimpel, A.; Siewerin, B.; Otto, M.; Stahl, K. Condition Monitoring of Slow-Speed Gear Wear Using a Transmission Error-Based Approach with Automated Feature Selection. *Int. J. Progn. Health Manag.* **2021**, *12*. [CrossRef]
11. Concli, F.; Pierri, L.; Sbarufatti, C. A Model-Based SHM Strategy for Gears—Development of a Hybrid FEM-Analytical Approach to Investigate the Effects of Surface Fatigue on the Vibrational Spectra of a Back-to-Back Test Rig. *Appl. Sci.* **2021**, *11*, 2026. [CrossRef]
12. Fromberger, M.L.; Kohn, B.; Utakapan, T.; Otto, M.; Stahl, K. Condition Monitoring by Position Encoders. In Proceedings of the INTER-NOISE and NOISE-CON Congress and Conference Proceedings, Hamburg, Germanny, 21–24 August 2016; Institute of Noise Control Engineering: Reston, VA, USA, 2016; Volume 253, pp. 7451–7459.
13. Feng, K.; Borghesani, P.; Smith, W.A.; Randall, R.B.; Chin, Z.Y.; Ren, J.; Peng, Z. Vibration-Based Updating of Wear Prediction for Spur Gears. *Wear* **2019**, *426–427*, 1410–1415. [CrossRef]
14. Sendlbeck, S.; Otto, M.; Stahl, K. A Gear Damage Estimation and Propagation Approach by Combining Simulated and Measured Transmission Error Data. In Proceedings of the International Conference on Advanced Vehicle Powertrains 2021, Beijing, China, 2–4 September 2021.
15. Fromberger, M.; Sendlbeck, S.; Rothemund, M.; Götz, J.; Otto, M.; Stahl, K. Comparing Data Sources for Condition Monitoring Suitability. *Forsch. Ingenieurwes.* **2019**, *83*, 521–527. [CrossRef]
16. Sendlbeck, S.; Fromberger, M.; Otto, M.; Stahl, K. Vibration-Based Gear Condition Monitoring Using an Improved Section-Specific Approach without the Need of Historic Reference Data. In Proceedings of the 27th International Congress on Sound and Vibration (ICSV2021), Online, 11–16 July 2021.
17. Fromberger, M.; Weinberger, U.; Kohn, B. Evaluating Signal Processing Methods for Use in Gearbox Condition Monitoring. In Proceedings of the 24th international congress on sound and vibration, London, UK, 23–27 July 2017.
18. Sensorintegrierende Maschinenelemente—Wegbereiter Der Flächendeckenden Digitalisierung. Available online: https://www.spp2305.de/ (accessed on 7 August 2022).
19. Holm-Hansen, B.T.; Gao, R.X. Time-Scale Analysis Adapted for Bearing Diagnostics. In Proceedings of the SPIE 3833, Intelligent Systems in Design and Manufacturing II, Boston, MA, USA, 20 August 1999; Volume 3833.
20. Holm-Hansen, B.T.; Gao, R.X. Monitoring of Loading Status Inside Rolling Element Bearings Through Electromechanical Sensor Integration. In Proceedings of the ASME International Mechanical Engineering Congress and Exposition, Proceedings (IMECE), Dallas, TX, USA, 16–21 November 1997; pp. 329–335. [CrossRef]
21. Holm-Hansen, B.T.; Gao, R.X. Multiple Defect Analysis of a Sensor Integrated Ball Bearing. In Proceedings of the ASME International Mechanical Engineering Congress and Exposition, Proceedings (IMECE), Anaheim, CA, USA, 15–20 November 1998; pp. 9–14. [CrossRef]
22. Holm-Hansen, B.T.; Gao, R.X. Smart Bearing Utilizing Embedded Sensors: Design Considerations. In Proceedings of the Smart Structures and Materials 1997: Smart Structures and Integrated Systems, San Diego, CA, USA, 3–6 March 1997; Volume 3041, pp. 602–610. [CrossRef]

23. Holm-Hansen, B.T.; Gao, R.X. Integrated Microsensor Module for a Smart Bearing with On-Line Fault Detection Capabilities. In Proceedings of the Conference Record—IEEE Instrumentation and Measurement Technology Conference, Ottawa, ON, Canada, 19–21 May 1997; Volume 2, pp. 1160–1163. [CrossRef]
24. Holm-Hansen, B.T.; Gao, R.X. Vibration Analysis of a Sensor-Integrated Ball Bearing. *J. Vib. Acoust.* **2000**, *122*, 384–392. [CrossRef]
25. Holm-Hansen, B.T.; Gao, R.X. Structural Design and Analysis for a Sensor-Integrated Ball Bearing. *Finite Elem. Anal. Des.* **2000**, *34*, 257–270. [CrossRef]
26. Gao, R.X.; Holm-Hansen, B.T.; Wang, C. Design of a Mechatronic Bearing through Sensor Integration. In Proceedings of the SPIE 3518, Sensors and Controls for Intelligent Machining, Agile Manufacturing, and Mechatronics, Boston, MA USA, 17 December 1998; Volume 3518.
27. Holm-Hansen, B.T.; Gao, R.X.; Zhang, L. Customized Wavelet for Bearing Defect Detection. *J. Dyn. Syst. Meas. Control* **2004**, *126*, 740–745. [CrossRef]
28. Schirra, T.; Martin, G.; Vogel, S.; Kirchner, E. Ball Bearings as Sensors for Systematical Combination of Load and Failure Monitoring. In Proceedings of the International Design Conference, Dubrovnik, Croatia, 21–24 May 2018; Volume 6, pp. 3011–3022. [CrossRef]
29. Schirra, T.; Martin, G.; Kirchner, E. Design of and with Sensing Macine Elements—Using the Example of a Sensing Rolling Bearing. *Proc. Des. Soc.* **2021**, *1*, 1063–1072. [CrossRef]
30. Vogel, S.; Martin, G.; Schirra, T.; Kirchner, E. Robust Design for Mechatronic Machine Elements—How Robust Design Enables the Application of Mechatronic Shaft-Hub Connection. In Proceedings of the International Design Conference, Dubrovnik, Croatia, 21–24 May 2018; Volume 6, pp. 3033–3040. [CrossRef]
31. Groche, P.; Brenneis, M. Manufacturing and Use of Novel Sensoric Fasteners for Monitoring Forming Processes. *Measurement* **2014**, *53*, 136–144. [CrossRef]
32. Gräbner, D.; Dödtmann, S.; Dumstorff, G.; Lucklum, F. 3-D-Printed Smart Screw: Functionalization during Additive Fabrication. *J. Sens. Sens. Syst.* **2018**, *7*, 143–151. [CrossRef]
33. Möhring, H.C.; Bertram, O. Integrated Autonomous Monitoring of Ball Screw Drives. *CIRP Ann.* **2012**, *61*, 355–358. [CrossRef]
34. Biehl, S.; Staufenbiel, S.; Recknagel, S.; Denkena, B.; Bertram, O. Thin film sensors for condition monitoring in ball screw drives. In Proceedings of the 1st Joint International Symposium on System-Integrated Intelligence, Hannover, Germany, 28–29 June 2012; pp. 27–29.
35. Biehl, S.; Staufenbiel, S.; Hauschild, F.; Albert, A. Novel Measurement and Monitoring System for Forming Processes Based on Piezoresistive Thin Film Systems. *Microsyst. Technol.* **2010**, *16*, 879–883. [CrossRef]
36. Peters, J.; Ott, L.; Gwosch, T.; Matthiesen, S. Requirements for Sensor Integrating Machine Elements: A Review of Wear and Vibration Characteristics of Gears. *Sens. Integr. Gears* **2020**. [CrossRef]
37. Peters, J.; Ott, L.; Dörr, M.; Gwosch, T.; Matthiesen, S. Design of Sensor Integrating Gears: Methodical Development, Integration and Verification of an in-Situ MEMS Sensor System. *Procedia CIRP* **2021**, *100*, 672–677. [CrossRef]
38. Peters, J.; Ott, L.; Dörr, M.; Gwosch, T.; Matthiesen, S. Sensor-Integrating Gears: Wear Detection by in-Situ MEMS Acceleration Sensors. *Forsch. Ingenieurwes.* **2022**, *86*, 421–432. [CrossRef]
39. Sridhar, V.; Chana, K. Development of a Novel Sensor for Gear Teeth Wear and Damage Detection. *Int. J. Progn. Health Manag.* **2021**, *12*. [CrossRef]
40. Binder, M.; Stapff, V.; Heinig, A.; Schmitt, M.; Seidel, C.; Reinhart, G. Additive Manufacturing of a Passive, Sensor-Monitored 16MnCr5 Steel Gear Incorporating a Wireless Signal Transmission System. *Procedia CIRP* **2022**, *107*, 505–510. [CrossRef]
41. Schmitt, M.; Kamps, T.; Siglmüller, F.; Winkler, J.; Schlick, G.; Seidel, C.; Tobie, T.; Stahl, K.; Reinhart, G. Laser-Based Powder Bed Fusion of 16MnCr5 and Resulting Material Properties. *Addit. Manuf.* **2020**, *35*, 101372. [CrossRef]
42. Bonaiti, L.; Concli, F.; Gorla, C.; Rosa, F. Bending Fatigue Behaviour of 17-4 PH Gears Produced via Selective Laser Melting. *Procedia Struct. Integr.* **2019**, *24*, 764–774. [CrossRef]
43. Concli, F.; Bonaiti, L.; Gerosa, R.; Cortese, L.; Nalli, F.; Rosa, F.; Gorla, C. Bending Fatigue Behavior of 17-4 PH Gears Produced by Additive Manufacturing. *Appl. Sci.* **2021**, *11*, 3019. [CrossRef]
44. Kadach, D. *FVA-Heft Nr. 1205 Einfluss Der Lastverteilung Auf Die Grübchentragfähigkeit von Einsatzgehärteten Stirnrädern*; Forschungsvereinigung Antriebstechnik e.V. (FVA): Frankfurt, Germany, 2017.
45. Kadach, D. *FVA-Heft Nr. 1164—Einfluss von Stillstandsmarkierungen Auf Die Flankentragfähigkeit von Zahnrädern*; Forschungsvereinigung Antriebstechnik e.V. (FVA): Frankfurt, Germany, 2015.
46. Kadach, D.; Tobie, T.; Stahl, K. Fretting Lines on Gears—Systematic Investigations on the Formation Conditions and Mechanisms. In Proceedings of the JSME international conference on motion and power transmissions, Kyoto, Japan, 28 February–3 March 2017; Volume 2017, pp. 4–5. [CrossRef]
47. Kadach, D.; Matt, P.; Tobie, T.; Stahl, K. Influences of the Facing Edge Condition on the Flank Load Carrying Capacity of Helical Gears. In Proceedings of the ASME 2015 International Design Engineering Technical Conferences and Computers and Information in Engineering Conference, Boston, MA, USA, 2–5 August 2015. [CrossRef]
48. Kadach, D. Stillstandsmarkierungen an Zahnrädern Und Deren Auswirkungen Auf Die Flankentragfähigkeit. Ph.D. Thesis, Technische Universität München, München, Germany, 2015.
49. *ISO 14635-1*; Gears—FZG Test Procedures—Part 1: FZG Test Method A/8,3/90 for Relative Scuffing Load-Carrying Capacity of Oils. International Organization for Standardization: Geneva, Switzerland, 2006.
50. Hoehn, B.-R.; Oster, P.; Tobie, T. Test Methods for Gear Lubricants. *Michaelis Metode Ispitivanja. Goriva Maz.* **2008**, *47*, 129–152.

51. *ISO 6336-1: 2006*; Calculation of Load Capacity of Spur and Helical Gears—Part 1: Basic Principles, Introduction and General Influence Factors. International Organization for Standardization: Geneva, Switzerland, 2006.
52. *ISO 6336-2:2019*; Calculation of Load Capacity of Spur and Helical Gears—Part 2: Calculation of Surface Durability (Pitting). International Organization for Standardization: Geneva, Switzerland, 2019.
53. *ISO 6336-3:2019*; Calculation of Load Capacity of Spur and Helical Gears—Part 3: Calculation of Tooth Bending Strength. International Organization for Standardization: Geneva, Switzerland, 2019.
54. *ISO 6336-5:2019*; Calculation of Load Capacity of Spur and Helical Gears—Part 5: Strength and Quality of Materials. International Organization for Standardization: Geneva, Switzerland, 2019.
55. *ISO 6336-6:2019*; Calculation of Load Capacity of Spur and Helical Gears—Part 6: Calculation of Service Life under Variable Load. International Organization for Standardization: Geneva, Switzerland, 2019.
56. Götz, J.; Sepp, S.; Otto, M.; Stahl, K. Low Excitation Spur Gears with Variable Tip Diameter. In *INTER-NOISE and NOISE-CON Congress and Conference Proceedings*; Institute of Noise Control Engineering: Washington, DC, USA, 2021; Volume 263, pp. 1275–1285.
57. Utakapan, T.; Kohn, B.; Fromberger, M.; Otto, M.; Stahl, K. Evaluation of Gear Noise Behaviour with Application Force Level: Conference Proceedings. *Forsch. Ingenieurwes.* **2017**, *81*, 59–64. [CrossRef]
58. Enthoven, J.C.; Cann, P.M.; Spikes, H.A. Temperature and Scuffing. *Tribol. Trans.* **1993**, *36*, 258–266. [CrossRef]
59. Bowman, W.F.; Stachowiak, G.W. A Review of Scuffing Models. *Tribol. Lett.* **1996**, *2*, 113–131. [CrossRef]
60. Niemann, G.; Winter, H.; Höhn, B.-R.; Stahl, K. Maschinenelemente 1. In *Maschinenelemente 1*; Springer: Berlin, Germany, 2019. [CrossRef]
61. Bibel, G.D.; Reddy, S.K.; Savage, M.; Handschuh, R.F. Effects of Rim Thickness on Spur Gear Bending Stress. *J. Mech. Des.* **1994**, *116*, 1157–1162. [CrossRef]
62. Marunić, G. Effects of Rim and Web Thickness on Gear Tooth Root, Rim and Web Stresses. *Key Eng. Mater.* **2008**, *385–387*, 117–120. [CrossRef]
63. Oda, S.; Nagamura, K.; Aoki, K. Stress Analysis of Thin Rim Spur Gears by Finite Element Method. *Bull. JSME* **1981**, *24*, 1273–1280. [CrossRef]
64. Li, S. Deformation and Bending Stress Analysis of a Three-Dimensional, Thin-Rimmed Gear. *J. Mech. Des.* **2002**, *124*, 129–135. [CrossRef]
65. Conrado, E.; Davoli, P. The "True" Bending Stress in Spur Gears. *Gear Technol.* **2007**, *8*, 52–57.
66. Kawalec, A.; Wiktor, J. Tooth-Root Stress Calculation of Internal Spur Gears. *Proc. Inst. Mech. Eng. B J. Eng. Manuf.* **2004**, *218*, 1153–1166. [CrossRef]
67. Politis, D.J.; Politis, N.J.; Lin, J. Review of Recent Developments in Manufacturing Lightweight Multi-Metal Gears. *Prod. Eng.* **2021**, *15*, 235–262. [CrossRef]
68. Ramadani, R.; Pal, S.; Kegl, M.; Predan, J.; Drstvenšek, I.; Pehan, S.; Belšak, A. Topology Optimization and Additive Manufacturing in Producing Lightweight and Low Vibration Gear Body. *Int. J. Adv. Manuf. Technol.* **2021**, *113*, 3389–3399. [CrossRef]
69. Mura, A.; Curà, F.; Pasculli, L. Optimisation Methodology for Lightweight Gears to Be Produced by Additive Manufacturing Techniques. *Proc. Inst. Mech. Eng. C J. Mech. Eng. Sci.* **2017**, *232*, 3512–3523. [CrossRef]
70. Truman, C.E.; Booker, J.D. Analysis of a Shrink-Fit Failure on a Gear Hub/Shaft Assembly. *Eng. Fail. Anal.* **2007**, *14*, 557–572. [CrossRef]
71. Strozzi, A.; Baldini, A.; Giacopini, M.; Bertocchi, E.; Bertocchi, L. Achievement of a Uniform Contact Pressure in a Shaft–Hub Press-Fit. *Proc. Inst. Mech. Eng. Part C J. Mech. Eng. Sci.* **2012**, *227*, 405–419. [CrossRef]
72. Leonhardt, C.; Otto, M.; Stahl, K. Potential of Lightweight Construction and Component Properties of Joined Spur Gears. In *Dritev-Getriebe in Fahrzeugen*; VDI Wissensforum GmbH: Dusseldorf, Germany, 2019.
73. Chu, S.J.; Jeong, T.K.; Jung, E.H. Effect of Radial Interference on Torque Capacity of Press- and Shrink-Fit Gears. *Int. J. Automot. Technol.* **2016**, *17*, 763–768. [CrossRef]
74. Bae, J.H.; Kim, J.S.; Hwang, B.C.; Bae, W.B.; Kim, M.S.; Kim, C. Prediction of the Dimensional Deformation of the Addendum and Dedendum after the Warm Shrink Fitting Process Using a Correction Coefficient. *Int. J. Automot. Technol.* **2012**, *13*, 285–291. [CrossRef]
75. Leonhardt, C. *FVA 797—Einfluss von Querpressverbänden Auf Die Zahnfußtragfähigkeit Außenverzahnter Stirnradverzahnungen*; Forschungsvereinigung Antriebstechnik e.V. (FVA): Frankfurt, Germany, 2021.
76. Güven, F. Effect of Design Parameters on Stresses Occurring at the Tooth Root in a Spur Gear Pressed on a Shaft. *Proc. Inst. Mech. Eng. Part E J. Process Mech. Eng.* **2021**, *235*, 1164–1174. [CrossRef]
77. Gorla, C.; Rosa, F.; Conrado, E.; Tesfahunegn, Y.A. Combined Effects of Lead Crowning and Assembly Deviations on Meshing Characteristics of Helical Gears. *Int. J. Appl. Eng. Res.* **2016**, *11*, 11681–11694.
78. Bonaiti, L.; Bayoumi, A.B.M.; Concli, F.; Rosa, F.; Gorla, C. Gear Root Bending Strength: A Comparison between Single Tooth Bending Fatigue Tests and Meshing Gears. *J. Mech. Des. Trans. ASME* **2021**, *143*, 103402. [CrossRef]
79. Broßmann, U. Über Den Einfluß Der Zahnfußausrundung Und Des Schrägungswinkels Auf Beanspruchung Und Festigkeit Schrägverzahnter Stirnräder. Ph.D. Thesis, Technische Universität München, München, Germany, 1979.
80. Papadopoulos, I.V.; Davoli, P.; Gorla, C.; Filippini, M.; Bernasconi, A. A Comparative Study of Multiaxial High-Cycle Fatigue Criteria for Metals. *Int. J. Fatigue* **1997**, *19*, 219–235. [CrossRef]

MDPI
St. Alban-Anlage 66
4052 Basel
Switzerland
Tel. +41 61 683 77 34
Fax +41 61 302 89 18
www.mdpi.com

Machines Editorial Office
E-mail: machines@mdpi.com
www.mdpi.com/journal/machines

www.ingramcontent.com/pod-product-compliance
Lightning Source LLC
LaVergne TN
LVHW071000170726
843515LV00006B/1038

* 9 7 8 3 0 3 6 5 7 4 7 6 9 *